2D MATERIALS FOR SURFACE PLASMON RESONANCE-BASED SENSORS

2D MATERIALS FOR SURFACE PLASMON RESONANCE-BASED SENSORS

Sanjeev Kumar Raghuwanshi
Santosh Kumar
Yadvendra Singh

CRC Press is an imprint of the
Taylor & Francis Group, an **informa** business

First edition published 2022
by CRC Press
6000 Broken Sound Parkway NW, Suite 300, Boca Raton, FL 33487-2742

and by CRC Press
2 Park Square, Milton Park, Abingdon, Oxon, OX14 4RN

Library of Congress Cataloging-in-Publication Data
Names: Raghuwanshi, Sanjeev Kumar, author. | Kumar, Santosh (Physics professor), author. | Singh, Yadvendra, author.
Title: 2D materials for surface plasmon resonance-based sensors / Sanjeev Kumar Raghuwanshi, Santosh Kumar, Yadvendra Singh.
Other titles: Two-dimensional materials for surface plasmon resonance-based sensors
Description: First edition. | Boca Raton, FL : CRC Press, 2022. | Series: Emerging materials and technologies | Includes bibliographical references and index.
Identifiers: LCCN 2021028863 (print) | LCCN 2021028864 (ebook) | ISBN 9781032041421 (hbk) | ISBN 9781032041469 (pbk) | ISBN 9781003190738 (ebk)
Subjects: LCSH: Optical detectors--Materials. | Two-dimensional materials. | Surface plasmon resonance.
Classification: LCC TK8360.O67 R34 2022 (print) |
LCC TK8360.O67 (ebook) | DDC 681/.250284--dc23
LC record available at https://lccn.loc.gov/2021028863
LC ebook record available at https://lccn.loc.gov/2021028864

ISBN: 978-1-032-04142-1 (hbk)
ISBN: 978-1-032-04146-9 (pbk)
ISBN: 978-1-003-19073-8 (ebk)

DOI: 10.1201/9781003190738

Typeset in Times
by MPS Limited, Dehradun

Contents

Preface

The idea behind writing this book is that few books are available in the field of optical and plasmonic sensors that cover the work on plasmonic-based fiber optic sensors. But it is difficult to find any dedicated book that generally covers the recent advancement of 2-dimensional (2D) materials in plasmonic-based sensors for chemical/biochemical/ biosensor applications. Due to the plasmonic properties, some nanomaterials such as 2D nanomaterials like graphene, black phosphorous (BP), and transition metal dichalcogenides (TMDCs), i.e., MoS_2, $MoSe_2$, WS_2, and WSe_2 have gained great attention for the sensitivity enhancement of SPR biosensors. It became a current topic of research due to the exclusive optical and electrical properties used in biosensing applications. Due to the unique properties of 2D nanomaterials, a robust framework for various system components can be developed for different device components.

The proposed book contains both the rigorous theory and the synthesis techniques of 2D materials for plasmonic sensing devices and a related variety of sensors. The introduction section of book covers the basic concept of plasmonic and conventional surface plasmon resonance-based sensors. The next subsequent chapters discussed the use of novel 2D materials in the field of plasmonic sensors with different fields of applications. This book provides the synthesis and characterization of latest 2D nanomaterials like graphene, reduced graphene oxide, black phosphorus (BP), transition metal dichalcogenides (TMDCs), etc. for the applications of SPR plasmonic sensors. This book also provides a practical approach to the many skills needed in the highly interdisciplinary field of chemical/biochemical/biosensor technology. Overall, this book offers an overview of current research in this area as well as pointers to its further directions. Including examples from daily lives and recent developments over plasmonic biosensing devices would help in creating readers' interest and they would love to learn the subject.

Plasmonic sensors/communications are necessary subjects that need to decipher by students and practitioners in Electrical, Electronics and Communication Engineering, and Physics. Linking plasmonic sensors with their applications in the chemical/ biomedical field forms the basis of our book. In last decade, plasmonic-optical fiber biosensors had a significant impact on both laboratory research and clinical diagnostics due to its rapid and reliable detection characteristics.

Sensitivity is an important performance index for evaluating surface plasmon resonance based sensors. Sensitivity enhancement has always been a hot topic. It is found that the different refractive indices of samples require different combinations of prism and metal film for better sensitivity. Furthermore, the sensitivity can be enhance by coating two-dimensional (2D) materials with appropriate layers on the metal film. At this time, it is necessary to choose the best film configuration to enhance sensitivity. With the emergence of more and more 2D materials, selecting the best configuration manually became more complicated. Compared with the traditional manual method of selecting materials and layers, this book proposes different study to quickly and effectively find the optimal film configuration that enhances sensitivity. By using this method, not only can the optimal number of layers of 2D materials be determined quickly, but also the optimal configuration can be conveniently found when many materials are available. Readers of this book will be able to understand and explore the possible application, and impact of 2D material in various structures of emerging trends of 2D materials in SPR based sensors.

This book deals with the recent advances in various plasmonics sensing techniques. Theoretical and experimentally, the detection schemes, their realization, configuration, components employed, and application of sensors with 2D materials are the main scope of this work. The total coverage of the book encompasses the plasmonic phenomenon and its application in optical fiber sensing, broadly distributed over eight chapters. Chapters elaborate on the possible configuration, and impact of the several 2D materials in the SPR sensors.

In Chapter 1, the Introduction of Plasmons & Plasmonics sensor discussed with its importance, techniques, principle, and performance in brief. In Chapter 2, theoretical design and analysis of light properties for different media, excitation of surface plasmons in different materials, several configurations of surface plasmon resonance (SPR) phenomenon is explained for SPR chemical sensors assisted by 2-D materials. Chapter 3 covers the theoretical nature and study of surface plasmon resonance (SPR) chemical sensors aided by two-dimensional materials. A brief introduction to SPR chemical sensors is covered, along with some 2-D materials and mathematical equations. The optical properties and application of Black Phosphorus (BP) and Phosphorene in SPR sensors is discussed in Chapter 4. Chapter 5 discusses MXene's suitability as a two-dimensional material for surface plasmon resonance sensors and it's significance in SPR sensors. In Chapter 6, theoretical and experimental analysis of SPR sensor assisted by Single-wall Carbon Nanotube's and its growth method for thin layer to vary the electronic and optical properties is discussed. In Chapter 7, another 2D material TMDC (Transition Metal Dichalcogenides) in SPR sensing is discussed. This chapter also includes the challenges, classification, and its applications. The SPR sensors for clinical diagnosis (detection of glucose, uric acid, ascorbic acid, protein, DNA/RNA, microorganisms) is covered in Chapter 8. In Chapter 9, we discussed the application of 2D materials in different field such as physical sensing, mining industries, biomedical industries, structural health monitoring.

By the nine chapters of the proposed book, we have provided the emerging trends of 2D materials for SPR based sensors, theoretically and experimentally. With the help of this book, readers can get an insight into SPR sensors and can be able to think about the new design for further research work and improvement in sensing performance of SPR sensors and for more practical and feasible applications. The chapter's results are well verified and it will be helpful for the industry and academic concerns for better understanding and prototype development.

The authors thank to the CRC press, in particular the editor who meticulously edited the book for the publication.

Sanjeev Kumar Raghuwanshi
Santosh Kumar
Yadvendra Singh

Contributors

Dr. Sanjeev Kumar Raghuwanshi (M'12, SM'19) received the Bachelor's degree in electronic and instrumentation engineering from S.G.S.I.T.S. Indore, Madhya Pradesh, India and the Master's degree in Solid State 7/9 Technology from Indian Institute of Technology, Kharagpur, in Aug. 1999 and Jan. 2002, respectively. Since July 2009, he has been obtained PhD degree in the field of optics from the Department of Electrical Communication Engineering of Indian Institute of Science, Bangalore India. He is an Associate Professor in Electronics Engineering Department of Indian Institute of Technology (ISM) Dhanbad, India. He was the Post-Doctoral Research Fellow during 2014-2015 at Instrumentation and Sensor Division, School of Engineering and Mathematical Sciences, City University London, Northampton Square, London. He has published more than 100 peer reviewed and indexed International SCI Journal papers in the last 10 years. He has been published 6 books on contemporary optical fiber domain. Since last 5 years 6 Indian patents have been filed and published. He has been sanctioned and executed several R&D projects from different central Government funding agencies including, Department of Atomic Energy, Government of India (GOI), Indian Space Research Organization (ISRO) GOI Council of scientific and Industrial research organization (CSIR) GOI etc. He served as reviewer of Journals like IEEE-Transactions on measurement and instrumentation, IEEE sensor journal, IEEE photonics technology letter, IEEE Quantum electronics to mention few. He is editorial board members and reviewers of several Indian journals. He receipts the Erusmus Mundus Scholarship for his Post Doc study. He is a Fellow of the Optical Society of India (OSI), Life member of IETE, senior member of IEEE (USA) and a Life Member of the International Academy of Physics Sciences.

Dr. Santosh Kumar received a Ph.D. degree from IIT (ISM) Dhanbad, Dhanbad, India. He is currently an Associate Professor at the School of Physics Science and Information Technology, Liaocheng University, Liaocheng, China. He has received an International Travel Grant from SERB-DST, Government of India in 2016. Till now, he has guided seven M.Tech. dissertations and six Ph.D. candidates. He has published more than 140 research articles in national and international SCI journals and conferences. He has presented many articles at conferences held in Belgium, China and USA. He has reviewed more than 530 SCI journals of IEEE, Elsevier, Springer, OSA, SPIE, and Nature. His current research interests include optical fiber sensors, nano and biophotonics, terahertz sensing and spectroscopy, and waveguide and interferometer. He is a Life Fellow Member of the Optical Society of India (OSI) and a Senior Member of IEEE, OSA and SPIE. He is also a Traveling Lecturer of OSA. Recently, he has nominated for the chief-elect of OSA Optical Biosensors technical group. He has delivered many invited talks and serves as the session chair in IEEE conferences. He is Associate Editor of the IEEE Sensors Journal, IEEE Access Journal and OSA Biomedical Optics Express. He is also a Senior Editor of IEEE Transactions on NanoBioscience.

Dr. Yadvendra Singh received his Ph.D. degree from Indian Institute of Technology (ISM) Dhanbad, Jharkhand, India. He has received master's degree (M.E.) from Thapar University, Patiala, Punjab, India in 2013. He is currently a Postdoc Fellow at the Department of Physics, Indian Institute of Technology Roorkee, India. He is a former Assistant Professor at Bipin Tripathi Kumaon Institute of Technology (An autonomous college of Uttarakhand Govt.), Dwarahat, Almora, India. He has published more than 20 research papers in National and International SCI Journals and Conferences, and 2 Indian Patents. He has presented many papers at conferences held in USA and Strasbourg, France. He has received a volunteer travel grant from SPIE, and attended SPIE Photonics West-2019 and SPIE Photonics West-2020 Conferences at California USA, during February 2019 and February 2020 respectively. His current research interests include optical sensors, FBG and TFBG based chemical/Biochemical sensors, optical and Plasmonic devices, semiconductor/metal nanostructures with emphasis on optical biosensors, and Nanotechnology. He is on the board of reviewers of IEEE Industrial Electronics. He is an active member of various societies like IEEE, OSA, and SPIE. He is a Life Fellow member of the Optical Society of India (OSI). He is also a Fellow (Life) Member of the Institution of Electronics and Telecommunication Engineers (IETE), India. He is an Associate (Life) Member of the Institution of Engineers India. He is an active member of the IEEE Photonics Society.

1 Introduction of Plasmons and Plasmonics

1.1 WHAT IS PLASMONICS?

Theoretically, Ritchie revealed the origin of surface plasmons on metal and dielectric interfaces. Thurbadar noticed during 1959 a shine reflectivity occurs through a large drop of thin coat of metal film; however, this phenomenon wasn't associated on the surface. In 1960, Powell and Swan found that electrons at the metal-dielectric interface could excite plasmons on the surface (Homola, Yee, and Gauglitz 1999). Stern and Ferrell revealed that the metal-assisted dielectric interface supports the existence of high-intensity electromagnetic waves on interface supported by specific field continuity conditions (Wakamatsu and Saito 2007). Later during 1968, Otto demonstrates that reflection property of metal thin film was due to excitation of surface plasmon. The history of surface plasmon resonance (SPR) is almost as old as 100 years. Recently it has attracted a great interest from researchers for multiple reasons. In 1902, Wood observed surface plasmon for the first time in history (Schasfoort 2017). According to his observation, when polarized light shone on grating embedded in a mirror, dark/bright fringes were seen in the back-reflected light. In 1907, Zenneck revealed theoretically that the electromagnetic waves of the radio frequency surface occur at the boundary of two media when the first medium is a lossless dielectric and the other is a lossy medium or metal. He concluded that binding of the electromagnetic waves to the interface is due to the imaginary part of permittivity constant of material. In 1909, Somerfield revealed that the surface wave field amplitudes present on the dielectric/metal interface changes inversely as the square root of the distance from the dipole source. In 1941, Fano theoretically concluded that Wood's reports on the anomaly of the excitation of surface waves on the grating embedded surface during 1902 were correct (Zheng et al. 2017). Theoretically, in 1957, Ritchie revealed the excitation of the metal surface plasmons. However, Otto's configuration is not very convenient for a real scenario, due to existence of a finite gap between the high refractive index (RI) prism and metal thin layer; rather it is more useful to perform the study of crystal surface. In 1971, Kretschmann tailored the configuration of Otto by directly depositing a thin metallic layer onto the surface of a high RI coupling prism while keeping in directly contact with the metallic thin layer of the dielectric medium (sample) whose RI is to be measured (Gupta and Verma 2009; Mishra, Mishra, and Gupta 2015; Politano and Chiarello 2014; Politano, Formoso, and Chiarello 2010).

If a transverse magnetic (TM) or p-polarized light is incident on prism/dielectric/metal structure, then some of the light will be transmitted and some of them will be

DOI: 10.1201/9781003190738-1

reflected with a function of the incident angle. However, one sharp dip occurs at some specific conditions of incidence angle which is called the resonance angle. The resonance angle is highly sensitive to the thickness of each layer and RI of surrounding medium. Hence, the dielectric constant of the unknown samples can be easily estimated by observing the resonance condition. This technique is called the angular interrogation method (Sharma and Gupta 2007). However, the prism-coupled SPR sensor has many flaws like bulky size of high RI prism, presence of many moving mechatronics components, precise positioning of a light source, detector, and a requirement of collimating sources with correct polarization properties to mention a few. Hence, such types of sensing chips are difficult to commercialize in a market. Furthermore, the prism-assisted SPR-sensing chips are unable to perform remote sensing. After introduction of SPR-based sensing devices based on Kretschmann geometry, a need arose for optical fiber technology-based SPR-sensing chips. In fact, an optical fiber technology-based SPR-sensing probe provides several advantages over a prism-based SPR-sensing probe, due to its miniaturized geometry, remote sensing capability, multiplexing capability, versatile sensing features, etc. (Liedberg, Nylander, and Lundström 1995). In addition, SPR fiber-optic sensing provides a simplified optical design and multiplex sensing capability. The metal layer is deposited directly on the unclad core of the optical fiber in a fiber-optic SPR sensor. In an SPR-based fiber-optic sensor, the SPR sensing is performed with a fixed angle of incidence while a multiple wavelength optical source is used to excite the SPR dip. This method is called a method of questioning wavelengths. This is because the spectral distribution of light with an optical fiber can be preserved. The sensor is also normally manufactured on a multimode optical fiber, where the rays are accepted in a range decided by an acceptance cone of optical fiber. Generally speaking, gold and silver metals are the preferred choice for SPR excitations. In fact, there are some merits and demerits of both metals. Silver has a low value of imaginary parts of permittivity constants, which gives a sharp resonance dip leading to better accuracy of detection. It is a matter of fact that silver can be oxidized easily when it comes in direct contact with air or water. Silver is not very stable compared to gold, which is very stable and not reactive (Bhatia and Gupta 2011; Feng et al. 2008; Gupta and Sharma 2005; Huang, Ho, Kong, and Kabashin 2012; Motogaito et al. 2015; Patskovsky, Kabashin, Meunier, and Luong 2003; Patskovsky, Bah, Meunier, and Kabashin 2006; Pyshkin 2013; Shalabney and Abdulhalim 2010; Urbonas et al. 2015; Verma and Gupta 2010; Wang et al. 2013). A gold-coated SPR sensor provides substantially high sensitivity compared to silver. It is desirable to have high sensitivity and detection accuracy, hence a single metal-coated SPR sensor is advantageous. Maharana and Jha (2012) demonstrated a prism-assisted SPR sensor with a bimetallic coating of silver and gold, which contains all the advantages of both individual metals. The same principle has been applied to fiber-optic SPR-sensing development. A lot of modifications have been provided in the SPR-sensing chip to enhance the sensitivity and figure of merit (FOM). Recently, symmetric versus asymmetric metal coating has been provided on SPR-sensing probes to improve the number of SPR resonance dips (Zhao, Zhang, Zhu, and Shi 2019). Also, tapered and dual-tapered optical fiber SPR-sensing chips have been realized to improve the sensing performance of gas and liquid (Herrera et al. 2011). Currently, the vapor phase detection by the THz SPR-sensing chip has been studied. The RI detection of gas is quite challenging hence mid-IR fiber-optic SPR-sensing chip has been under research at present. Continuing this, a large number of theoretical and experimental research studies are being carried out over time to improve the performance of SPR-sensing chips.

Plasmonics is the study of the interaction between electromagnetic field and photon due to metal-metal/dielectric interface under controlled circumstances. These density waves are developed at optical wavelengths and very precise controllers by outer circumstances. A plasmon is a collective oscillation of electrons on a metal/dielectric interface. The coupling of plasmon with a lightwave creates other forms of quasiparticles, so-called plasma polaritons. Plasmon excitation requires a precise condition of metal film thickness and interface condition with dielectric materials. Hence, free electrons of metal require the interaction with photons on an interface to create plasmonic oscillations. The SPR condition is satisfied at some particular incident angle and an electromagnetic field penetrates to the outer layer of the composite structure (Kedenburg, Vieweg, Gissibl, and Giessen 2012; Lin et al. 2006; Maharana, Jha, and Palei 2014; Ouyang et al. 2016; Wu et al. 2016; Zeng, Baillargeat, Ho, and Yong 2014).

1.2 SURFACE PLASMON RESONANCE (SPR)

SPR is supposed to be a coherent vibration of the electrons at the boundary of a metal/dielectric while the composite structure is excited by the incident light. The precise incident condition generates a surface plasmon wave (SPW) with some peculiar optical phenomena and properties. In general, plasmon is a collective oscillation of an electron on a metal/dielectric interface. Excitation of SPR depends on the type of contact between two mediums like a positive dielectric constant with a negative dielectric constant medium.

It is known that the dielectric constant $\mathcal{E} = (\boldsymbol{n} + \boldsymbol{i}\mathbf{k})^2$, where RI and the absorption constant are represented by $\boldsymbol{n}$ and $\mathbf{k}$, is also excepted for the second surface $\boldsymbol{n} \ll \mathbf{k}$. This is valid for the case of metals. It is also important that the total internal reflection (TIR) condition must be satisfied for the incident medium. Now, for finding the SPR frequency, we start from Maxwell's equation and use the boundary condition as follows, discussing the importance of the wave vector, *k,* in the process:

$$\nabla . D = \rho \tag{1.1}$$

$$\nabla . B = 0 \tag{1.2}$$

$$\nabla \times E = -\frac{\partial B}{\partial T} \tag{1.3}$$

$$\nabla \times H = J + \frac{\partial D}{\partial t} \tag{1.4}$$

Wave equations:

$$\nabla \times E = -\mu\mu_0 \frac{\partial H}{\partial t} \tag{1.5}$$

$$\nabla \times H = \varepsilon\varepsilon_0 \frac{\partial E}{\partial t} \tag{1.6}$$

Using Eq. (1.5) and Eq. (1.6), we have

$$\nabla^2 E = \mu\mu_0 \sigma \frac{\partial E}{\partial t} + \varepsilon\varepsilon_0 \mu\mu_0 \frac{\partial^2 E}{\partial^2 t} \tag{1.7}$$

$$\nabla^2 H = \mu\mu_0 \sigma \frac{\partial H}{\partial t} + \varepsilon\varepsilon_0 \mu\mu_0 \frac{\partial^2 H}{\partial^2 t} \tag{1.8}$$

From Eq. (1.7), if we consider a plane wave form

$$E = E^\circ e^{-i(kr - wt)} \tag{1.9}$$

where E° is the polarization vector, k is the wave number, and r is the position vector.

So, we get

$$k^2 = i\mu\mu_0 \sigma w + \varepsilon\varepsilon_0 \mu\mu_0 w^2 \tag{1.10}$$

The k factor is an important parameter regarding the propagation of the plane wave. Its relation with that of the RI of the medium through which it will propagate is of immense importance and discussion is provided as follows:

$$k = \frac{2\pi}{\lambda} = \frac{w}{v} = \frac{nw}{c} \tag{1.11}$$

As of Eq. (1.11), k depends on n (RI of the medium), which again depends on the material property as ε.

If we consider the material to have a complex RI of metal, then

$$k = \frac{Nw}{c} \tag{1.12}$$

where N is complex RI.

$$\text{So,} \quad N = (n + ik) \tag{1.13}$$

where n = RI, k = absorption coefficient.

Eq. (1.12) is rewritten as

$$k = \frac{(\boldsymbol{n} + \boldsymbol{i}\mathbf{k})\, w}{c} \tag{1.14}$$

as RI is the square root of material permittivity,

$$N = \sqrt{\epsilon_1 + i\epsilon_2} \tag{1.15}$$

Squaring Eq. (1.13) and comparing to Eq. (1.15), we have

$$\epsilon_1 = n^2 - \mathrm{k}^2 \tag{1.16}$$

$$\epsilon_2 = 2n\mathrm{k} \tag{1.17}$$

This part discusses the dependency of the RI, which will be an important parameter for plasma frequency.

1.2.1 Different Types of Plasmon and Polariton

1.2.1.1 Polariton

Polariton is due to coupling between an elementary excitation and a photon or light-matter interaction.

1.2.1.2 Types of Polaritons

Plasmon polariton: This is due to a state of coupling between a plasmon and a photon.

Phonon polariton: This is due to a state of coupling between a phonon and a photon.

1.3 WHAT IS SURFACE PLASMON POLARITONS (SPP)?

The joint excitation of a surface plasmon and a photon is called Surface Plasmon Polariton (SPP). However, excitation of a Surface Plasmon would not be possible unless a specific condition fulfills both energy and momentum conservations. In generalized terms, the metal-insulator or metal-dielectric interfaces are the normally chosen metals, which are noble metals. Most of the noble metals, unlike most base metals, have good corrosion resistance. Hence, they do not suffer from oxidation problems. However, noble metals are very precious due to their mining process under earth crust. Some of the noble metals generally found in Earth's crust are ruthenium, rhodium, palladium, silver, osmium, iridium, platinum, and gold. For the study of surface plasmonics, optical properties of noble metals must be studied first. For this, we have to study many models. Two of them are listed below in detail.

1.3.1 Drude Model

The classical theory of the Drude model explains how free electrons found in metals interact with an external electromagnetic field at wavelength λ. In short, we have to somehow find out the frequency-dependent complex permittivity of metals. Oscillations of free electron gas have some relative phase shift in operation of the driving EM field as per the principle of the free electron Drude model. This is the reason why most of the metals have exhibit a negative dielectric constant at optical wavelength, which is one of the desirable conditions for the existence of SPR phenomena. These oscillations can be related to driven, damped harmonic oscillations, which can be derived below.

Time dependence of electric field having harmonic is represented by the following expression:

$$E(r, t) = Re(E_0 \, e_x e^{-iwt}) \tag{1.18}$$

Driving force is given by:

$$-eE_x = -eE_0 \, e^{-iwt} \tag{1.19}$$

Damping force is given by:

$$-m_e \Gamma v = -m_e \Gamma \frac{dx}{dt}$$

where damping rate, Γ, which depends on the electron mean free path resultant of scattering effects, m_e is the effective mass, and e is the electric charge of electrons. It is also apparent that x is a displacement of electron and v is the velocity. The Drude model of the free-electron gas relates as follows:

$$m_e \frac{d^2x}{d^2t} + m_e \Gamma \frac{dx}{dt} = -eE_0 \, e^{-iwt}$$

$$\frac{d^2x}{d^2t} + \Gamma \frac{dx}{dt} = -\frac{eE_0 \, e^{-iwt}}{m_e}$$

Let,

$$x' = u \tag{1.20}$$

$$u' + \Gamma u = -\frac{eE_0 \, e^{-iwt}}{m_e}$$

Integrating Factor (IF) = $e^{\Gamma t}$

$$u(t) = -\frac{1}{e^{\Gamma t}} \int e^{\Gamma t} \frac{eE_0 e^{-iwt}}{m_e} dt$$

$$u(t) = -\frac{1}{e^{\Gamma t}} \frac{eE_0}{m_e} \int e^{(\Gamma - iw)t} dt$$

$$u(t) = -\frac{1}{e^{\Gamma t}} \frac{eE_0}{m_e} \frac{e^{(\Gamma - iw)t}}{(\Gamma - iw)} + c$$

where c = integration constant.

But according to Eq. (1.3):

$$x(t) = \int u(t) dt$$

$$x(t) = \int - \frac{eE_0}{m_e} \frac{e^{(\Gamma - iw - \Gamma)t}}{(\Gamma - iw)} \, \mathrm{dt}$$

For the time being, we are assuming initial conditions equals to zero, i.e., c = 0.

$$x(t) = \frac{eE_0}{m_e} \frac{1}{(w^2 + i\Gamma w)} e^{-iwt}$$

In generalized form:

$$\begin{aligned} x(t) &= A\, e^{-iwt} \\ A &= \frac{eE_0}{m_e} \frac{1}{(w^2 + i\Gamma w)} = x = Electron \;\; Displacement \end{aligned} \tag{1.21}$$

According to standard formulas, dipole momentum per unit volume or resonant macroscopic polarization of the given medium may be represented as,

$$P = -Nexe_x$$

By using Eq. (1.21), we get

$$P = -Ne \frac{eE_0}{m_e} \frac{1}{(w^2 + i\Gamma w)} \frac{E}{E_0}$$

By using Eq. (1.19), we get

We can also write it as:

$$P = \frac{Ne^2}{m_e} \frac{E}{(-w^2 - i\Gamma w)}$$

Relative permittivity is defined for the case of isotropic material as follows:

$$D = \varepsilon_0 E + P = \varepsilon_0 \varepsilon_r E \tag{1.22}$$

$$\begin{aligned} D &= \varepsilon_0 E + \frac{Ne^2}{m_e} \frac{E}{(-w^2 - i\Gamma w)} \\ D &= E\left[1 - \frac{Ne^2}{m_e} \frac{E}{(w^2 + i\Gamma w)\varepsilon_0}\right] \end{aligned} \tag{1.23}$$

On comparing Eq. (1.22) and (1.23), we get

We find out that relative permittivity is a function of frequency as given by

$$\begin{aligned} \varepsilon_r(w) &= \left[1 - \frac{Ne^2}{m_e} \frac{1}{(w^2 + i\Gamma w)\varepsilon_0}\right] \\ \varepsilon_r(w) &= 1 - \frac{w_p^2}{(w^2 + i\Gamma w)} \end{aligned} \tag{1.24}$$

where, $w_p = \sqrt{\frac{Ne^2}{\varepsilon_0 m_e}}$ is called plasma frequency related to volume. Also, N stands for a total number of free electrons that exists in unit volume. Upon rationalizing Eq. (1.24), we get a real and complex part of permittivity as given by

$$\varepsilon_r' = 1 - \frac{w_p^2}{(w^2 + \Gamma^2)} \tag{1.25}$$

and

$$\varepsilon_r'' = \frac{w_p^2 \Gamma}{w(w^2 + \Gamma^2)} \tag{1.26}$$

The complex value of RI "n" related to the relative permeability, μ_r, and the relative permittivity, ε_r, of the material is as follows:

$$n = \sqrt{\varepsilon_r \mu_r} = n_r + ik \tag{1.27}$$

The real part denotes the quotient of the phase velocity in free space (vacuum) in the given material. However, the imaginary part represents the absorption coefficient of the material.

1.3.1.1 Conclusion from the Above Expression

When the frequency is below w_p, the metal retains the metallic characteristics ($\varepsilon_r' < 0$). It is a fact that the imaginary part of the RI would be large if the real part of the permittivity is negative enough. It turns out that the light can only penetrate to a small extent in metal. For large frequencies that are close to $w_p(w_p > \Gamma)$, $w \gg \Gamma$; hence the damping factor Γ is insignificant. Eq. (1.24) can be simplified to:

$$\varepsilon_r(w) = 1 - \left(\frac{w_p}{w}\right)^2 \tag{1.28}$$

This defines the real part of dielectric constant of an undamped free-electron plasma. For the case of $\omega > \omega_p$, it means that metal behaves as a dielectric because the permittivity is positive.

1.3.2 Lorentz Model

In the infrared regime, the Drude model gives quite accurate results for optical properties of metals; however, for the case of a shorter wavelength regime ≤ 650 *nm*, the Drude model does not give satisfactory results. The reason behind this anomaly is due to the lower energy shell bound electrons that try to interact with the incident light. Also, a higher-energy shell excited photon can excite lower energy band electrons to the higher energy level conduction band.

If we study the Lorentz Dipole oscillator classical model, it reveals how the electron, bound to atoms, interacts with an external field at the optical wavelength $\lambda = \frac{2\pi c}{\omega}$.

The coherent oscillation of electric field drives the electrons to their atom in harmonic motion. If "λ" coincides with the natural resonance wavelengths of the atom, the resonance condition of ion pairs is fulfilled. In that case, atoms can therefore substantially absorb energy from the field and the medium exhibits absorption. However, in another way, the medium behaves optically transparent, if the "λ" or "ω" does not coincide with any natural resonance wavelength or frequency. As a result, the medium would not reveal any absorption phenomena. The interaction between atoms and an electromagnetic wave would sustain a single natural resonance wavelength, say "λ_0." Various types of forces working on the oscillator are given as follows:

Driving Force: $-eE_x = -eE_0\, e^{-iwt}$
Spring Force: $-K_s x$
Damping Force: $-m_b \Upsilon v = -m_b \Upsilon \frac{dx}{dt}$

where "Υ" is the radiating damping coefficient in the case of tightly bound electrons, and "K_s" is the spring kind of constant with $w_0 = \sqrt{K_s/m_b}$, m_b is the bound electron's effective mass, which is not equal to the effective mass of a free electron.

Since then, the equation of motion is:

$$m_b \frac{d^2x}{d^2t} + m_b \Upsilon \frac{dx}{dt} + m_b {w_0}^2 = -eE_0 e^{-iwt} \tag{1.29}$$

$$\frac{d^2x}{d^2t} + \Upsilon \frac{dx}{dt} + {w_0}^2 = -\frac{eE_0 e^{-iwt}}{m_b} \tag{1.30}$$

It's better to solve this problem separately by finding a general solution and a particular solution.

$$D^2x + \Upsilon Dx + {w_0}^2 x = -\frac{eE_0 e^{-iwt}}{m_b}$$

where, $\frac{d}{dt} = D$.

$$x(t) = \frac{1}{D^2 + \Upsilon D + {w_0}^2}\left(-\frac{eE_0 e^{-iwt}}{m_b}\right)$$
$$x(t) = \frac{-eE_0}{m_b}\frac{1}{(-jw)^2 + \Upsilon(-jw) + {w_0}^2} \tag{1.31}$$
$$x(t) = \frac{-eE_0}{m_b}\frac{1}{{w_0}^2 - w^2 - i\Upsilon w}$$

Now, again, with the help of standard formula, the resonant macroscopic polarization, i.e., Dipole momentum per unit volume, is given by:

$$P = -N'ex e_x \tag{1.32}$$

Now, by using Eq. (1.31) in Eq. (1.32), we get:

$$P = -N'e\left(\frac{-eE_0}{m_b}\frac{1}{{w_0}^2 - w^2 - i\Upsilon w}\right)\frac{E}{E_0} \tag{1.33}$$

$$P = \frac{N'e^2}{m_b}\left(\frac{1}{{w_0}^2 - w^2 - i\Upsilon w}\right)E \tag{1.34}$$

For isotropic materials, the relative permittivity is defined as

$$D = \varepsilon_0 E + P = \varepsilon_0 \varepsilon_r E$$

By using Eq. (1.34), we get:

$$D = \varepsilon_0 E\left(1 + \frac{N'e^2}{m_b} \frac{1}{{w_0}^2 - w^2 - i\Upsilon w}\right)$$

Now from Lorentz model, dielectric function can be expressed contributed with bound electrons as follows:

$$\varepsilon_r(w) = 1 + \frac{N'e^2}{m_b} \frac{1}{{w_0}^2 - w^2 - i\Upsilon w}$$
$$\varepsilon_r(w) = 1 + \frac{w_p'^2}{{w_0}^2 - w^2 - i\Upsilon w} \tag{1.35}$$

where in Drude model $w_{p'} = \sqrt{\frac{N'e^2}{\varepsilon_0 m_b}}$, which has some other physical meaning compared to the plasma frequency. Also, N' is the bound electrons' density. The real and imaginary parts of the dielectric constant can be obtained by rationalizing Eq. (1.24) and finally we get:

$$\varepsilon_r' = 1 + \frac{w_p'^2({w_0}^2 - w^2)}{({w_0}^2 - w^2)^2 + \Upsilon^2 w^2} \tag{1.36}$$

and,

$$\varepsilon_r'' = \frac{w_p'^2 w\, \Upsilon}{({w_0}^2 - w^2)^2 + \Upsilon^2 w^2} \tag{1.37}$$

For the systems with many oscillators, the Lorentz model is a superposition of more than one of the above resonances. In that case, the dielectric constant may be expressed as follows.

The optical constant can be written as:

$$\varepsilon_r(w) = 1 + \sum_{j=1}^{m} \frac{f_j w_{pj}^2}{{w_j}^2 - w^2 - i\Upsilon j w} \tag{1.38}$$

where f_j stands for the oscillator strength. For the case of silver, the dielectric constant is better fitted with the Lorentz–Drude model, which includes both bound electron effects and free-electron effects.

1.4 SURFACE PLASMON POLARITONS (SPPS) AT METAL/ INSULATOR INTERFACES

The four Maxwell's equations can be written as per Eq. (1.1–1.4), where, for the case of dielectric optical fiber, usually $\rho = 0$ and $j = 0$ are assumed. Let us discuss single-dimensional SPPs on a single metal-dielectric interface. Let the surface

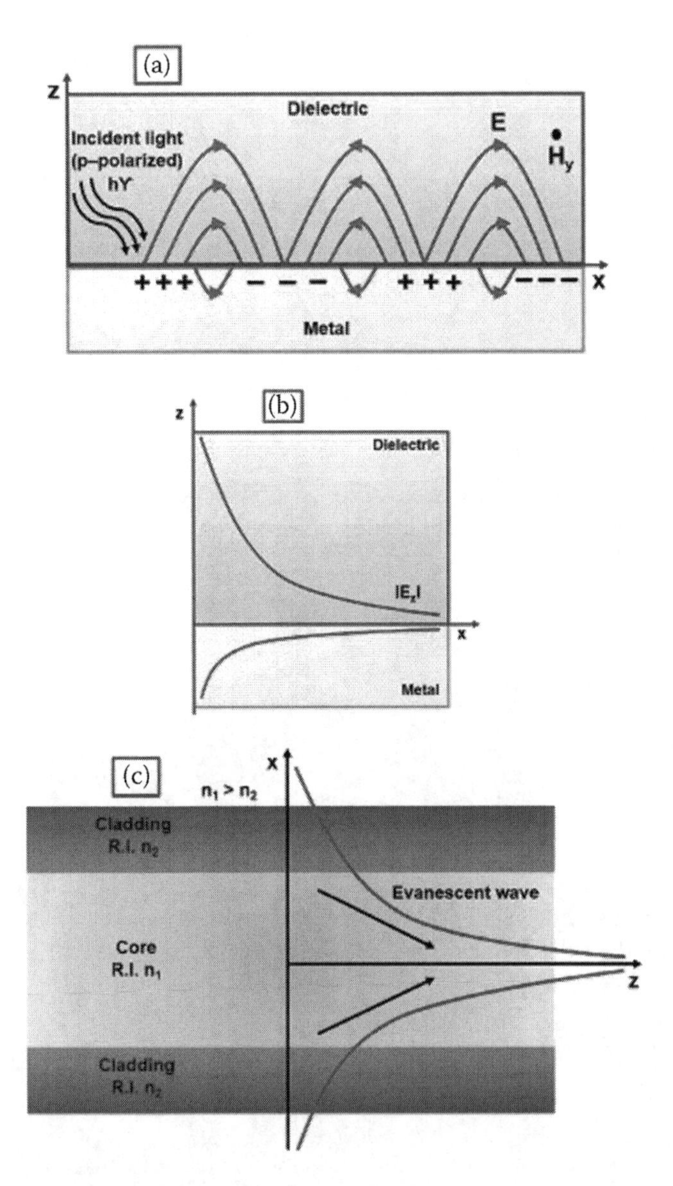

FIGURE 1.1 Schematic illustration of Surface Plasmon Polaritons (SPPs) on a single interface of metal-dielectric. Reprinted with permission from Sensors and Actuators B: Chemical. Copyright, 2020, Elsevier (Tabassum and Kant 2020).

plasmon waves propagate in the X-direction, as shown in Fig. 1.1(a). Also assume that the geometry is invariant in the Y-direction. It is assumed that the X – Y plane has a $\varepsilon_1(w) = \varepsilon_r' + \varepsilon_r''$ related to Eq. (1.36) and Eq. (1.37) and a dielectric (z > 0) constant ε_2(w). Also, it is apparent that the case of metal (z < 0) has a negative real part of the dielectric constant and region (z > 0) with the positive dielectric constant for time harmonic waves ($\frac{d}{dt} = -iw$) propagating along the X-direction ($\frac{d}{dx} = i\beta$) with no spatial variation along the Y-direction ($\frac{d}{dy} = 0$).

For the TM mode, from Maxwell's last two curl equations, we get:

$$\begin{vmatrix} i & j & k \\ \frac{d}{dx} & \frac{d}{dy} = 0 & \frac{d}{dz} \\ E_x & 0 & E_z \end{vmatrix} = -iwu_0(H_y j)$$

The above matrix reduces to

$$\frac{dE_x}{dz} - i\beta E_z = iwu_0 H_y \tag{1.39}$$

$$\frac{dH_y}{dz} = iw\varepsilon_0 \varepsilon E_x \tag{1.40}$$

$$i\beta H_y = iw\varepsilon_0 \varepsilon E_Z \tag{1.41}$$

On differentiating Eq. (1.40), we get:

$$\frac{d^2 H_y}{d^2 z} = iw\varepsilon_0 \varepsilon \frac{dE_x}{dz} \tag{1.42}$$

On putting Eq. (1.39) in Eq. (1.42), we get:

$$\frac{d^2 H_y}{d^2 z} = iw\varepsilon_0 \varepsilon (iwu_0 H_y + i\beta E_z) \tag{1.43}$$

Also, from Eq. (1.40) and Eq. (1.42), we get:

$$\frac{d^2 H_y}{d^2 z} = -w^2 u_0 \varepsilon_0 \varepsilon H_y + \beta^2 H_y \tag{1.44}$$

$$\frac{d^2 H_y}{d^2 z} + (w^2 u_0 \varepsilon_0 \varepsilon + \beta^2) H_y = 0 \tag{1.45}$$

Since as we know that:

$w\sqrt{u_0 \varepsilon_0} = K$= Wave Number

So, we can write Eq. (1.45) as:

$$\frac{d^2H_y}{d^2z} + (K_0^2\varepsilon - \beta^2)H_y = 0 \tag{1.46}$$

Eq. (1.45) is the wave equation for TM modes.

Note: TE modes do not exist for a plasmonic wave at metal/insulator interface.

Proof for above note:

For the TE mode, if we take Maxwell's second curl equation and solve similarl to the above, we get:

$$\frac{d^2E_y}{d^2z} + (K_0^2\varepsilon - \beta^2)E_y = 0 \tag{1.47}$$

The field component can be easily found from the solution of Eq. (1.47), which can be solved by a differential equation solution technique with time variations as:

For $z > 0$ (i.e., dielectric material):

$$\frac{d}{dz} = D$$

By Eq. (1.46), we get:

$$\begin{aligned} D^2 &= K_2^2 \\ H_y(z) &= A_2 e^{i\beta x} e^{-K_2 z} \end{aligned} \tag{1.48}$$

Also, from Eq. (1.40) and (1.41), we get:

$$\begin{aligned} E_x(z) &= \frac{1}{iw\varepsilon_0\varepsilon_2}\frac{dH_y}{dz} \\ E_x(z) &= \frac{1}{iw\varepsilon_0\varepsilon_2}\frac{d(A_2 e^{i\beta x} e^{-K_2 z})}{dz} \\ E_x(z) &= iA_2\frac{1}{w\varepsilon_0\varepsilon_2}K_2 e^{i\beta x} e^{-K_2 z} \end{aligned} \tag{1.49}$$

Similarly, in the same fashion:

$$\begin{aligned} E_z(z) &= -\frac{i\beta H_y}{iw\varepsilon_0\varepsilon_2} \\ E_z(z) &= -\frac{i\beta A_2 e^{i\beta x} e^{-K_2 z}}{iw\varepsilon_0\varepsilon_2} \\ E_z(z) &= -A_2\frac{\beta}{w\varepsilon_0\varepsilon_2} e^{i\beta x} e^{-K_2 z} \end{aligned} \tag{1.50}$$

For $z < 0$ (i.e., metal material):

$$H_y(z) = A_1 e^{i\beta x} e^{K_1 z} \tag{1.51}$$

$$E_x(z) = iA_1 \frac{1}{w\varepsilon_0 \varepsilon_1} K_1 e^{i\beta x} e^{K_1 z} \tag{1.52}$$

$$E_z(z) = -A_1 \frac{\beta}{w\varepsilon_0 \varepsilon_1} e^{i\beta x} e^{K_1 z} \tag{1.53}$$

In which

$$K_i^2 = \beta^2 - K_0^2 \varepsilon_i \; (i = 1,\ 2) \tag{1.54}$$

With the same procedure as stated above, the field component can be easily found from the solution of Eq. (1.23), which can be solved by a differential equation solution technique with time variations given the following set of six equations.

For z > 0 (i.e., dielectric material):

$$E_y(z) = A_2 e^{i\beta x} e^{-K_2 z} \tag{1.55}$$

$$H_x(z) = -iA_2 \frac{1}{wu_0} K_2 e^{i\beta x} e^{-K_2 z} \tag{1.56}$$

$$H_z(z) = A_2 \frac{\beta}{wu_0} e^{i\beta x} e^{-K_2 z} \tag{1.57}$$

For z < 0 (i.e., metal material):

$$E_y(z) = A_1 e^{i\beta x} e^{K_1 z} \tag{1.58}$$

$$H_x(z) = iA_1 \frac{1}{wu_0} K_1 e^{i\beta x} e^{K_1 z} \tag{1.59}$$

$$H_z(z) = A_1 \frac{\beta}{wu_0} e^{i\beta x} e^{K_1 z} \tag{1.60}$$

Continuity of the $E_y(z)$ and $H_x(z)$ at the interface, e.g., at z = 0, leads to the condition:

$$A_1(K_1 + K_2) = 0$$

Since field confinement to the surface requires Re $[K_1] > 0$ and Re $[K_2] > 0$, this condition must be only fulfilled if $A_1 = 0$, so that $A_2 = A_1 = 0$. Hence, SPPs would only exist for TM polarization. This proves that no surface wave exists for TE polarization. Field propagation perpendicular to interface in the metal region has a penetration depth or skin depth (δ_m) and for the dielectric region, field confinement (δ_d) of the electromagnetic wave would be defined by:

$$\delta_m = 1/|K_1| \text{ and } \delta_d = 1/|K_2|$$

In fact, the magnitude of penetration depth (δ_m) is defined by:

$$\delta_m = \frac{\lambda}{2\pi n_1\sqrt{sin^2\theta - \left(\frac{n_2}{n_1}\right)^2}}$$

where n_1 and n_2 are the RI of high and low RI medium, θ is the angle of an incident from normal to interface, while λ is the wavelength of light in free space. Continuity of H_y and E_x at the interface, i.e., at z = 0, results in

$$A_1 = A_2 \text{ and } \frac{K_2}{K_1} = -\frac{\varepsilon_2}{\varepsilon_1} \tag{1.61}$$

Combining Eq. (1.54) and Eq. (1.61), we get:

$$\begin{aligned} K_i^2 &= \beta^2 - K_0^2\varepsilon_i \\ &= > \beta^2 = K_1^2 + K_0^2\varepsilon_1 \\ &= > \beta^2 = K_2^2 + K_0^2\varepsilon_2 \end{aligned} \tag{1.62}$$

By Eq. (1.38), we get:

$$K_2 = -\frac{\varepsilon_2}{\varepsilon_1}K_1 \tag{1.63}$$

Putting Eq. (1.63) in Eq. (1.62), we get:

$$\beta^2 = \left(-\frac{\varepsilon_2}{\varepsilon_1}K_1\right)^2 + K_0^2\varepsilon_2$$

$$\beta^2 = \left(-\frac{\varepsilon_2}{\varepsilon_1}\right)^2(\beta^2 - K_0^2\varepsilon_1) + K_0^2\varepsilon_2$$

On further solving, we get:

$$\beta^2 = \frac{w^2u_0\varepsilon_0\varepsilon_1\varepsilon_2}{\varepsilon_1 + \varepsilon_2}$$

$$w^2 u_0 \varepsilon_0 = K^2 = \left(\frac{w}{c_0}\right)^2$$

From above, finally we obtained a relation as:

$$\beta = \frac{w}{c_0}\sqrt{\frac{\varepsilon_1 \varepsilon_2}{\varepsilon_1 + \varepsilon_2}} \quad (1.64)$$

which is the eigenvalue or dispersion equation of SPPs on a single metal/dielectric interface. Fig. 1.1(a) reveals the electric field lines of excited SPPs via the TM mode on a metal/dielectric interface, whereas Fig. 1.1(b) shows the exponential decay of the field intensity irrespective of the point of the dielectric/metal interface. Fig. 1.1(c) shows the high multimode optical fiber structure where a high RI core is sandwiched by two low RI cladding layers. Hence, the light will be trapped inside the core region as long as the TIR condition is satisfied. However, each time the ray hits (reflect and refract of rays) the core-cladding interface, an evanescent wave is generated. Fig. 1.8(a) shows the corresponding dispersion curve of SPPs for the case of direct light. As an extension, Fig. 1.8(c) reflects the graph for a dispersion curve of evanescent waves at various interfaces in the Kretschmann configuration (Tabassum and Kant 2020).

1.5 DISCUSSION USING FRESNEL'S EQUATION

After applying boundary conditions to a plane wave, Eq. (1.9), passing from one medium to another and simplifying we get Fresnel's equation, as specified in Fig 1.2.

$$rp = \frac{n_1 \cos \theta_2 - n_2 \cos \theta_1}{n_1 \cos \theta_2 + n_2 \cos \theta_1} \quad (1.65)$$

$$t_p = \frac{2n_1 \cos \theta_1}{n_1 \cos \theta_2 + n_2 \cos \theta_1} \quad (1.66)$$

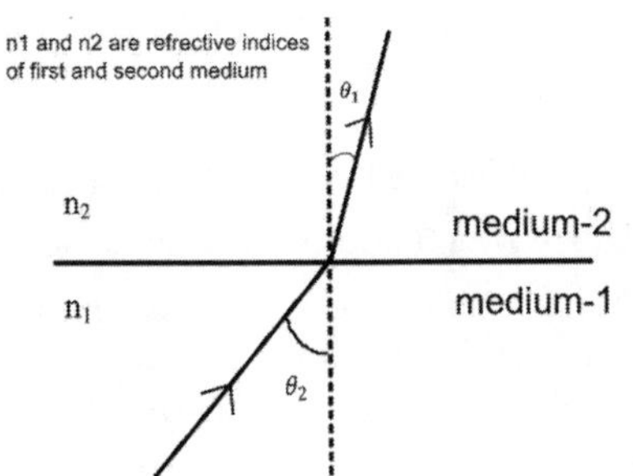

FIGURE 1.2 Two-layer model in which medium 1 and medium 2 have n_1, θ_1 and n_2, θ_2 as refractive indices and corresponding angles, respectively.

$$r_s = \frac{n_1 \cos \theta_1 - n_2 \cos \theta_2}{n_1 \cos \theta_1 + n_2 \cos \theta_2} \quad (1.67)$$

$$t_s = \frac{2n_1 \cos \theta_1}{n_1 \cos \theta_1 + n_2 \cos \theta_2} \quad (1.68)$$

In Eqs. (1.65–1.68), the "p" and "s" stand for p-type and s-type polarization, "r" and "t" for reflection and transmission coefficient, and n_1 and n_2 are refractive indices of first and second medium with corresponding incident and refracted angle of θ_1 and θ_2. These equations are further used in finding the reflectance and transmittance of n number of layers, which has been discussed using the concept of propagation matrices and matching matrices for a matrix form.

1.6 EXCITATION TECHNIQUES

For excitation purposes it is required that *kx* value must be coupled with the wave number vector of surface plasmons that is present at the interface. Energy would transfer between photon and electrons due to optical power coupling to surface plasmons. Since the allowed wave propagation constant of SPW is always greater than the wave propagation constant associated with the incident light traveling in free space. Hence, surface plasmon would not be possible to excite by directly shining light on metal/dielectric interface until some high-RI prism is incorporated to excite the SPR. However, there are several techniques to excite a surface plasmon wave. The most versatile excitation technique is to use a prism to provide phase matching between the plasmon wave and the incident wave. Some techniques are

1. Kretschmann
2. Otto
3. Diffraction grating
4. Waveguide coupling

Out of these, the most common is the Kretschmann's configuration, in which a high-RI prism is deployed and coated by a thin metal film. In this proposed scheme, light travels through a high-RI medium to a lower RI medium; hence, it will be fully reflected at the interface if and only if the TIR condition is satisfied. Simultaneously, TIR generates an evanescent field along with the interface whose amplitudes exponentially decrease with the distance from the interface. This phenomenon is called skin depth. The amplitude of the evanescent field is reduced by $\frac{1}{e}$, called the penetration depth, which is about a unit wavelength. More discussion on this topic can be done if we consider the situation that the light is reflected from a metal surface covered with a dielectric medium (a glass prism (pr) ($\epsilon_{pr} > 1$)). Here, in the prism, x and z represent the components of a wave vector:

$$k = \frac{2\pi}{\lambda} = \frac{w}{v} = \frac{nw}{c} = \frac{w\sqrt{\epsilon}}{c} \tag{1.69}$$

Applying for kx and kz components,

$$k_x^{pr} = k_{pr} sin\theta_{pr} \tag{1.70}$$

$$k_z^{pr} = k_{pr} cos\theta_{pr} \tag{1.71}$$

Using Eq. (1.69) in Eq. (1.70) and Eq. (1.71),

$$k_x^{pr} = \sqrt{\varepsilon_{pr}}\left(\frac{w}{c}\right) sin\theta_{pr} = n_{pr}\left(\frac{w}{c}\right) sin\theta_{pr} \tag{1.72}$$

$$k_z^{pr} = \sqrt{\varepsilon_{pr}}\left(\frac{w}{c}\right) cos\theta_{pr} = n_{pr}\left(\frac{w}{c}\right) cos\theta_{pr} \tag{1.73}$$

For the prism base Kretschmann configuration, the SPR condition of the light at a metal (1) and air (2) interface is given by taking Eq. (1.72) for k_x^{sp}

$$k_x^{pr} = k_x^{sp} \tag{1.74}$$

$$\sqrt{\epsilon_{pr}}\left(\frac{w}{c}\right) sin\theta_{pr} = \left(\frac{w}{c}\right)\sqrt{\frac{\epsilon_2 \times \epsilon_1}{\epsilon_2 + \epsilon_1}} \tag{1.75}$$

1.7 MATRIX METHOD FOR REFLECTANCE OF MULTILAYERS

In this context, the multiple layers concept is taken as follows using the matrix method. The matrix method has been widely used to analyze the multilayer structure of SPR sensors. In this section, the transmission matrix has been presented to compute the reflection coefficient. Fig. 1.3 reveals the multilayer structure when the p-polarized light is incident on the first interface, resulting in the reflection and refraction at each interface.

From Snell's Law, the reflection and refraction angles are related by

$$sin\theta_0 = n_i sin\theta_i = n_t sin\theta_t \tag{1.76}$$

for i = 1, 2, 3,.........M.

Wave propagation constant β in each layer is related by following expression:

$$\beta_i = k_i l_i \tag{1.77}$$

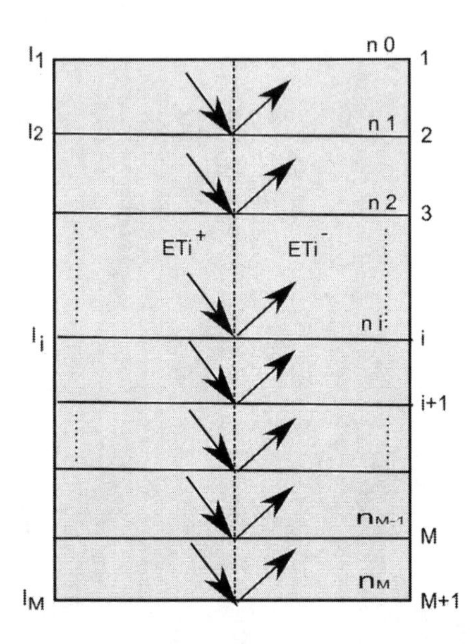

FIGURE 1.3 Multilayer reflection and transmission by thin period of thin films.

where $k_i = k_0 n_i cos\theta_i$

$k_0 = \frac{w}{c_0} = \frac{2\pi}{\lambda}$ is the free-space wave number.

By using Snell's Law, the wave propagation constant at each region can be expressed as follows:

$$\begin{aligned}
\beta_i &= k_i l_i = k_0 n_i cos\theta_i l_i \\
&= \frac{w}{c_0} n_i cos\theta_i l_i \\
&= \frac{2\pi}{\lambda} n_i cos\theta_i l_i \\
&= \frac{2\pi}{\lambda} n_i l_i \sqrt{1 - sin^2\theta_i} \\
&= \frac{2\pi}{\lambda} n_i l_i \sqrt{1 - \frac{n_0^2 sin^2\theta_o}{n_i^2}} for \ i = 1, \ 2, \ldots\ldots\ldots M
\end{aligned} \tag{1.78}$$

From Fresnel's Eq. (1.64),

$$\begin{aligned}
r_p &= \frac{n_1 \cos \theta_2 - n_2 \cos \theta_1}{n_1 \cos \theta_2 + n_2 \cos \theta_1} \\
&= \frac{\frac{n_1}{\cos \theta_1} - \frac{n_2}{\cos \theta_2}}{\frac{n_1}{\cos \theta_1} + \frac{n_2}{\cos \theta_2}}
\end{aligned} \tag{1.79}$$

Also from Eq. (1.67),

$$r_s = \frac{n_1 \cos \theta_1 - n_2 \cos \theta_2}{n_1 \cos \theta_1 + n_2 \cos \theta_2} \tag{1.80}$$

Taking into account Eq. (1.56) and Eq. (1.57), the generalized form can be written as

$$r_{T,i} = \frac{n_{T,i-1} - n_{T,i}}{n_{T,i-1} + n_{T,i}} \; for \; i = 1, \; 2, \; 3, \ldots\ldots M + 1 \tag{1.81}$$

$$n_{T,i} = \begin{cases} \dfrac{n_i}{cos\theta_i} \; TM, \\ n_i . \; cos\theta_i \; TE \end{cases}$$

for TM, parallel, p-polarized, TE, perpendicular, s-polarized and i = 0, 1, 2…M, as $T + R = 1$, for T represents transmittance and R represents reflectance we can have

$$t_{T,i} = \sqrt{T_{T,i}} \tag{1.82}$$

and $T_{T,i} = 1 - R_{T,i}$

$$\Rightarrow T_{T,i} = 1 - (r_{T,i})^2 \tag{1.83}$$

Using Eq. (1.81)

$$\begin{aligned} \Rightarrow T_{T,i} &= 1 - \left(\frac{n_{T,i-1} - n_{T,i}}{n_{T,i-1} + n_{T,i}} \right)^2 \\ \Rightarrow T_{T,i} &= \frac{(n_{T,i-1} + n_{T,i})^2 - (n_{T,i-1} - n_{T,i})^2}{(n_{T,i-1} + n_{T,i})^2} \\ \Rightarrow T_{T,i} &= \frac{(n_{T,i-1} + n_{T,i})^2 - (n_{T,i-1} - n_{T,i})^2}{(n_{T,i-1} + n_{T,i})^2} \\ \Rightarrow T_{T,i} &= \frac{(\, 2n_{T,i-1} \,)^2}{(n_{T,i-1} + n_{T,i})^2} \end{aligned} \tag{1.84}$$

So from Eq. (1.82) and Eq. (1.84),

$$\begin{aligned} t_{T,i} &= \sqrt{T_{T,i}} \\ &= \frac{2n_{T,i-1}}{n_{T,i-1} + n_{T,i}} \; for \; i = 0, \;\; 1, \;\; 2, \;\; 3 \ldots\ldots\ldots. M + 1 \end{aligned} \tag{1.85}$$

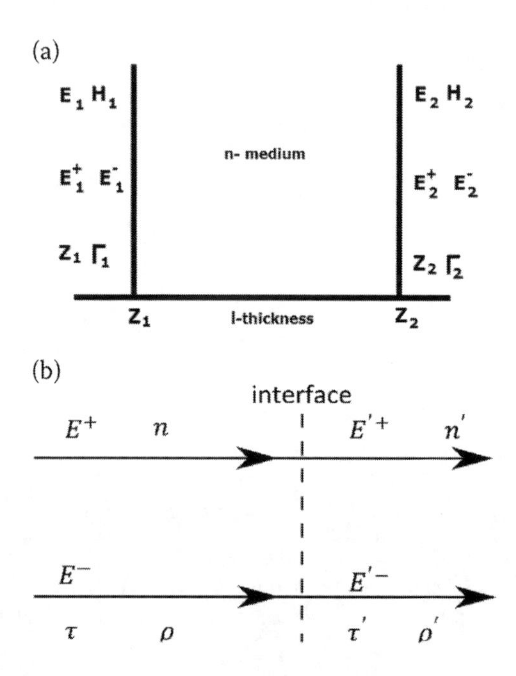

FIGURE 1.4 a) Field quantities propagated between two positions in space. b) Fields across interface (consider simple straight transmit without any bend). Reprinted with permission from IEEE Sensors Journal. Copyright, 2016, IEEE (Raghuwanshi, Kumar, and Athokpam 2016).

The reflectivity and transmittance of incident light from the multilayer structure may be calculated from the matrix multiplication of an individual matching matrix by putting in appropriate boundary conditions. The matching matrix consists of a forward/backward propagating field on each side of the interface, representing the relationship in terms of individual matrix multiplications. Field continuity conditions must be satisfied at each interface of a multilayer structure. An allowed Eigen matrix provides information on forward/backward propagation of an electromagnetic field from one interface to the other.

A transfer matrix or transition matrix is a combination of a matching matrix of relating the electromagnetic fields across the various interfaces. The concept of allowed propagation of all matching matrices is being provided here in brief in Fig. 1.4(a). For propagation considering a normal incident of wave

$$E(z) = E_O^+ e^{-jkz} + E_O^- e^{jkz} = E^+(z) + E^-(z)$$

And, $H(z) = \frac{1}{n}[E^+(z) - E^-(z)]$

Now,

$$E_1^+ = E_O^+ e^{-jkz1} = E_O^+ e^{-jk(z2-l)} = E_2^+ e^{jkl} \quad (1.86)$$

$$E_2^+ = E_O^+ e^{-jkz2}$$

Similarly,

$$E_1^- = E_2^- e^{-jkl} \tag{1.87}$$

Using Eq. (1.86) and Eq. (1.87),

$$\begin{matrix} E_1^+ \\ E_1^- \end{matrix} = \begin{bmatrix} e^{jkl} & 0 \\ 0 & e^{-jkl} \end{bmatrix} \begin{bmatrix} E_2^+ \\ E_2^- \end{bmatrix} \tag{1.88}$$

If we assume a planar interface residing at an x-y plane and propagation direction is z, the matching matrix method can be derived while matching two dielectrics and a conductive medium with characteristic impedance n and n' as represented in Fig. 1.4(b). Suppose the field is incident normally and tangential to the interface plane. The boundary condition must require that the total electric and magnetic field be continuous across the two sides of the interface. Hence

$$E = E' \Rightarrow E^+ + E^- = E'^+ + E'^- \tag{1.89}$$

and, $H = H'$

Also

$$\frac{1}{n}[E^+ - E^-] = \frac{1}{n'}[E'^+ + E'^-] \tag{1.90}$$

Now, adding Eq. (1.89) and Eq. (1.90),

$$\begin{aligned} E^+ + E^- + \frac{1}{n}[E^+ - E^-] &= E'^+ + E'^- \frac{1}{n'}[E'^+ + E'^-] \\ \Rightarrow 2E^+ &= E'^+\left(1 + \frac{n}{n'}\right) + E'^-\left(1 + \frac{n}{n'}\right) \\ \Rightarrow E^+ &= \frac{n' + n}{2n'}E'^+ + \frac{n' - n}{2n'}E'^- \end{aligned} \tag{1.91}$$

Similarly, subtracting Eq. (1.89) from Eq. (1.90),

$$\begin{aligned} 2E^- &= E'^+ + E'^- - \frac{n}{n'}(E'^+ - E'^-) \\ &= \frac{n'}{n'}(E'^+ + E'^-) - \frac{n}{n'}(E'^+ - E'^-) \end{aligned}$$

So,

$$E^- = \frac{n' - n}{2n'} E'^+ + \frac{n' + n}{2n'} E'^- \tag{1.92}$$

since $E^\pm$ (on the right) and $E'^\pm$ (on the left), respectively, are related in a matrix form as follows

$$\begin{bmatrix} E^+ \\ E^- \end{bmatrix} = \begin{bmatrix} \frac{n'+n}{2n'} & \frac{n'-n}{2n'} \\ \frac{n'-n}{2n'} & \frac{n'+n}{2n'} \end{bmatrix} \begin{bmatrix} E'^+ \\ E'^- \end{bmatrix} \tag{1.93}$$

This can be verified by Fresnel's equations; hence, new parameters are defined as

$$\tau = \frac{2n'}{n + n'}$$

$$\rho = \frac{n' - n}{n + n'}$$

So, Eq. (1.93) can be modified as

$$\begin{aligned} \begin{bmatrix} E^+ \\ E^- \end{bmatrix} &= \begin{bmatrix} \frac{1}{\tau} & \frac{\rho}{\tau} \\ \frac{\rho}{\tau} & \frac{1}{\tau} \end{bmatrix} \begin{bmatrix} E'^+ \\ E'^- \end{bmatrix} \\ \Rightarrow \begin{bmatrix} E^+ \\ E^- \end{bmatrix} &= \frac{1}{\tau} \begin{bmatrix} 1 & \rho \\ \rho & 1 \end{bmatrix} \begin{bmatrix} E'^+ \\ E'^- \end{bmatrix} \end{aligned} \tag{1.94}$$

Inversely considering, we have

$$\begin{bmatrix} E'^+ \\ E'^- \end{bmatrix} = \frac{1}{\tau'} \begin{bmatrix} 1 & \rho' \\ \rho' & 1 \end{bmatrix} \begin{bmatrix} E^+ \\ E^- \end{bmatrix} \tag{1.95}$$

where in this case

$$\tau' = \frac{2n}{n + n'}$$

$$\rho' = \frac{n - n'}{n + n'}$$

From Eq. (1.92) and Eq. (1.94) considering the case of Fig. 1.3 generalizing the case for i^{th} layers,

$$\begin{bmatrix} E_{T,i}^{+} \\ E_{T,i}^{-} \end{bmatrix} = matching\ \ matrices \times propagation\ \ matrices \times \begin{bmatrix} E_{T,i+1}^{+} \\ E_{T,i+1}^{-} \end{bmatrix}$$
$$\begin{bmatrix} E_{T,i}^{+} \\ E_{T,i}^{-} \end{bmatrix} = \frac{1}{t_{T\ i}} \begin{bmatrix} 1 & r_{T,i} \\ r_{T,i} & 1 \end{bmatrix} \begin{bmatrix} e^{i\beta_i} & 0 \\ 0 & e^{-i\beta_i} \end{bmatrix} \begin{bmatrix} E_{T,i+1}^{+} \\ E_{T,i+1}^{-} \end{bmatrix} \tag{1.96}$$

The amplitudes ratio of reflection coefficient, which is a complex quantity of a transverse electric field, is given by

$$r_{TE} = \left| \frac{E_{T,i}^{-}}{E_{T,I}^{+}} \right| \tag{1.97}$$

And so, the reflectance is given as

$$R_{TE} = |\ r_{TE}|^2 \tag{1.98}$$

The total reflectivity or reflectance R_{TM} is denoted in the case of the TM mode. The reflection coefficient, R, of an SPR sensor is given by the ratio of the reflectance as:

$$R = \frac{R_{TM}}{R_{TE}} \tag{1.99}$$

Taking a case of multiple layers with the data as provided below, the reflectance is plotted as a function of an angle for the set of the parameters. It is to be noted that in an SPR experiment, reflectance is the main parameter to be measured. From derivation, it is apparent that reflectance is dependent on which materials the stack is composed of, the thickness of the materials, as well as the angle of illumination. The reflectivity is minimum at a specific angle, which is called the SPR dip. The SPR is correlated to the material adsorbed onto the thin metal film. From Eq. (1.41), taking medium 2 as air and medium 1 as metal (gold), then

$$\epsilon_1 = \epsilon_1{}^{*} + i\epsilon_1{}^{**} \tag{1.100}$$

$$\varepsilon 2 = \epsilon 2 \tag{1.101}$$

For the SPR condition, assume

$$\epsilon_1{}^{**} < |\epsilon_1{}^{*}| \tag{1.102}$$

Put Eq. (1.64) into Eq. (1.62) to get

$$kz_1{}^2 = \left(\frac{\epsilon_1{}^2}{\epsilon_1 + \epsilon_2}\right)\left(\frac{w}{c}\right)^2 \tag{1.103}$$

$$kz_2{}^2 = \left(\frac{\epsilon_2{}^2}{\epsilon_1 + \epsilon_2}\right)\left(\frac{w}{c}\right)^2 \tag{1.104}$$

If the value of Eq. (1.100) and Eq. (1.101) are put in Eq. (1.64) and identified, the real part is

$$\mathrm{Re}(kx) = \frac{w}{c}\sqrt{\left(\frac{\epsilon_1{}^* \times \epsilon_2}{\epsilon_1{}^* + \epsilon_2}\right)} \tag{1.105}$$

Decay of the surface plasmon is along the x-direction, a large *kx* denominator of Eq. (1.105) becomes zero

$$\epsilon_1{}^* + \epsilon_2 = 0 \tag{1.106}$$

The dielectric constant of a thin metal film is given by the **Drude model** as

$$\varepsilon \equiv (n_r + in_i)^2 \equiv 1 - \frac{w_p{}^2}{w(w + i\gamma)} \tag{1.107}$$

where n_r, n_i are the real and imaginary part of RI, w is the angular frequency of the incident light, and wp and γ are the plasmon frequency and the reciprocal of relaxation time ($\approx 10^{-14}s$) of a thin metal film, respectively. Now, the real part of (1.107)

$$n_r{}^2 - n_i{}^2 = 1 - \frac{w_p{}^2}{w^2 + \gamma^2} \tag{1.108}$$

and imaginary part is

$$2n_r n_i = \frac{\gamma .\, w_p{}^2}{w(w^2 + \gamma^2)} \tag{1.109}$$

So, comparing Eq. (1.108) with Eq. (1.100),

$$\epsilon_1{}^* = 1 - \frac{w_p{}^2}{w^2 + \gamma^2} \tag{1.110}$$

At a high frequency (no damping of electrons considered) neglecting γ, i.e., $w \gg \gamma$ we get (from (1.107))

$$n^2 = 1 - \left(\frac{w_p}{w}\right)^2 \tag{1.111}$$

also

$$\epsilon_1 = n^2 = 1 - \left(\frac{w_p}{w}\right)^2 \tag{1.112}$$

The factor (w_p/w) is taken into consideration and Eq. (1.112) can be interpreted using the graph. Electromagnetic waves should have been propagating without damping when (w_p/w) is positive and real; however, it should be totally reflected when (w_p/w) is negative. Some other interpretation that can be derived from Eq. (1.89) has also been sorted out and listed below. From the observation, at optical frequencies, the dielectric constant is approximately real and negative.

$$\epsilon = 1 - \left(\frac{w_p}{w}\right)^2 \tag{1.113}$$

For $w < w_p$, the EM wave is attenuated as it propagates in the metal.

For $w > w_p$, the EM wave propagates in the metal.

For $w = w_p$, no transverse EM wave can be supported in the metal.

The electron sea oscillates longitudinally at its natural frequency. Such oscillations cannot be coupled to a propagating EM wave. So, for simple metals, the dielectric constant is given by plasma frequency as

$$\epsilon_1{}^* = 1 - \left(\frac{w_p}{w}\right)^2 \tag{1.114}$$

Now, from (1.106) and Eq. (1.114), surface plasmon frequency w_{sp} is obtained as

$$\begin{aligned} -\epsilon_2 &= 1 - \left(\frac{w_p}{w_{sp}}\right)^2 \\ \Rightarrow \epsilon_2 &= \left(\frac{w_p}{w_{sp}}\right)^2 - 1 \\ \Rightarrow 1 + \epsilon_2 &= \left(\frac{w_p}{w_{sp}}\right)^2 \\ \Rightarrow w_{sp} &= \frac{w_p}{\sqrt{1 + \epsilon_2}} \end{aligned} \tag{1.115}$$

The term wp, plasma frequency, which has been used, is given by

$$w_p = \sqrt{\frac{n_e \times e^2}{\varepsilon_0 \times m_e}}\ (in\ S.\ I.\ Unit) \tag{1.116}$$

or

$$w_p = \sqrt{\frac{4\pi n_e \times e^2}{m_e}} \quad (in\ C.\ G.\ S.\ unit)$$

where n_e is free electron density
e is electron charge
m_e is electron mass

1.8 SENSING PRINCIPLE OF SPR

At the metal-dielectric interface, the excitation of surface plasmons results in energy transfer to surface plasmons through an incident wave. The reflected light intensity reduces at some specific incident angle, better known as the resonance angle. In this connection, Fig. 1.5 reveals the reflectance versus angle of incident. In Fig. 1.5, point "A" reveals the edge of TIR. TIR occurs at a particular angle, which depends on the dielectric constants of power coupling prism and cover (sensing) medium. In fact, this angle gives a clear idea about the correct alignment or misalignment of a sample. Point "B" reveals the resonance angle of SPR. Point "B" may not often touch the baseline. The SPR dip critically depends on the thickness of metal. For the case of gold, the maximum dip occurs at 50 nm thickness. The shift of angle gives the information about the RI of analyte or a sample.

Point "C" reveals the FWHM (full width at half maximum) of SPR for a non-substrate; it is a function of imaginary value of the permittivity of the metal. In this connection, Fig. 1.6(a) shows the multilayer fiber-optic SPR sensor concept with etched cladding features. The multimode fiber has been taken for developing the

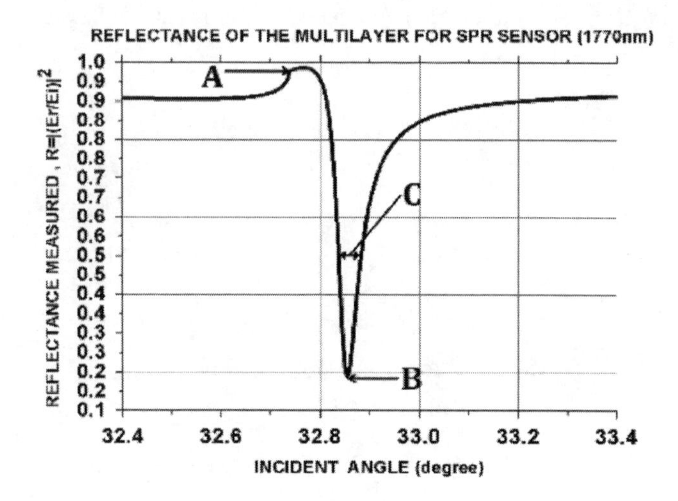

FIGURE 1.5 Reflectance (R) versus incidence (θ) at the prism/metal interface. Reprinted with permission from IEEE Sensors Journal. Copyright, 2016, IEEE (Raghuwanshi, Kumar, and Athokpam 2016).

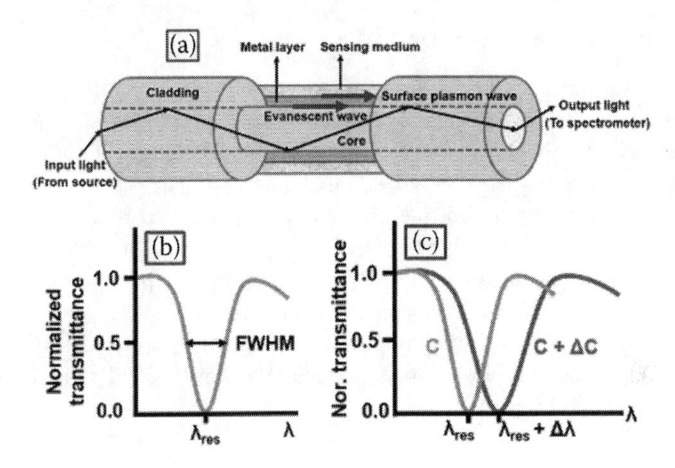

FIGURE 1.6 a) Multimode fiber-based SPR sensor with partially etched cladding features, b) their FWHM concept with c) shift of resonance wavelength due to different sensing medium refractive index. Reprinted with permission from Sensors and Actuators B: Chemical. Copyright, 2020, Elsevier (Tabassum and Kant 2020).

SPR sensor while cladding has been partially etched down by a 40% HF solution, followed by a coating of a thin metal film. Light is coupled from a white light source, followed by detection through a spectrometer. Fig. 1.6(b) shows the FWHM concept, while FWHM has a relation with width of plasmon and detection accuracy. Hence, one can determine the RI of the sensing medium by monitoring the resonance wavelength ($\Delta\lambda$) from the reflected light intensity spectrum, as revealed in Fig. 1.6(c). The spectral interrogation method is much better to cover a wider range of the sensing medium refractive. One of the main applications of the SPR phenomena is its usage in sensing purposes as SPR sensors. Surface RI changes can be precisely estimated using the intense electromagnetic field as an outcome of SPR phenomena, which is the basis of developing the highly sensitive SPR sensor devices. SPR resonance wavelength is very much sensitive to a very small perturbation of sensor surfaces RI.

The allowed eigenvalue of the wave propagation constant is estimated by the RI constant of the dielectric/metal of the given structure supporting the SP wave. A slight change of RI of a dielectric material or metal will turn out to modify the allowed propagation constant value, which will eventually change the resonance wavelength peak or angle in the reflectivity spectrum. An SPR biosensor may be defined if the outer sensing layer of an SPR sensing chip is functionalized by antibodies or any other appropriate biorecognition substances. In that process, it is quite possible to recognize and interact with specific analyte concentrations. When the right concentration (in ppm) of analyte is introduced into an SPR biosensor chip, the effective permittivity of the composite structure would be changed due to the interaction of binding between biorecognition and the analyte molecules. The change of RI would result in a change of resonance wavelength, which can be detected by a high-resolution spectrophotometer.

1.8.1 Dual-mode SPR Sensor

Two separate sensing layers with separate flow cells have been introduced in a dual-channel system of an SPR sensor. A low or high RI layer is coated on the first or second channel to shift the resonance coupling wavelength to a shorter or longer wavelength. Sometimes the second channel might be used as a reference compensation of the first channel for the biosensing experiments. Either of the channels might be functionalized with bio-molecular recognition to interact with a specific molecule. The proposed dual-mode SPR sensor chip is capable enough to distinguish a specific type of response from a nonspecific substance and bulk effects. Unlike the typical SPR sensors, dual-mode SPR sensors are supported by two different modes, namely short-range surface plasmons (SRSPs) and long-range surface plasmons (LRSPs).

1.8.2 Main Concept

Some additional layers with appropriate RI might be placed adjacent to the metal film to develop an early symmetric RI profile, which supports both anti-symmetric [short-range (SR)] modes and symmetric [long-range (LR)]. Both of the modes have some different fluid penetration capabilities in the solution; hence, these modes might be used to distinguish between interfering bulk RI changes and surface interaction of interest. In the design of the dual-mode SPR sensors, special metals like gold or silver are chosen, preferably gold. The surface plasmon waves generate at the interface of two materials with opposite dielectric permittivity. Also, two materials should have a real dielectric constant of opposite sign; hence, one should choose a metal with negative real relative RI, such as silver or gold. Also, gold is a preferred metal for an SPR sensor due to its environmental stability. One more thing in dual-channel SPR sensors, a Teflon AF layer may be introduced under the gold layer to generate an asymmetric RI profile. Since the Teflon (AF) has a dielectric constant close to water, 1.31 at a 632 nm wavelength, it is suitable for a variety of applications like petrochemical industries and biosensing applications to mention a few.

The basic concept of usage of this additional layer lies in the fact that surface plasmon modes are excited at a metal/dielectric interface. The mode can be coupled if the thin metal film is subjected to a medium with a similar dielectric constant (symmetric structure). As a result of this coupling, it leads to the short and long-range (or symmetric and anti-symmetric) surface plasmon modes. Moreover, the selection of (gold) and Teflon for optical properties as well as their thickness is crucial for velocity and attenuation of the SPR modes. The excitation technique and condition is important to determine the appropriate thickness of gold thin film and Teflon layer.

1.8.3 Sensing Response of Dual-mode SPR Sensors

The sensing response has been considered for self-referencing wavelength interrogated SPR sensors. Some of the differences between long-range surface plasmon (LRSP) and short-range surface plasmon (SRSP) are provided in Table 1.1.

TABLE 1.1
Differences Between the Long Range and Short Range Surface Plasmon

S. No.	LRSP	SRSP
1.	Its long range	Its short range
2.	It has lower absorption loss	It has higher absorption loss
3.	Symmetric mode	Anti-symmetric mode
4.	Penetrate deeply into a dielectric	More concentrated at a metal surface. Influenced more by surface binding.

From the above table, it is clear that the allowed propagation constant would be influenced by surface binding effects; hence, the short-range mode will be affected more strongly compared to the long-range mode. Self-referencing SPR sensors work on this principle of the difference in LRSP and SRSP. For dual-mode **wavelength interrogation,** SPR sensor case, short-range (SR), and long-range (LR), surface plasmon resonance excitation will result in changes of different resonance wavelength shifts due to both surface binding and background index changes. Fig. 1.7(a,b) shows one such analysis of a multilayer SPR sensor where the long-range SPR mode is quite distinguishable from the short-range SPR sensor (Raghuwanshi et al. 2018). Hence, one can utilize the two resonances to discriminate between the two phenomena.

The case of the SPR sensor range is approximately linear to both surface binding and bulk index change, then for both LR and SR, the SPR wavelength shifts $\Delta\lambda_{LR}$ and $\Delta\lambda_{SR}$, respectively given by

$$\Delta\lambda_{LR} = S_{S-LR}\Delta d + S_{B-LR}\Delta n_B \tag{1.117}$$

$$\Delta\lambda_{SR} = S_{S-SR}\Delta d + S_{B-SR}\Delta n_B \tag{1.118}$$

where S_{S-LR} and S_{S-SR} are the surface sensitivities in (nm/nm) (wavelength/thickness) for the long- and short-range modes, and S_{B-LR} and S_{B-SR} are the sensitivities due to the bulk RI in (nm/RI unit) (wavelength/RIU). The notations Δn_B and Δd are the bulk RI change and binding layer thickness change, respectively. After simulation, it is easy to get the resonance wavelength shifts and sensitivities; hence, the bulk index changes and surface layer thickness can be estimated by using Eq. (1.117) as follows:

$$\begin{gathered}\Delta\lambda_{LR} - S_{S-LR}\Delta d = S_{B-LR}\Delta n_B \\ \Rightarrow \Delta n_B = \frac{\Delta\lambda_{LR} - S_{S-LR}\Delta d}{S_{B-LR}}\end{gathered} \tag{1.119}$$

Putting this Δn_B value in Eq. (1.118),

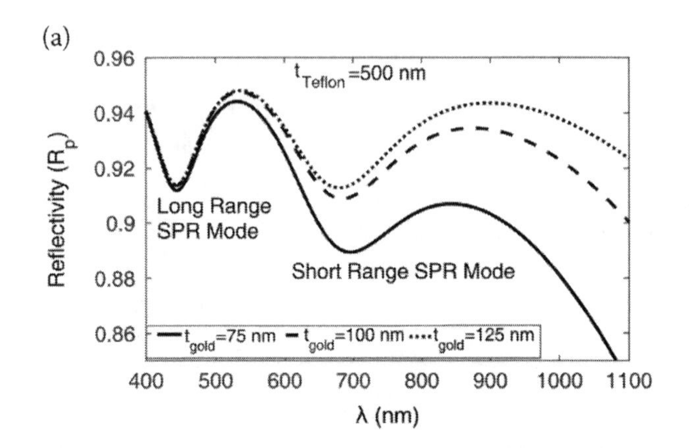

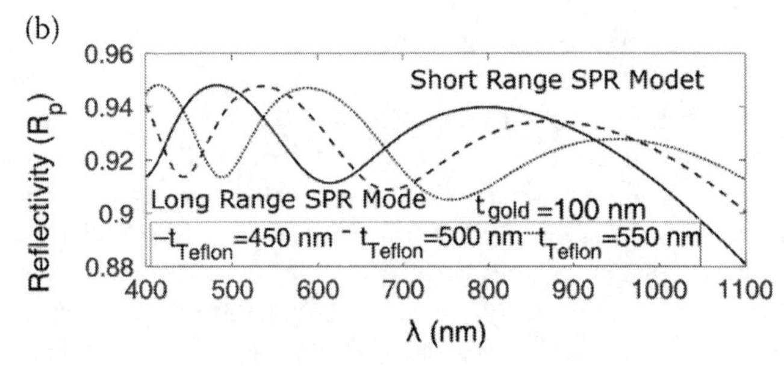

FIGURE 1.7 a, b) Variation in reflectivity (R_P) versus λ due to various thickness of gold and Teflon layers for the wavelength interrogation case to the multilayer SPR sensor system. Reprinted with permission from IEEE Transactions on Instrumentation and Measurement. Copyright, 2018, IEEE (Raghuwanshi and Kumar 2018).

$$\begin{aligned}
\Delta\lambda_{SR} &= S_{S-SR}\Delta d + S_{B-SR}\Delta n_B\left(\frac{\Delta\lambda_{LR} - S_{S-LR}\Delta d}{S_{B-LR}}\right)\\
\Delta\lambda_{SR} &= \Delta d\left[\frac{S_{S-SR}S_{B-LR} - S_{B-SR}S_{S-LR}}{S_{B-LR}}\right] + \frac{S_{B-SR}}{S_{B-LR}}\Delta\lambda_{LR}\\
&\Rightarrow \Delta\lambda_{SR} - \frac{S_{B-SR}}{S_{B-LR}}\Delta\lambda_{LR} = \Delta d\left[\frac{S_{S-SR}S_{B-LR} - S_{B-SR}S_{S-LR}}{S_{B-LR}}\right]\\
&\Rightarrow S_{B-LR}\Delta\lambda_{SR} - S_{B-SR}\Delta\lambda_{LR} = \Delta d\left[S_{S-SR}S_{B-LR} - S_{B-SR}S_{S-LR}\right]\\
&\Rightarrow \Delta d = \frac{S_{B-LR}\Delta\lambda_{SR} - S_{B-SR}\Delta\lambda_{LR}}{S_{S-SR}S_{B-LR} - S_{B-SR}S_{S-LR}}
\end{aligned} \tag{1.120}$$

Similarly, from Eq. (1.94),

$$\begin{aligned}
&\Delta\lambda_{LR} - S_{S-LR}\Delta d = S_{B-LR}\Delta n_B\\
&\Rightarrow \Delta\lambda_{LR} - S_{B-LR}\Delta n_B = S_{S-LR}\Delta d\\
&\Rightarrow \frac{\Delta\lambda_{LR} - S_{B-LR}\Delta n_B}{S_{S-LR}} = \Delta d
\end{aligned}$$

Putting Δd in Eq. (1.95),

$$\begin{aligned}\Delta\lambda_{SR} &= S_{S-SR}\left[\frac{\Delta\lambda_{LR} - S_{B-LR}\Delta n_B}{S_{S-LR}}\right] + S_{B-SR}\Delta n_B \\ &\Rightarrow \Delta\lambda_{SR}S_{S-LR} = S_{S-SR}\Delta\lambda_{LR} - S_{B-LR}\Delta n_B + S_{S-LR}S_{B-SR}\Delta n_B \\ &\Rightarrow \Delta\lambda_{SR}S_{S-LR} = S_{S-SR}\Delta\lambda_{LR} - S_{S-SR}S_{B-LR}\Delta n_B + S_{S-LR}S_{B-SR}\Delta n_B \\ &\Rightarrow \Delta\lambda_{SR}S_{S-LR} - S_{S-SR}\Delta\lambda_{LR} = -S_{S-SR}S_{B-LR}\Delta n_B + S_{S-LR}S_{B-SR}\Delta n_B \\ &\Rightarrow \Delta n_B = \frac{S_{S-LR}\Delta\lambda_{SR} - S_{S-SR}\Delta\lambda_{LR}}{S_{B-SR}\,S_{S-LR} - S_{S-SR}S_{B-LR}}\end{aligned} \tag{1.121}$$

Hence, by knowing two different resonance wavelengths, one can separately determine the two unknowns of interest: equivalently fractional coverage or surface layer thickness change and bulk index change. It is a matter of fact that, in some experiments, the nonlinear effect cannot be ignored where there are bulk index changes in large quantities or surface concentrations. In such cases, resonance wavelengths are the function of the bound layer property and background index changes and we must consider the more complicated relationship as follows:

$$\Delta\lambda_{LR} = S_{S-LR}\Delta d + S_{B-LR}\Delta n_B + S_{SB-LR}\Delta d\Delta n_B \tag{1.122}$$

$$\Delta\lambda_{SR} = S_{S-SR}\Delta d + S_{B-SR}\Delta n_B + S_{SB-SR}\Delta d\Delta n_B \tag{1.123}$$

The surface and bulk sensitivities in the last equation will not be the same as those found for the linear model because the basic functions (Δd, Δn_B) and (Δd, Δn_B) are not orthogonal. Similarly, for the case of the **angular interrogation** scheme, the resonance angle shift as a function of the sensitivities $\Delta\theta_{LR}$ and $\Delta\theta_{SR}$ can be expressed respectively as

$$\Delta\theta_{LR} = S_{S-LR}\Delta d + S_{B-LR}\Delta n_B \tag{1.124}$$

$$\Delta\theta_{SR} = S_{S-SR}\Delta d + S_{B-SR}\Delta n_B \tag{1.125}$$

From Eq. (1.101),

$$\begin{aligned}\Delta\theta_{LR} - S_{S-LR}\Delta d &= S_{B-LR}\Delta n_B \\ \Delta n_B &= \frac{\Delta\theta_{LR} - S_{S-LR}\Delta d}{S_{B-LR}}\end{aligned} \tag{1.126}$$

Putting Eq. (1.126) into Eq. (1.125),

$$\begin{aligned}\Delta\theta_{SR} &= S_{S-SR}\Delta d + S_{B-SR}\left(\frac{\Delta\theta_{LR} - S_{S-LR}\Delta d}{S_{B-LR}}\right)\\ &\Rightarrow \Delta\theta_{SR}S_{B-LR} = S_{S-SR}S_{B-LR}\Delta d + S_{B-SR}\Delta\theta_{LR} - S_{S-LR}S_{B-SR}\Delta d\\ &\Rightarrow \Delta\theta_{SR}S_{B-LR} - S_{B-SR}\Delta\theta_{LR} = S_{S-SR}S_{B-LR}\Delta d - S_{S-LR}S_{B-SR}\Delta d\\ &\Rightarrow \Delta d = \frac{\Delta\theta_{SR}S_{B-LR} - S_{B-SR}\Delta\theta_{LR}}{S_{S-SR}S_{B-LR} - S_{S-LR}S_{B-SR}}\end{aligned} \tag{1.127}$$

It can be also written as

$$\Delta d = \frac{\frac{\Delta\theta_{LR}}{S_{B-LR}} - \frac{\Delta\theta_{SR}}{S_{B-SR}}}{\frac{S_{S-LR}}{S_{B-LR}} - \frac{S_{B-SR}}{S_{S-SR}}} \tag{1.128}$$

Similarly for the Δn_B,

$$\Delta n_B = \frac{\frac{\Delta\theta_{LR}}{S_{S-LR}} - \frac{\Delta\theta_{SR}}{S_{S-SR}}}{\frac{S_{B-LR}}{S_{S-LR}} - \frac{S_{B-SR}}{S_{S-SR}}} \tag{1.129}$$

The bulk index change and surface layer thickness change may be easily determined by knowing the resonance wavelength or angle shift of two SPRs.

1.9 PERFORMANCE PARAMETERS OF THE SPR SENSOR

In the case of fiber-optic SPR sensors, as shown in Fig. 1.6(a), the broadband source has been used to excite the multimode structure at a particular angle of incident. Hence, the output light detected at the spectrophotometer is a function of wavelength. This case is called the wavelength interrogation. Contrary to this, Fig. 1.8(b) shows the prism-based SPR sensor where a single wavelength light is incident to a high refractive prism and collected light is a function of the incident angle, which is called the angular interrogation of the SPR sensor, while the resonance angle (θ_{res}) is highly sensitive to the surrounding medium RI. A slight change of the surrounding medium RI δn_d changes the resonance from θ_{res_1} to θ_{res_2}, as revealed in Fig. 1.8(d,e). For the wavelength interrogation case, this shift corresponds to $\Delta\lambda$, as shown in Fig. 1.6(c). Sensitivity is one of the important parameters for an SPR-based sensor for either of the interrogation cases. It should be as high as possible.

Thus, an SPR sensor's sensitivity depends on the resonance angle shift or the resonance wavelength due to a change in the surrounding medium's RI. Sensitivity would be better for a significant shift in resonance wavelength. Resonance wavelength shift irrespective to change of surrounding or sensing medium RI is shown in Fig 1.6(c) and Fig. 1.8(d), respectively.

Two SPR curves are plotted for two different values of the RI of the sensing medium to define the sensitivity of an SPR-based sensor, as shown in Fig. 1.8(d).

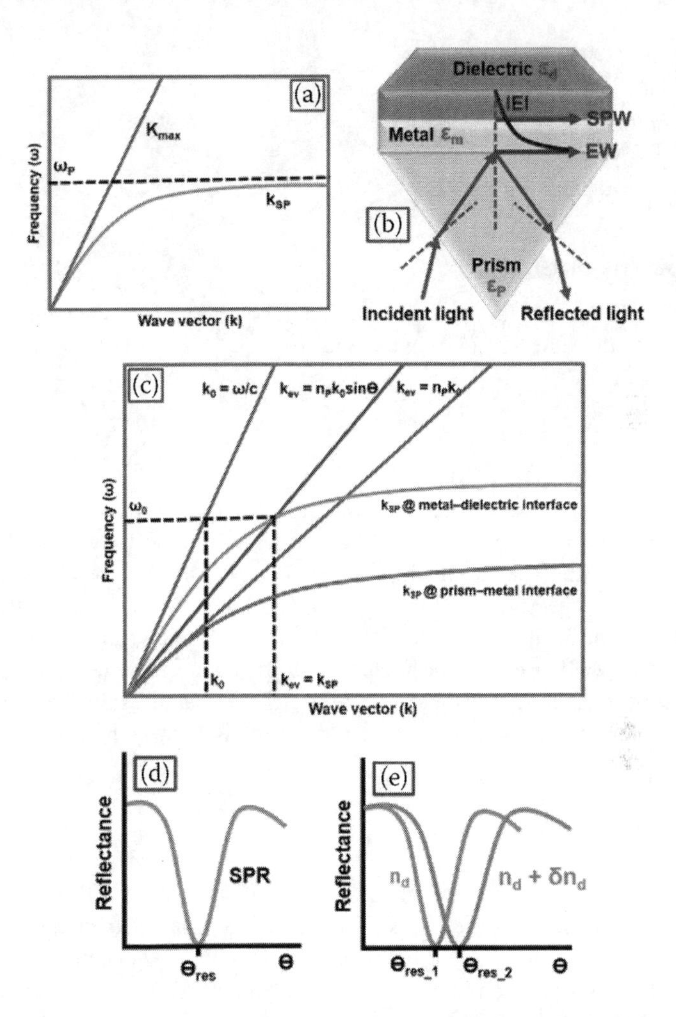

FIGURE 1.8 a) Dispersion graph for direct light illumination of SPPs. b) Prism excitation of SPPs in Kretschmanngeometry, c) dispersion graph of evanescent and SPPs waves at various interfaces d) and e) shift of resonance angle ($\delta\theta_{res}$) due to a change in concentration of sensing medium. Reprinted with permission from Sensors and Actuators B: Chemical. Copyright, 2020, Elsevier (Tabassum and Kant 2020).

On changing the RI of the sensing medium by an amount δn_d, the resonance angle also shifts by an amount $\delta\theta_{res} = \theta_{res\_2} - \theta_{res\_1}$. Hence, the sensitivity of an SPR-based fiber-optic sensor is defined as:

$$S = \frac{\delta\theta_{res}}{\delta n_d} \tag{1.130}$$

Some of the performance characteristics of SPR sensors worth mentioning are:

1. Sensitivity
2. Linearity
3. Resolution
4. Accuracy
5. Limit of detection (LOD)

1.9.1 Sensor Sensitivity

Suppose the change in sensor output is dY, corresponding to a change in measuring quantities dX, then the sensor S is defined by

$$S = \frac{dY}{dX} \tag{1.131}$$

The sensitivity can be calculated in terms of intensity change of an SPR spectrum measured in the form (Amplitude/RIU). Similarly, for the case of a phase interrogation scheme, the change of phase of amplitude reflection coefficient has been calculated with respect to the change of sensing layer RI measured in the form of (Degree/RIU), this scheme is quite different from the angular interrogation case. In fact, the effective RI theory can be applied to calculate the change of mode field pattern with a change of sensing layer dielectric constant. The particular incident value for which the dip is maximum or reflectance is minimum is considered to be the *coupling angle.*

1.9.2 Linearity

It defines the linear nature between the sensor output and the actual quantity to be measured. The linearity nature is taken to simplify the understanding of the sensor response while neglecting the non-linear nature. However, to get a more elaborate and accurate calibration, the non-linear nature also should be taken into consideration.

If the linear model of the sensor response is taken into account, then the sensitivity can be found by using a slope of the linear curve.

1.9.3 Resolution

It is the ratio of maximum change in output signal measured to the minute change of the input quantity measurand. The sensor resolution is the smallest change that can be predictably detected in the quantity measured. The main factor determining the sensor resolution is the noise in the sensing output. In an SPR sensor, it is generally the lowest change in the sensing layer RI, which can be an appreciable shift in the output.

1.9.4 Accuracy

Measurement is an essential thing to estimate the usefulness of SPR sensors in various fields like petrochemical, bio-sensing, pharmaceutical, and chemical industries to mention a few. In general, accuracy is a statistical quantity that denotes the closeness of estimates to the true values. In other words, an SPR sensor can measure the value accurately, known as accuracy. There is no guarantee that a high-resolution SPR-sensing chip may have better or poorer accuracy in specific experiments.

1.9.5 Limit of Detection

A slight alteration in bulk RI can be usually detected by using a high-resolution SPR-sensing chip. In another way, on the other hand, the limit of detection (LOD) is again a statistical quantity that depends on the sensor response in some experiments performed many times. To be precise, it should be repeated a minimum of three times, for a standard deviation of sensor response to a fixed concentration of an analyte. It is a matter of fact that the LOD can be more precisely determined if and only if the parameters of an analyte, mass transport phenomena on sensor surface, and biorecognition elements are known. In mathematical form, the LOD can be defined by knowing the sensor response for the blank sample and the three standard deviations of a blank sample represented by

$$Y_{LOD} = Y_{blank} + 3SD_{blank} \tag{1.132}$$

where Y_{blank} is the response of the sensor for a blank sample or analyte. If we assume that the blank signal is negligible, then the LOD will be better represented to

$$Y_{LOD} = 3SD_{blank} \tag{1.133}$$

So, the LOD can also be taken as a minimum concentration of a sample that can be estimated. The LOD should be low enough for any optimized SPR sensor. SPR sensors are popularly used due to several advantages, compared to the old or traditional biosensing process. Some prominent ones are listed below:

1. An SPR sensor is highly efficient as an RI transduction element to justify the changes on a sensor surface.
2. The antibody-antigen binding effect on a sensor surface has been successfully studied with high sensitivity due to an enhanced electric field associated with an interaction region.
3. An SPR-sensing chip has been designed and tested for real-time detection and in situ monitoring purposes with tremendous speed.

Despite the presence of such an advantage, one main problem associated with standard SPR sensors comes in the form of their inability to distinguish between a bulk refractive index change from a specific chemical or biological interaction on the surface of the sensor. The desirable one is the surface interaction. In order to tackle this problem, one of the research areas of SPR sensors came up with a self-referencing dual-mode SPR sensing system that can distinguish surface interaction from interfering bulk RI changes, which is discussed in this chapter.

1.10 OUTLINE OF BOOK

This book deals with the recent advances in various plasmonics sensing techniques by theoretical and experimental detection schemes, their realization, configuration, components employed, and application of sensors with 2D materials. The total coverage of the book encompasses the plasmonic phenomenon and its application in optical fiber-based sensing, broadly distributed over eight chapters. Chapters distributed the possible configuration, and impact of the several 2D materials in the SPR sensors.

In Chapter 1, the introduction of plasmons and plasmonics sensors is illustrated with their importance, techniques, principle, and performance in brief. In Chapter 2, the fundamental properties of 2D materials like band gap structure, characterization, and surface morphology, etc., are discussed in detail. Chapter 3 of this book deals with the theoretical design and analysis of light properties for different media. Basically, this chapter exploded surface plasmon resonance (SPR) chemical sensors with various configurations assisted by 2D materials followed by discussion on excitation of surface plasmons in different 2D materials. In Chapter 4, black phosphorus (BP) and phosphorene introduction, history, characterization, optical properties, and application of BP in surface plasmon resonance (SPR) sensors is explained. In Chapter 5, another 2D material, MXene, is introduced, along with its importance and practical application for the surface plasmon resonance sensing. In Chapter 6, theoretical and experimental analysis of an SPR sensor assisted by single-wall carbon nanotubes (SWCNT) is discussed, followed by their introduction, growth method of thin layer, electronic and optical properties of SWCNT structure, and applications and advantages over other material based sensor. Chapter 7 discusses recent developments of 2D material in the context of recent trends of SPR sensors, specifically targeting TMDC (Transition Metal Dichalcogenides) material. This chapter discusses their history, challenges, classification, and application. In Chapter 8, application of SPR sensors for clinical diagnostics (detection of glucose, uric acid, ascorbic acid, protein, DNA/RNA, microorganisms) is explained. Chapter 9 deals with the future trends for the SPR technology, like physical sensing biomedical industries, health monitoring, and mining industries is discussed herein.

This book contains a total of nine chapters and explains the emerging trends of some novel 2D materials in the context of surface plasmon resonance based sensors. It provides the opportunity to cover up-to-date knowledge of theoretical and experimental means of recent advancements of 2D materials in a very lucid manner. Indeed, this book covers the discussion of a broad range of 2D materials, providing the opportunity to readers to think about the new novel designs of 2D materials for further research work and for more practical and feasible applications. This chapter's results are well verified and it will be helpful in industry and academic areas for better understanding.

REFERENCES

Bhatia, Priya, and Banshi D. Gupta. 2011. "Surface-Plasmon-Resonance-Based Fiber-Optic Refractive Index Sensor: Sensitivity Enhancement." *Applied Optics* 50 (14): 2032–6.

Feng, Wei-Yi, Nan-Fu Chiu, Hui-Hsin Lu, Hsueh-Ching Shih, Dongfang Yang, and Chii-Wann Lin. 2008. "Surface Plasmon Resonance Biochip Based on ZnO Thin Film for Nitric Oxide Sensing." *Journal of Bionanoscience* 2 (1): 62–6.

Gupta, Banshi D., and Anuj K. Sharma. 2005. "Sensitivity Evaluation of a Multi-Layered Surface Plasmon Resonance-Based Fiber Optic Sensor: A Theoretical Study." *Sensors and Actuators B: Chemical* 107 (1): 40–6.

Gupta, Banshi D., and Rajneesh Kumar Verma. 2009. "Surface Plasmon Resonance-Based Fiber Optic Sensors: Principle, Probe Designs, and Some Applications." *Journal of Sensors*: 1–2.

Herrera, Natalia D'iaz, Óscar Esteban, Mar'ia-Cruz Navarrete, Agust'in González-Cano, Elena Benito-Peña, and Guillermo Orellana. 2011. "Improved Performance of SPR Sensors by a Chemical Etching of Tapered Optical Fibers." *Optics and Lasers in Engineering* 49 (8): 1065–8.

Homola, Jivrí, Sinclair S. Yee, and Günter Gauglitz. 1999. "Surface Plasmon Resonance Sensors." *Sensors and Actuators B: Chemical* 54 (1–2): 3–15.

Huang, Y. H., Ho Pui Ho, Siu Kai Kong, and Andrei V. Kabashin. 2012. "Phase-sensitive surface plasmon resonance biosensors: methodology, instrumentation and applications." *Annalen de Physik (Berlin)* 524 (11): 637–662.

Kedenburg, Stefan, Marius Vieweg, Timo Gissibl, and Harald Giessen. 2012. "Linear Refractive Index and Absorption Measurements of Nonlinear Optical Liquids in the Visible and Near-Infrared Spectral Region." *Optical Materials Express* 2 (11): 1588–611.

Liedberg, Bo, Claes Nylander, and Ingemar Lundström. 1995. "Biosensing with Surface Plasmon Resonance—How It All Started." *Biosensors and Bioelectronics* 10 (8): i–ix.

Lin, Chii-Wann, Kuo-Ping Chen, Min-Chi Su, Tze-Chien Hsiao, Sue-Sheng Lee, Shiming Lin, Xue-jing Shi, and Chih-Kung Lee. 2006. "Admittance Loci Design Method for Multilayer Surface Plasmon Resonance Devices." *Sensors and Actuators B: Chemical* 117 (1): 219–29.

Maharana, Pradeep Kumar, and Rajan Jha. 2012. "Chalcogenide Prism and Graphene Multilayer Based Surface Plasmon Resonance Affinity Biosensor for High Performance." *Sensors and Actuators B: Chemical* 169: 161–6.

Maharana, Pradeep Kumar, Rajan Jha, and Srikanta Palei. 2014. "Sensitivity Enhancement by Air Mediated Graphene Multilayer Based Surface Plasmon Resonance Biosensor for near Infrared." *Sensors and Actuators B: Chemical* 190: 494–501.

Mishra, Akhilesh K, Satyendra K Mishra, and Banshi D Gupta. 2015. "Gas-Clad Two-Way Fiber Optic SPR Sensor: A Novel Approach for Refractive Index Sensing." *Plasmonics* 10 (5): 1071–6.

Motogaito, Atsushi, Shohei Nakamura, Jyun Miyazaki, Hideto Miyake, and Kazumasa Hiramatsu. 2015. "Using Surface-Plasmon Polariton at the GaP-Au Interface in Order to Detect Chemical Species in High-Refractive-Index Media." *Optics Communications* 341: 64–8.

Ouyang, Qingling, Shuwen Zeng, Li Jiang, Liying Hong, Gaixia Xu, Xuan-Quyen Dinh, Jun Qian, et al. 2016. "Sensitivity Enhancement of Transition Metal Dichalcogenides/ Silicon Nanostructure-Based Surface Plasmon Resonance Biosensor." *Scientific Reports* 6 (1): 1–13.

Patskovsky, Sergiy, Souleymane Bah, Michel Meunier, and Andrei V Kabashin. 2006. "Characterization of High Refractive Index Semiconductor Films by Surface Plasmon Resonance." *Applied Optics* 45 (25): 6640–5.

Patskovsky, Sergiy, Andrei V. Kabashin, Michel Meunier, and John H. T. Luong. 2003. "Silicon-Based Surface Plasmon Resonance Sensing with Two Surface Plasmon Polariton Modes." *Applied Optics* 42 (34): 6905–9.

Politano, Antonio, and Gennaro Chiarello. 2014. "Plasmon Modes in Graphene: Status and Prospect." *Nanoscale* 6 (19): 10927–40.

Politano, A., V. Formoso, and G. Chiarello. 2010. "Plasmonic Modes Confined in Nanoscale Thin Silver Films Deposited onto Metallic Substrates." *Journal of Nanoscience and Nanotechnology* 10 (2): 1313–21.

Pyshkin, S. L. 2013. "Gallium Phosphide – New Prospect for Optoelectronics." *Advances in Optoelectronic Materials* 1 (4): 59–66.

Raghuwanshi, Sanjeev K., Manish Kumar, and Bidhanshel Singh Athokpam. 2016. "Analysis of Novel Class of Surface Plasmon Phenomena Having a Metamaterial Layer between Two Different Metals for Sensor Application." *IEEE Sensors Journal* 16 (17): 6617–24.

Raghuwanshi, Sanjeev Kumar, and Manish Kumar. 2018. "Highly Dispersion Tailored Property of Novel Class of Multimode Surface Plasmon Resonance Biosensor Assisted by Teflon and Metamaterial Layers." *IEEE Transactions on Instrumentation and Measurement* 68 (8): 2954–63.

Schasfoort, Richard B. M. 2017. *Handbook of Surface Plasmon Resonance*. Burlington House, Piccadilly, London: Royal Society of Chemistry.

Shalabney, Atef, and Ibrahim Abdulhalim. 2010. "Electromagnetic Fields Distribution in Multilayer Thin Film Structures and the Origin of Sensitivity Enhancement in Surface Plasmon Resonance Sensors." *Sensors and Actuators A: Physical* 159 (1): 24–32.

Sharma, Anuj K., and B. D. Gupta. 2007. "On the Performance of Different Bimetallic Combinations in Surface Plasmon Resonance Based Fiber Optic Sensors." *Journal of Applied Physics* 101 (9): 93111.

Tabassum, Rana, and Ravi Kant. 2020. "Recent Trends in Surface Plasmon Resonance Based Fiber-Optic Gas Sensors Utilizing Metal Oxides and Carbon Nanomaterials as Functional Entities." *Sensors and Actuators B: Chemical* 310: 127813.

Urbonas, Darius, Armandas Balčytis, Martynas Gabalis, Konstantinas Vaškevičius, Greta Naujokait.e, Saulius Juodkazis, and Raimondas Petruškevičius. 2015. "Ultra-Wide Free Spectral Range, Enhanced Sensitivity, and Removed Mode Splitting SOI Optical Ring Resonator with Dispersive Metal Nanodisks." *Optics Letters* 40 (13): 2977–80.

Verma, Rajneesh K., and Banshi D. Gupta. 2010. "Surface Plasmon Resonance Based Fiber Optic Sensor for the IR Region Using a Conducting Metal Oxide Film." *Journal of the Optical Society of America A* 27 (4): 846–51.

Wakamatsu, Takashi, and Kazuhiro Saito. 2007. "Interpretation of Attenuated-Total-Reflection Dips Observed in Surface Plasmon Resonance." *Journal of the Optical Society of America B* 24 (9): 2307–13.

Wang, Shuang, Benjamin D. Weil, Yanbin Li, Ken Xingze Wang, Erik Garnett, Shanhui Fan, and Yi Cui. 2013. "Large-Area Free-Standing Ultrathin Single-Crystal Silicon as Processable Materials." *Nano Letters* 13 (9): 4393–8.

Wu, Leiming, Yue Jia, Leyong Jiang, Jun Guo, Xiaoyu Dai, Yuanjiang Xiang, and Dianyuan Fan. 2016. "Sensitivity Improved SPR Biosensor Based on the MoS2/Graphene–Aluminum Hybrid Structure." *Journal of Lightwave Technology* 35 (1): 82–7.

Zeng, Shuwen, Dominique Baillargeat, Ho-Pui Ho, and Ken-Tye Yong. 2014. "Nanomaterials Enhanced Surface Plasmon Resonance for Biological and Chemical Sensing Applications." *Chemical Society Reviews* 43 (10): 3426–52.

Zhao, Xiao, Xian Zhang, Xiao-Song Zhu, and Yi-Wei Shi. 2019. "Long-Range Surface Plasmon Resonance Sensor Based on the GK570/Ag Coated Hollow Fiber with an Asymmetric Layer Structure." *Optics Express* 27 (7): 9550–60.

Zheng, Gaige, Xiujuan Zou, Yunyun Chen, Linhua Xu, and Weifeng Rao. 2017. "Fano Resonance in Graphene-MoS2 Heterostructure-Based Surface Plasmon Resonance Biosensor and Its Potential Applications." *Optical Materials* 66: 171–8.

2 Fundamental Optical Properties of 2D Materials

2.1 INTRODUCTION

Two-dimensional (2D) materials are crystalline solids that are made up of a single layer of atoms. 2D materials have length and width in centimeters, but their thickness is insignificant in comparison to their length and width, so they are referred to as 2D materials. Examples of 2D materials (Jayakumar, Surendranath, and Mohanan 2018) are transition metal dichalcogenides (TMDCs), graphene, layered double hydroxides, laponite clay, gC_3N_4, black phosphorous (BP), and hexagonal boron nitride (h-BN), as shown in Fig. 2.1.

These 2D materials are important in today's optical research, such as sensor growth, nano-tube development, and various types of biochemical sensing devices. With peculiar electrical, mechanical, thermal, and optical properties that vary from their bulk counterparts, 2D materials have become the most advanced area of research during the last few years (Jayakumar, Surendranath, and Mohanan 2018). In graphene, charge carriers are defined as the massless Dirac fermions, which have a mobility of 10,000 V cm^{-2} s^{-1}, which is quite high (Zhang and Cheung 2016). In the monolayer state, spin-valley coupling transforms TMDCs from indirect to direct band gap semiconductors (Late, Bhat, and Rout 2019; You, Bongu, Bao, and Panoiu 2018). Nanosheets of titanium oxide ($Ti_{0.87}O_2$) have a high electronic permittivity and dielectric constant, making them ideal for development of today's required electronics equipment. In 2D thin layers, the quantum confinement effect is major, and are highly dependent on composition and layer thickness. Furthermore, external fields, strain engineering, and chemical doping can all be utilized to fine-tune the 2D material properties, allowing for precise monitoring of their properties.

Silicon and silica optical systems have been the materials of choice for visible and near-infrared optical applications. However, these materials lack a direct band gap, and their crystal symmetry limits second-order optical nonlinearity. In order to incorporate these properties into next-generation system-on-a-chip systems, other materials must be incorporated into the system. To remove propagating wave perturbations in the current method, these materials should be made as small as possible, effectively minimizing light-matter interaction. New classes of 2D materials have recently been discovered with exceptional optical properties, such as exceptional nonlinear susceptibility and atomic thickness, making them excellent candidate materials to integrate into optical systems lacking these properties.

When compared to traditional bulk materials, 2D materials have gotten a lot of attention because of their exceptional electrical properties and outstanding physical

DOI: 10.1201/9781003190738-2

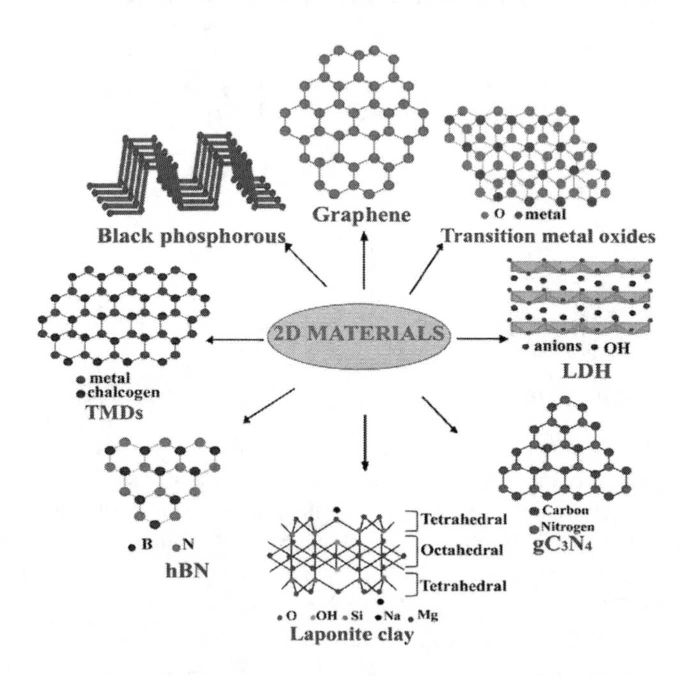

FIGURE 2.1 Different 2D material structures and their optical properties lay, hBN, and black phosphorous have various structures. Reprinted with permission from International Journal of Pharmaceutics Journal. Copyright, 2018, Elsevier (Jayakumar, Surendranath, and Mohanan 2018).

properties (such as high Young's modulus, ultralow weight, and high strength) (Zhang and Cheung 2016). The most widely studied 2D material in recent years has been graphene, which has the highest calculated Young's modulus (1 TPa). Graphene-filled polymer matrices have been shown in studies to greatly improve the composites' mechanical properties. Whether, pristine graphene limited with bandgap, that also which restricts use in area that require a semiconducting content. TMDCs and BP have an intrinsic bandgap that has the potential to replace graphene as a potential replacement material for optoelectronics and electronics applications and widen the scope of 2D materials research. Furthermore, the presence of a more sensitive piezoresistive effect in TMDCs under mechanical deformation in comparison to graphene makes them more appealing for novel applications such as tactile strain sensors nanogenerators and advanced nano-electro-mechanical systems (NEMS). Since 2D materials have fewer crystal defects and interlayer stacking faults, their Young's modulus is higher than that of the corresponding bulk materials.

Other recently discovered 2D materials, such as gallium selenide (GaSe), BP, h-BN, TMDCs such as WS_2, WSe_2, MoS_2, $MoSe_2$, $MoTe_2$, TiS_2, and perovskites have broadened the basic functionalities and properties. Due to their ultrafast response, and high optical nonlinearity, these materials are used in saturable absorbers and all-optical modulators used in wavelength converters, optical limiters, and Q-switching and passive mode-locking. These 2D materials in nonlinear optics have broadband and tunable optical absorption, ultrafast nonlinear optical response, ultrafast recovery time, large thermal and optical damage threshold, low fabrication cost, and high mechanical and chemical stability (You, Bongu, Bao, and Panoiu 2018).

Graphene is currently a very common 2D material in the research field due to its flexible properties. It is an important material of the 2D material family, and it is widely used in sensor systems, experimental validation of condensed matter physics problems, and innovative electronics. It appears like a honeycomb lattice of carbon with a single-atom thickness, that further gives the appearance shown in Fig. 2.2 (Late, Bhat, and Rout 2019).

A 2 × 2 matrix structure used to describe it is analogous to the massless Dirac equation (You et al. 2018). As 2D materials crystallize into graphite-like structures, however, they exhibit significant anisotropy in mechanical and electrical properties.

The monolayer TMDCs have wide applications in computers due to a direct bandgap nature. The electrostatic tunability of excitons in 2D and high binding energy led to the study of excitons and their subsequent function in many-body phenomena and strong Coulomb interaction (You, Bongu, Bao, and Panoiu 2018).

Density functional theory (DFT) calculations give a clear band difference in the monolayer of WSe_2 and MoS_2 at the hexagonal BZ's corners. The symmetry is the main distinction between bulk crystals and monolayers. The band structure changes can be observed by angle-resolved photoemission spectroscopy in various 2D materials, as shown in Fig. 2.3.

Bulk materials need to be thinned down to tune the bandgap for improved electronic and optical properties. Instead of modifying synthetic paths, which raises the risk of impurities, this was accomplished in 2D materials by monitoring external influences. The interplanar interaction enable the bandgap to be opened by turning the electric field, which is impossible in bulk materials. In the case of bilayer graphene, researchers theorized that changes in the electric field or the introduction of dopants will open a direct bandgap of value 250 meV. However, when the theoretical and experimental findings were compared, this did not quite match the forecast. The optical bandgap was estimated to be about 250 meV, whereas the electrical bandgap was found to be 130 meV (Zhang et al. 2009).

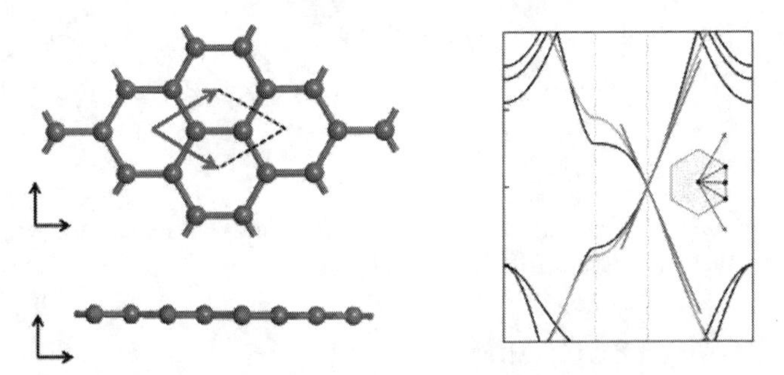

FIGURE 2.2 Atomic structure (A) and energy bands (B) of graphene. Reprinted with permission from Fundamentals and Sensing Applications of 2D Materials. Copyright, 2019, Elsevier (Late, Bhat, and Rout 2019).

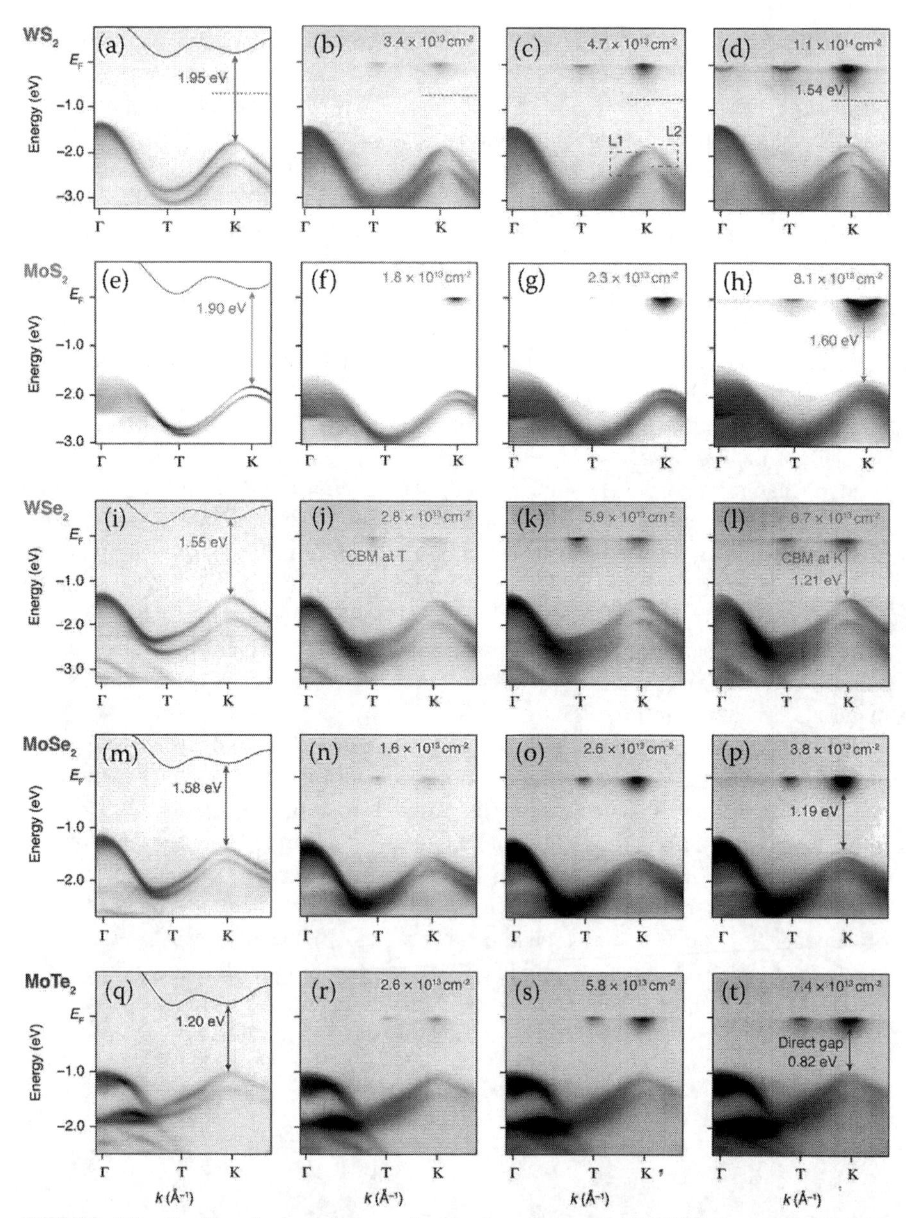

FIGURE 2.3 Surface doping changes the band structure of 2H TMDCs. Reprinted with permission from Fundamentals and Sensing Applications of 2D Materials. Copyright, 2019, Elsevier (Late, Bhat, and Rout 2019).

2.2 BANDGAP STRUCTURES

The basis of many essential features like electronics, computing, communication, optoelectronics, and sensing are semiconductors. The introduction of the point-contact transistor in 1947 can be traced to modern semiconductor technology. The demonstration opened the way to creating integrated and discrete semiconductor equipment, and systems that help the semiconductors are all-embracing elements of day-to-day life. This demonstration of the bandgap is a crucial characteristic to establish the electrical and optical semiconductor properties (Chaves et al. 2020). Beyond graphene, 2D materials have recently been identified with semi-conductive bandgaps in two-layer graphene and BP from terahertz and mid-infrared, as visible in the metal transition dichalcogenide, to ultraviolet h-BN. Almost all modern semiconductor devices use band-structure configuration by using heterostructures (Kroemer 2001), superlattices (Esaki and Tsu 1970), or other effects. More importantly, due to the very high surface-to-volume ratios of 2D material layers, its belt structure is very responsive to changes the external disturbances that changes the electronic structure and band gaps. Materials with a periodic dielectricity profile are photonic bandgap (PBG), which prevents light from spreading in one, two, or a number of polarization directions within materials from specific frequencies or wavelengths. This frequency range has a similar characteristic to an electronic bandgap (Tao 2005).

As shown in Fig. 2.4, PBG materials can be 1D, 2D, or 3D. The 2D PBGs have periodicity and homogeneity along the third coordinate axes. They can be

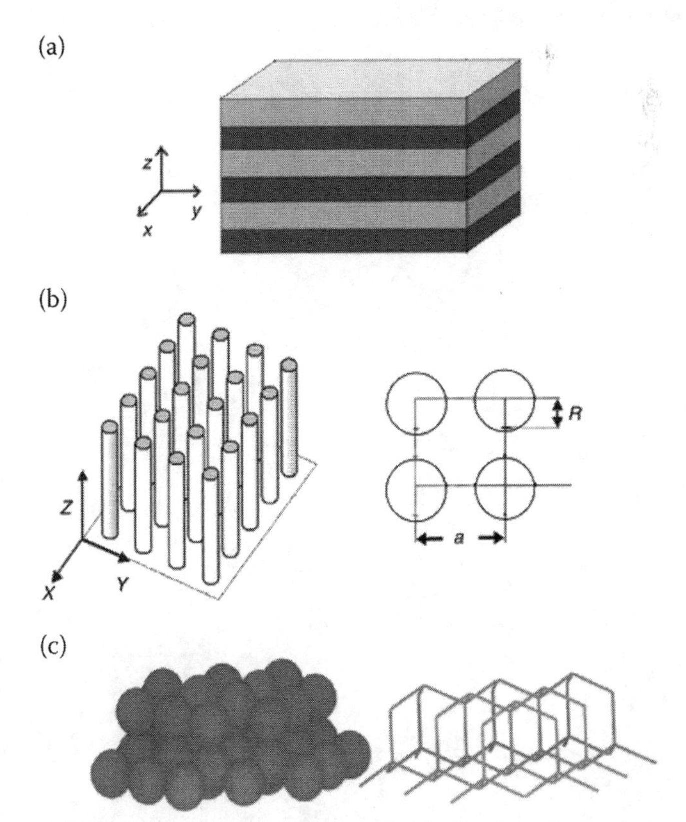

FIGURE 2.4 (a) 1D, (b) 2D, and (c) 3D photonic bandgap structures. Reprinted with permission from Wearable Electronics and Photonics. Copyright, 2005, Elsevier (Tao 2005).

produced by reactive ions or wet electrochemical etching using dry etching. The first way gives exact control over the size and setup of a hole (nm accuracy), but has a maximum etching depth capped. Electrical wet gravure can produce very deep holes and is appropriate for manufacturing, as high aspect ratio structures are not so predictable for the scale of the graved holes. The materials are selected based on their experimental significance for proof with a view of their crystal structures. The arrangement is according to the lower wavelength/bandgap, while the bar below the structure shows bandgap changes from bulk to mono form. Bulk bandgap is normally smaller than its monolayer represented as black bars, but exceptions exist, represented as red bars. 2D left-hand materials are shown by a gray box and are null or close to null, metal or semi-metallic. Fig. 2.5 illustrates the bandgap range of 2D materials from monolayer to bulk that display energy gaps in the terahertz and UV-visible electromagnetic spectrum (Chaves et al. 2020).

The 2H TMDC semiconductors (e.g., MoS_2) have 1–2-eV gaps and gained considerable attention because of their fascinating circular valley dichroism and exciton physics (Wang et al. 2018). Let us begin with the dependency on bandages in a much-studied elementary half manufacturer, namely BP. This material is a stable phosphorus allotrope, which can be exfoliated into several puckered wax bars (Liu et al. 2014).

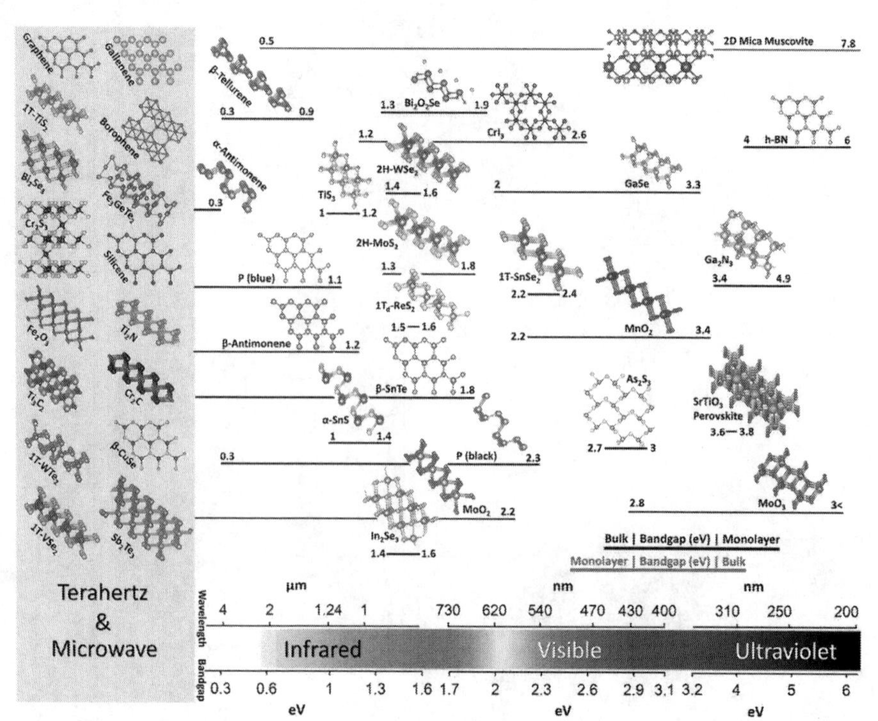

FIGURE 2.5 Bandgap elected a family of 2D materials and their bandgaps (Chaves et al. 2020).

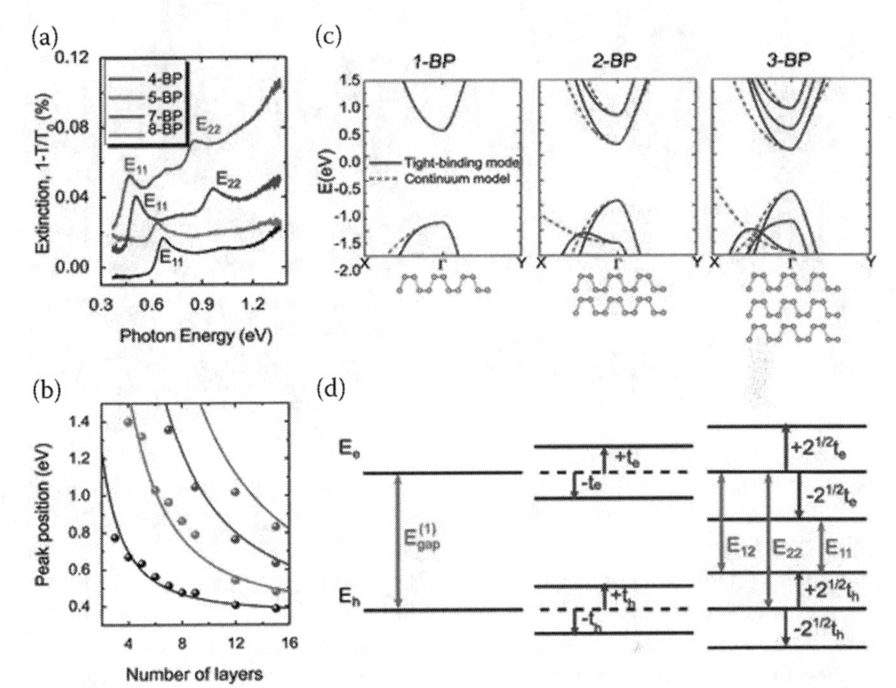

FIGURE 2.6 Bandgap diagrams by stacking the BP layers (Zhang, Huang, et al. 2017).

Absorption spectra are dependent on the number of layers of BP and it has been discovered in various experimental results. Fig. 2.6 shows the optical gaps (range from 1.66 eV to 0.30 eV in monolayer (Li et al. 2017; Zhang, Huang, et al. 2017; Zhang, Chaves, et al. 2018). Fig. 2.6 shows the absorption curve of the N layer BP; the curve shows the same index electron and hole transition (Low et al. 2014).

Fig. 2.6 collects the energies of these transitional subband peaks in Fig. 2.6 that are classified as E_{ii}, depending on their number. The peak features within the N-BP absorption field are clearly described in a simple theoretical model, i.e., solid lines in Fig. 2.6 (De Sousa et al. 2017).

2.3 OPTICAL BAND AND CHARACTERIZATION

Since 1914, BP has been synthesizing from white phosphorous in the presence of high temperature and pressure. The 2D form of BP is well known as a phosphorene and has been explored recently as a monolayer material that quite resembles graphene. Phosphorene is a prototype of BP that has earned a great value due to its premium properties, such as direct bandgap, tunable anisotropic, etc. Recently BP has received a lot of attention in various applications like sensing, optomechanical devices, optothermal devices, and low-power ultra-large-scale integration (ULSI) devices to mention a few. Phosphorus has been placed in the 15th position in the periodic table and is found abundantly in the Earth's crust. It is also found in human RNA, ATP, and DNA and plays a great role in the living cells of human blood. It is available in various allotropic forms, as shown in Fig. 2.7. Phosphorene has some

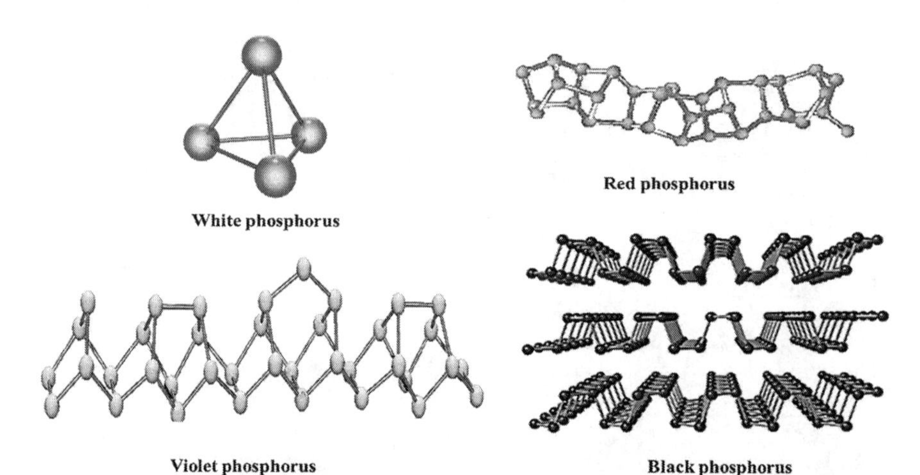

FIGURE 2.7 hosphorus has different forms of allotropes. Reprinted with permission from Journal of Industrial and Engineering Chemistry. Copyright, 2018, Elsevier (Irshad et al. 2018).

very unique properties, compared to other 2D materials. Among them, one such unique property is its high carrier mobility (600–1,000 cm^2 $(V\ s)^{-1}$) at ambient temperature. This unique property greatly depends on several layers in nanosheets based on their thickness. Carrier mobility is inversely proportional to thickness of the BP monolayer. Direct bandgap followed by a tunable energy gap property makes it a potential candidate for future applications of sensing as well. A large bandgap is attributed to a lack of hybridization between interlayers of BP. Since bandgap can be better controlled by modifying the layer-dependent property, it is an emerging material for various optoelectronic applications. Due to excellent optical and electrical properties of BP, such as a large surface-to-volume ratio, stability, high flexibility, and good thermal conductivity to name a few, it has been a good contender to other 2D materials like graphene and other TMDs. Conductance of BP is greatly changed due to the adsorption of biomolecules, which is the basic principle of highly sensitive BP-based SPR sensors. Also, adsorption of gas molecules depends on the composition of gases, which makes BP a selective gas-sensing material. Also, BP is a P-type material, so it has a great affinity to electron charge molecules, eventually enhancing its electrical conductivity. It will not be an exaggeration to say that charge conductivity would not exist without BP-based devices. Recently BP is seen as a very selective sensor for gas molecules containing nitrogen, which will lead to creating a hole in a BP layer, resulting in conductivity modulation to BP-based sensing devices.

In 2004, graphene was successfully separated from graphite, demonstrated by the existence of 2D material in an isolating single crystal form. Shortly after that, single-layer graphene was successfully laminated by other materials, leading to a vast application in sensing and electrochemical applications due to their excellent optical, electrical, magnetic, mechanical, and thermal properties. The linear optical property and electronic structure of 2D TMDC-based MOX_2 (where X = Te, S, Se) have been recently explored by many researchers. Reflectivity, optical absorption coefficient, real/imaginary part of permittivity (dielectric constant), and energy loss function have great importance for the applied electric field in two directions. As compared to zero bandgap graphene, with 2D materials like ZnO, AIN, BN, GaN, SiC, and TMDCs (dichalcogenides), they differ in terms of their sizable bandgap and linear range of spectrum, which has a wide application in field emission transistors (FETs) and energy harvestings. In most of the cases for low energy value, the dielectric functions would reveal highly isotropic behavior and for above it would reveal anisotropic behavior. The most fascinating features of 2D MOX_2 materials are the high absorption coefficient along with broad absorption spectrum, qualifying them for optoelectronic and solar applications. Bandgap engineering can be greatly altered via optical absorption phenomena while choosing a suitable thickness of 2D material at nanoscale dimensions. TMDCs sustain the incident bandgap (in their bulk form) and transit to direct bandgap upon scaling down to a single layer. Due to their inherent modulation property, they sustain a negligible short channel effect, high thermal stability, and high current on/off ratio in nanoscale semiconductor devices. MOS_2 has a similar hexagonal multi-layered structure, as shown in Fig. 2.7.

Fig. 2.8(a) reveals the bonding of Mo with two S atoms. It is a matter of fact that each layer of Mo atom is attracted by six S atoms, as revealed in Fig. 2.8 (b, c and d);

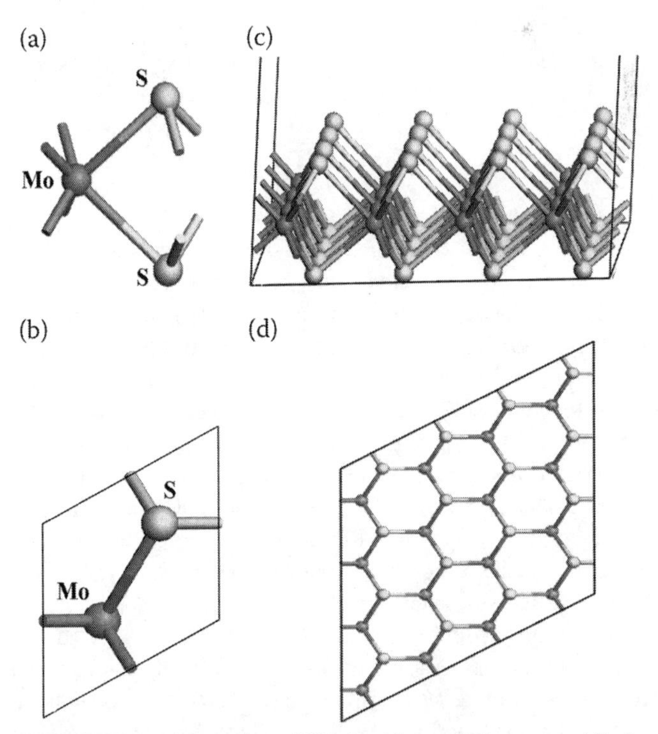

FIGURE 2.8 a) Bonding of "Mo" with two "S" atoms, while b, c, and d show that each layer of the "Mo" atom is attracted by six "S" atoms. Reprinted with permission from Physica E: Low-dimensional Systems and Nanostructures. Copyright, 2021, Elsevier (Beiranvand 2021).

TABLE 2.1
Computed and Experimental Results of Dielectric Function and Bandgap of Single Layer MoX_2. Reprinted with Permission from Physica E: Low-Dimensional Systems and Nanostructures. Copyright, 2021, Elsevier (Beiranvand 2021)

Material	Bandgap (eV)			ε (0)	
	Our results (GGA)	Experiment	Others	E ll x	E ll z
MoS_2	1.83	1.88 1.85 1.90	1.679, 1.866, 2.89, 1.60 (PBE), 2.05 (HSE), 2.82 (GW)	5.74	2.70
$MoSe_2$	1.52	1.57	1.444, 1.613, 1.35 (PBE), 1.75 (HSE), 2.41 (GW)	8.01	4.51
$MoTe_2$	1.11	–	0.95 (PBE), 1.30 (HSE), 1.75 (GW)	8.00	4.33

however, each S atom is bonded to three Mo atoms. It has been demonstrated experimentally and theoretically that monolayer MoS_2 exhibits a direct bandgap and offers a strong photoluminescence property. It is expected that other TMDCs also exhibit similar optical and electrical properties as MoS_2 or other MoX_2. Optical and electrical bandgaps and other physical properties of MoX_2 material depend on the complex dielectric function of MoS_2, $MoSe_2$, and $MoTe_2$ to mention a few. Precise determination of an optical bandgap property is attributed by the photoluminescence spectroscopy, photoabsorption, and photoemission techniques. Table 2.1 reveals the computed and experimental results of the dielectric function (which is a tensor quantity) and bandgap of a single-layer MoX_2. There are peculiar properties of materials like loss function, dielectric tensor, optical conductivity, reflectivity, and extinction coefficient that are still under investigation worldwide (Beiranvand 2021).

The Kohan sham DFT (Beiranvand 2021) method has been adopted for the bandgap calculation of 2D materials. Fig. 2.9 reveals band structure and bandgap engineering properties offered by the MoS_2 monolayer computed by the Kohan sham DFT method. The bandgap is substantially high for the case of MoS_2.

Graphene has been demonstrated as an excellent application in the electrochemistry point of view. The concept of van der Waals (vdW) hetero-structures was recently seen when two dissimilar 2D materials combined in a peculiar manner lead to some remarkable electronic properties (Dryfe 2019). The hetero-structure

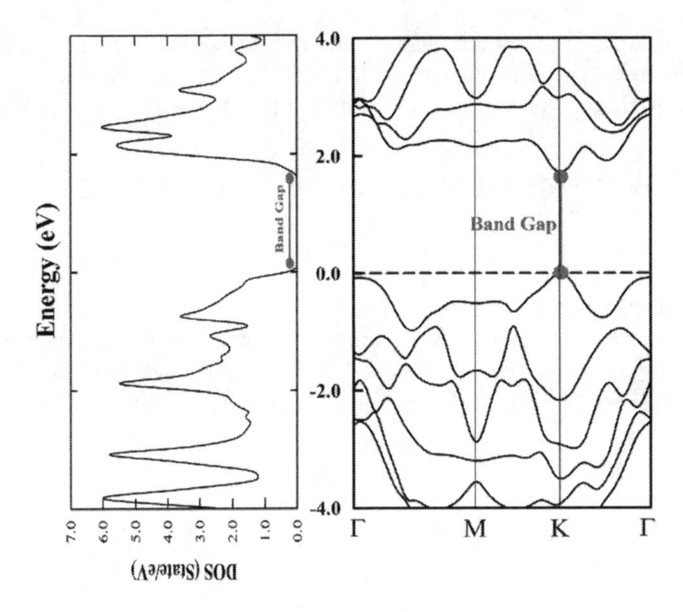

FIGURE 2.9 a) Bandgap energy and b) bandgap engineering property offered by the MoS_2 monolayer computed by the Kohan sham DFT method. Reprinted with permission from Physica E: Low-dimensional Systems and Nanostructures. Copyright, 2021, Elsevier (Beiranvand 2021).

concept based on 2D material in electrochemistry is a relatively new idea and has a vast application in the future needs of society. MoS_2 has depicted very strongly bonded 2D S-Mo-S atoms, which is strongly connected together via vdW forces. Fig. 2.10 reveals the 2D heterostructure formed by coupling between two distinct 2D materials via vdW forces.

Electron-hole interactions would have played a significant role in estimating optical and absorption spectrum. To computer the optical property, one needs to integrate over the densely packed Brillouin zone (BZ) in k-space. Convergence of solution needs to be addressed by the large sampling point of k-space (Beiranvand

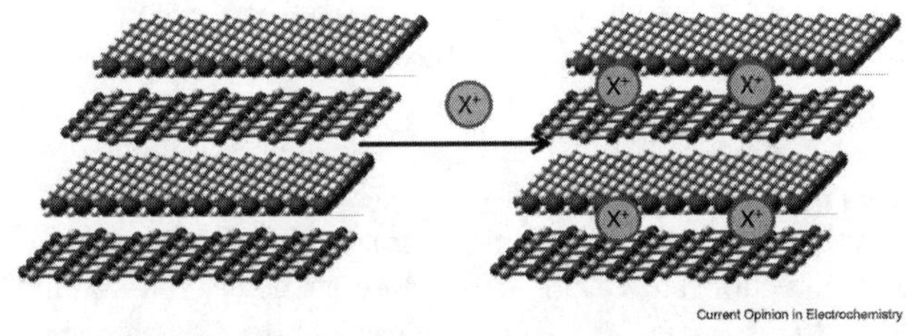

FIGURE 2.10 Demonstration of 2D heterostructure formation and electrochemical reaction by stacking layer over layer of two distinct materials. Reprinted with permission from Current Opinion in Electrochemistry. Copyright, 2019, Elsevier (Dryfe 2019).

2021). It is a matter of the fact that the Kramers-Kronig approximation converges well in the calculation of $\varepsilon_2(\omega)$ up to 50 eV. However, the imaginary part can be computed by the random phase approximation (RPA) method as follows:

$$\epsilon_2(\omega) = \frac{Ve^2}{2\pi\hbar m^2\omega^2}\int d^3k \sum_{nn'} |<kn|p|kn'>|^2 f(kn) \mathrm{x} (1 - f(kn'))\delta(E_{(kn)} - E(kn') - \hbar\omega \tag{2.1}$$

where $\hbar\omega$ denotes the incident photon energy, p is the momentum, and meaning of the various terms is as usual. The complex dielectric function is a very important parameter to determine the interactions between matter and light in all these 2D materials. It is mentioned by (Beiranvand 2021):

$$\epsilon^{\perp}(\omega) = \frac{\epsilon^{xx}(\omega) + \epsilon^{yy}(\omega)}{2} \tag{2.2}$$

and

$$\epsilon^{||}(\omega) = \epsilon^{zz}(\omega) \tag{2.3}$$

The band structure gives information about the imaginary part of the dielectric function and the Kramers-Kronig approximation provides the information about the imaginary part of the dielectric function. It has been found that it shows the highly anisotropic behavior for an energy level ≤9 eV and would become isotropic for an energy level ≥9 eV. Table 2.1 summarizes the polarization property at $\omega = 0$ (static dielectric constant). Graphene is almost transparent and lightweight, with a good barrier to pass gas and water molecules (Bouzidi, Barhoumi, and Said 2021). Also, electronic mobility is quite high; higher than silicon. In the case of planar structure, graphene atoms are so mobile that they act like photons, practically massless. Hence, they can travel a much longer distance with negligible loss. Some graphene is not a semiconductor; hence its application in electronics remains questionable. To circumvent this issue, a new material has been synthesized to fulfill the need. Some of them are h-BN, phosphorene, stanine, silicone, TMDCs, etc. However, all of these materials are quite unstable and oxidize much easier. To substantiate this, a new family discovered, named oxyhalides, is based on exceptional properties. These materials are still not fully explored and continuous efforts are being made worldwide to synthesize them. In fact, oxyhalides have been found to have various applications in LEDs, solar cells, storage, and telecommunication antennas. Fig. 2.11 shows geometrical features of different oxyhalides.

Recently, Wurtzite 2D and 3D materials of group VI such as AlN, GeC, SiC, ZnS in the (0 0 0 1) direction are found to be more stable with a similar hexagonal structure like TMDCs or halides. Also, 3D-ZnS has found a great application in short-wavelength LED devices (Lashgari et al. 2016). It is basically a binary compound, as shown in Fig. 2.12, which has a dual behavior in low temperatures at ambient pressure (Zinc-Blende) and in high temperatures and pressure (Wurtzite phase).

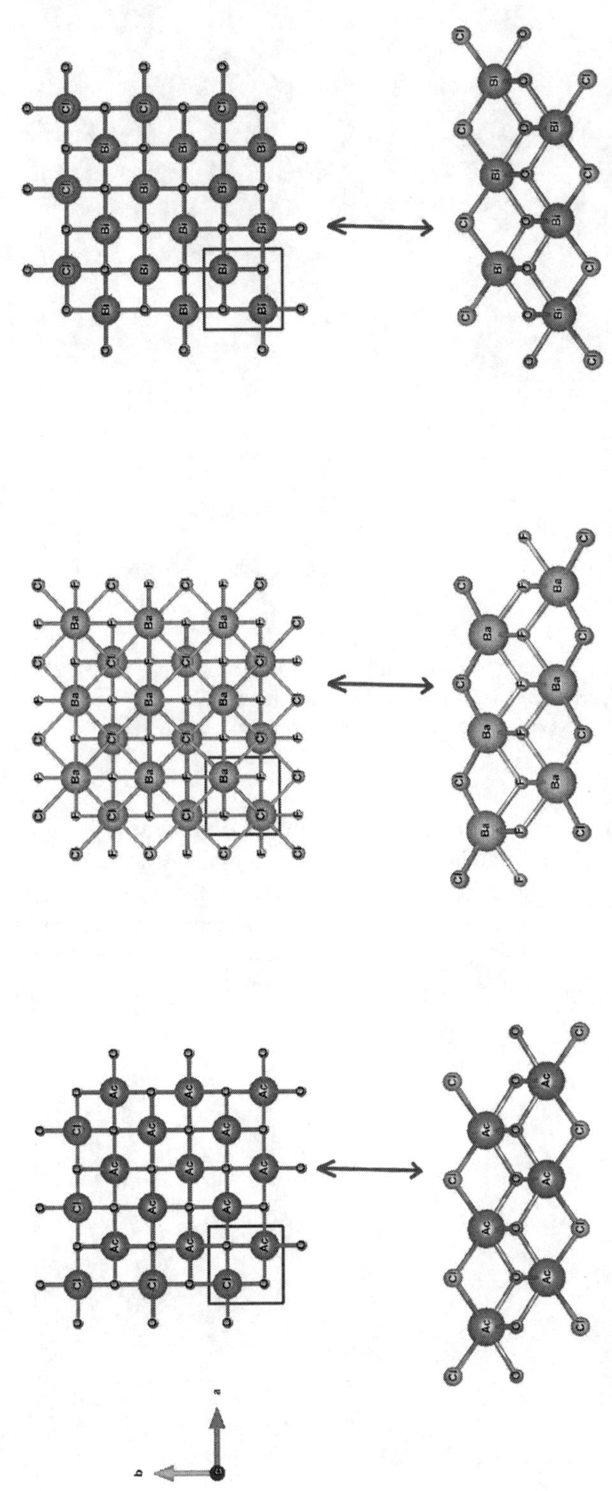

FIGURE 2.11 Geometrical features of 2D AlOCl, BaFCl, and BiOCl monolayers. Reprinted with permission from Optik. Copyright, 2021, Elsevier (Bouzidi, Barhoumi, and Said 2021).

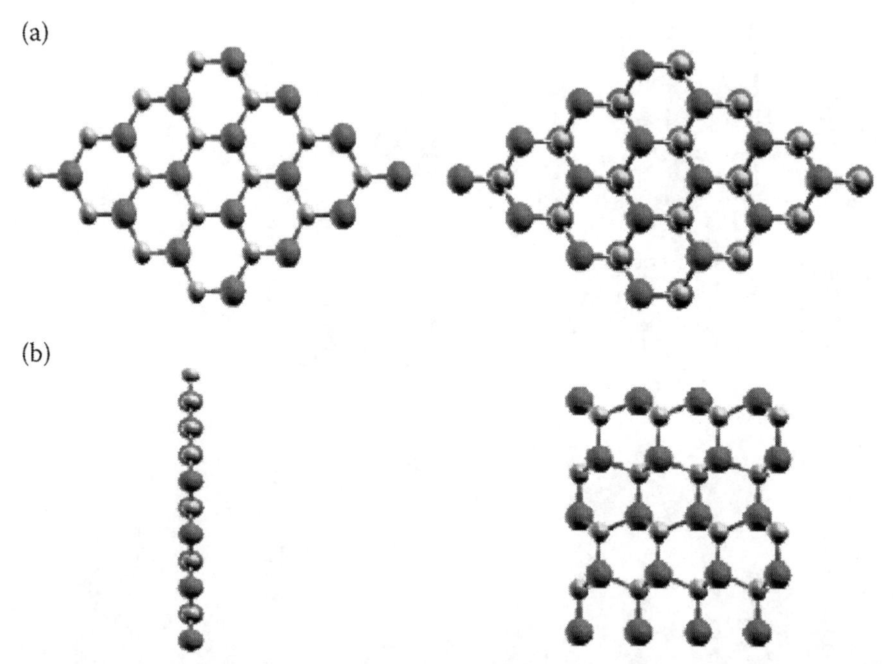

FIGURE 2.12 Dual nature of ZnS: a) top view and b) side view, as a graphene-like 2D-ZnS (left) behavior and 3D-ZnS (right) in the Wurtzite phase. Reprinted with permission from Applied Surface Science. Copyright, 2016, Elsevier (Lashgari et al. 2016).

Nowadays, Group V, 2D monolayer material called blue phosphorene (BP) came into light for their most stable and remarkable properties in optoelectronics applications. BP exhibits a wide, indirect bandgap more superior than graphene and black phosphorene. BP has a similar honeycomb type of appearance, as shown in Fig. 2.13 (Shaikh et al. 2020). Fig. 2.14 reveals the bandgap property and total density states of BP. It is evident that BP has an indirect bandgap located in between

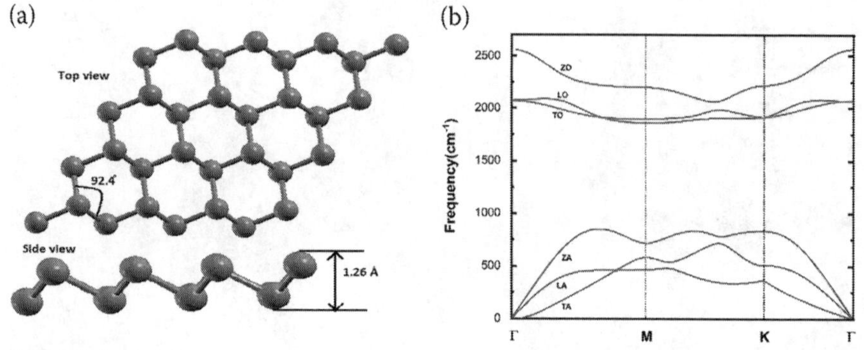

FIGURE 2.13 Geometrical features of Blue phosphorene. Reprinted with permission from Materials Today: Proceedings. Copyright, 2020, Elsevier (Shaikh et al. 2020).

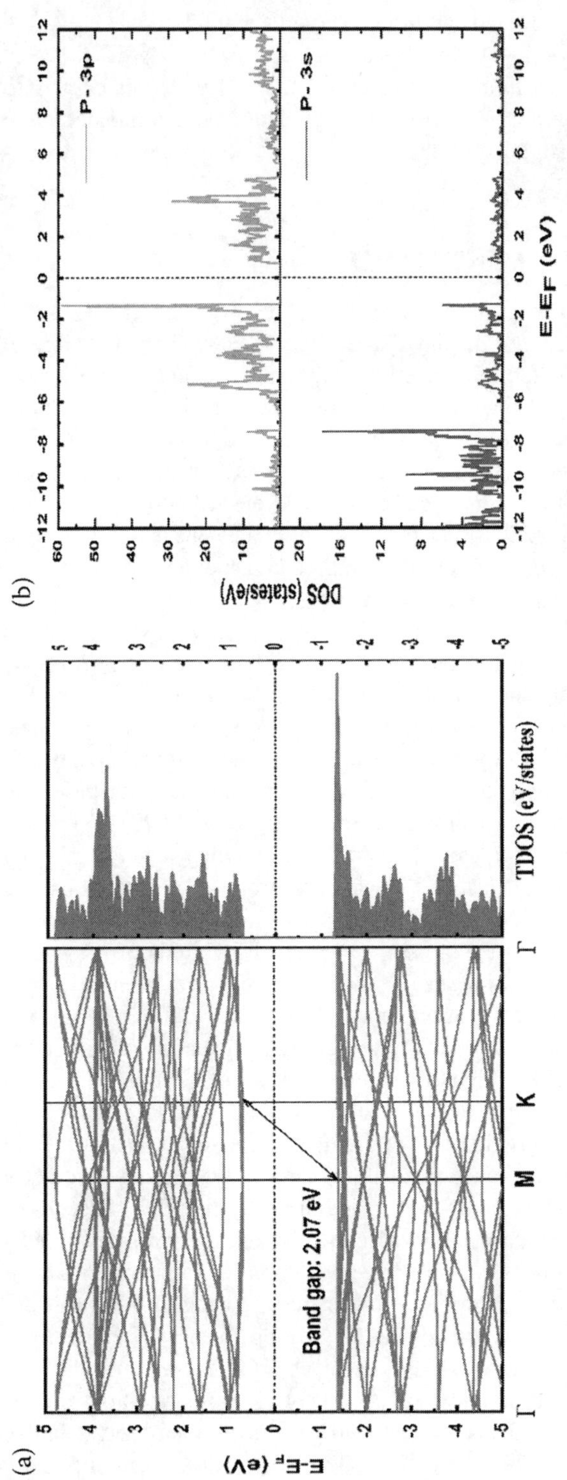

FIGURE 2.14 Bandgap property and total density states of blue phosphorene. Reprinted with permission from Materials Today: Proceedings. Copyright, 2020, Elsevier (Shaikh et al. 2020).

the peak point of K (conduction band) and the minimum point M (valence band), calculated along with $\Gamma - M - K - \Gamma$ directions.

Very recently, another class of emerging layered oxide 2D materials came into the picture. Due to their ultrathin property of 2D metal oxide, most of the atoms run over the surfaces, which distinguishes them from their traditional bulk oxides properties (Kumbhakar et al. 2021).

2.4 EXCITONS, CONFINEMENT EFFECT

Nowadays, flexible nanotechnology has a wide scope in several applications like sensors, radio-frequency, actuators, displays, and photovoltaics (Sun, Li, Ullrich, and Yang 2020). The flexible electronic materials help in the fabrication of low-cost devices, and are lightweight, durable, and stretchable due to their novel functionalities. Out of those flexible materials families, 2D-layered nanosheets such as MoS_2 and BP are mostly used to fabricate flexible devices that are scalable and mechanically flexible due to their significant optical and electronic properties. An inhomogeneous strain field in 2D MoS_2 materials produces the excitons found around the surrounding area of high-tensile strain and generates the excitonic funnel. This funnel effect overcomes the Shockley-Queisser limit and helps in harvesting and manipulating light. Further, this effect was found in TMDCs (MoS_2) as well as BP material. Here, it has been found that the funnel effect in MoS_2 is opposite as a comparison of BP, and it means excitons are pushed away from highly strained regions, i.e., called the inverse funnel effect. Thus, these two materials (MoS_2 and BP) have the special property of excitonic funnel effects. 2D-layered TMDCs materials have wide applications in developing many devices based on their peculiar photoelectric and physical properties (Chang et al. 2020). They also exhibit the spin-valley coupling characteristics, its carrier mobility, and electro- and photo-catalytic activities are high; thus, it is ideal for heterojunction solar cells, gas sensing, optoelectronics, valley electronics, etc. Apart from this, one more unique feature is the high exciton binding energy (in the order of hundreds of meV) that helps in optical transitions. Here, exciton, i.e., electron-hole pairs are tightly coupled and phonon (lattice vibrations) exhibits photoelectric properties in MoS_2 and $MoSe_2$ materials due to reduced dielectric screening and quantum confinement. This exciton-phonon interaction exhibits the up-conversion of excitonic luminescence that helps in the optical refrigeration of semiconductors. A special tool, resonant Raman spectroscopy, is used for determining the exciton transition of 2D materials and to find the exciton-phonon coupling and its scattering symmetry (Chang et al. 2020). Here, the exciton process inside the WS_2 material is discussed. The tube furnace can be used at ambient pressure to carry out the synthesis of WS_2 films, as shown in Fig. 2.15. For this, first WO_3 powder and sulfur powder can be used as precursors for WS_2 films. During the experiment, 24 mg of WO_3 powder was taken in the ceramic zone that is placed near the heating zone. The ceramic boat is covered with two silicon wafers; first is placed above the WO_3 powder and second is placed at the top position and a space is kept to allow the vapor out, as shown in Fig. 2.15. Raman measurements can also perform to measure the lattice vibration characteristics, as shown in Fig. 2.16. Due to stronger exciton-phonon

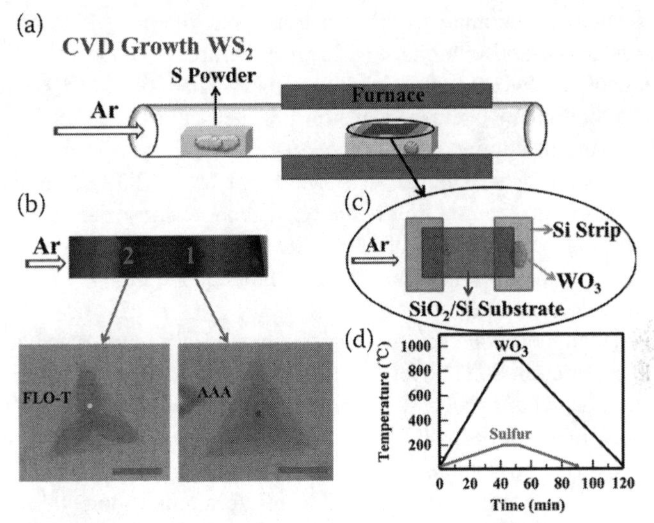

FIGURE 2.15 a) Process of WS_2 synthesis, b) SiO_2/Si substrate growth, c) WO_3 precursor, d) temperature curves. Reprinted with permission from Journal of Luminescence. Copyright, 2020, Elsevier (Chang et al. 2020).

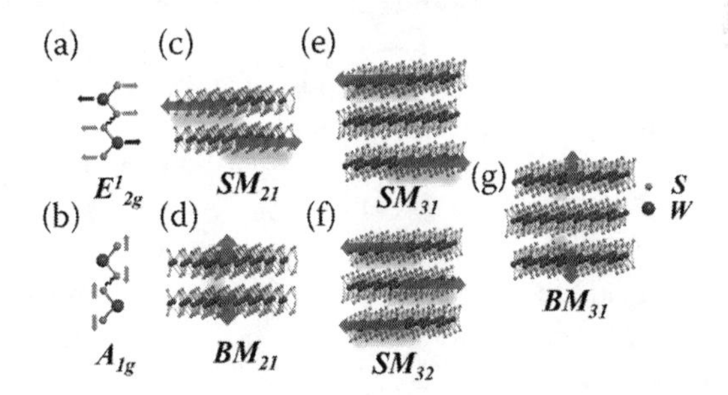

FIGURE 2.16 a–b) Raman modes show atomic displacement pattern, c–d) vibrational schematics in bilayer WS_2, e–g) vibrational schematics in trilayer WS_2, here arrows show vibrations in the layers. Reprinted with permission from Journal of Luminescence. Copyright, 2020, Elsevier (Chang et al. 2020).

coupling at a low frequency, there is a chance of an increment of exciton Bohr radius and scattering cross-section and backscattering forbidden phonon is dominant in the case of WS_2.

Similarly, one more 2D-quantum dots (2D-QDs) based in MoS_2 has been reported by Golovynskyi et al. (Golovynskyi, Bosi, Seravalli, and Li 2021). For this, first the MoS_2 2D QDs kept over the SiO_2/Si material with the help of chemical vapor deposition method; thereafter, various properties like exciton and trion photoluminescence can be analyzed. The average size of MoS_2 2D QDs is 40 nm and its shape is round and exhibits a tri-layer structure. It found that this type of structure produces the two photoluminescence bands that are A and B excitons. In the 2D QDs, reduction in lateral dimension produces the blueshift emission that generates the quantum confinement. The A exciton band is asymmetric in nature and superimposed the trion band. It also found that MoS_2 2D QDs emission efficiency is six times more than the MoS_2 flakes. Its charge carrier localization is also high and offers huge applications of MoS_2 2D QDs by varying its thickness and lateral sizes in nanodevices that involve the efficient modulation of optical properties and multicolor engineering.

Confinement can be achieved through a geometrically separated and definite atmosphere (Tareen et al. 2021). Here, a different scaffold as a confining boundary helps in the electrocatalysis process. 2D materials utilize the properties of vdW gaps of metal-organic frameworks (MOFs), porous sites (0D), channels in nanotubes (1D), and 2D cover. Confinements are classified in 1D, 2D, and 3D confinement in 2D materials. Here, confining words came from the atoms (nonmetals or metals), compensation from the lattice, or association to the 2D materials that are due to vigorous covalent bonds between each doping (Tareen et al. 2021). Here, the confined atoms generate active sites with electrocatalytic natures due to robust electronic interplay between the compact atoms and 2D materials that create the current electronic states. A single layer of TMDCs crystal phase transition takes place from a semiconductor to metallic phase due to the degree of lithiation, as shown in Fig. 2.17.

The bulk space also helps in the generation of novel nano-scale chemical surroundings. It also helps in the formation of nano-confinement by using the kinetics and energetic properties of electrocatalytic reactions. Several scaffolds are reported in support of confinement. For this, porous surfaces, nanotubes, and vdW gaps of 2D materials have a major role. Nowadays, space confinements of 2D materials are emerging trends for nanosized electrocatalysts. Thus, this confined electrocatalysis of 2D materials is called single atom confinement in the vdW gaps between two separate layers, and space in a 2D cover as shown in Fig. 2.18.

Similarly, confinement in tin monoxide (SnO) has been discussed by Chen et al. (2020). As SnO is a p-type and has transparent conducting oxides due to this reason, it is widely used in transparent thin-film transistors. In the bulk SnO, layers are arranged in the crystallographic direction; these layers are interconnected through the vdW. SnO is also a type of 2D material that appears like graphene, WS_2, and SnS_2; thus, larger sheets of SnO monolayers can be prepared that can exhibit the novel properties. It has been also proved through the experiments that the band structure of SnO is tunable and can be possible through the variation of SnO layers by proper consideration of its thickness. Due to its tunability nature, it can be used in many applications such as transparent electronics, photocatalysis, photovoltaics, etc. Here, morphology of thin films decides the performance of the device.

Further, to studies of quantum confinement effects, the thickness of SnO films should be optimized and its infinite extension should be in the plane of the film. But,

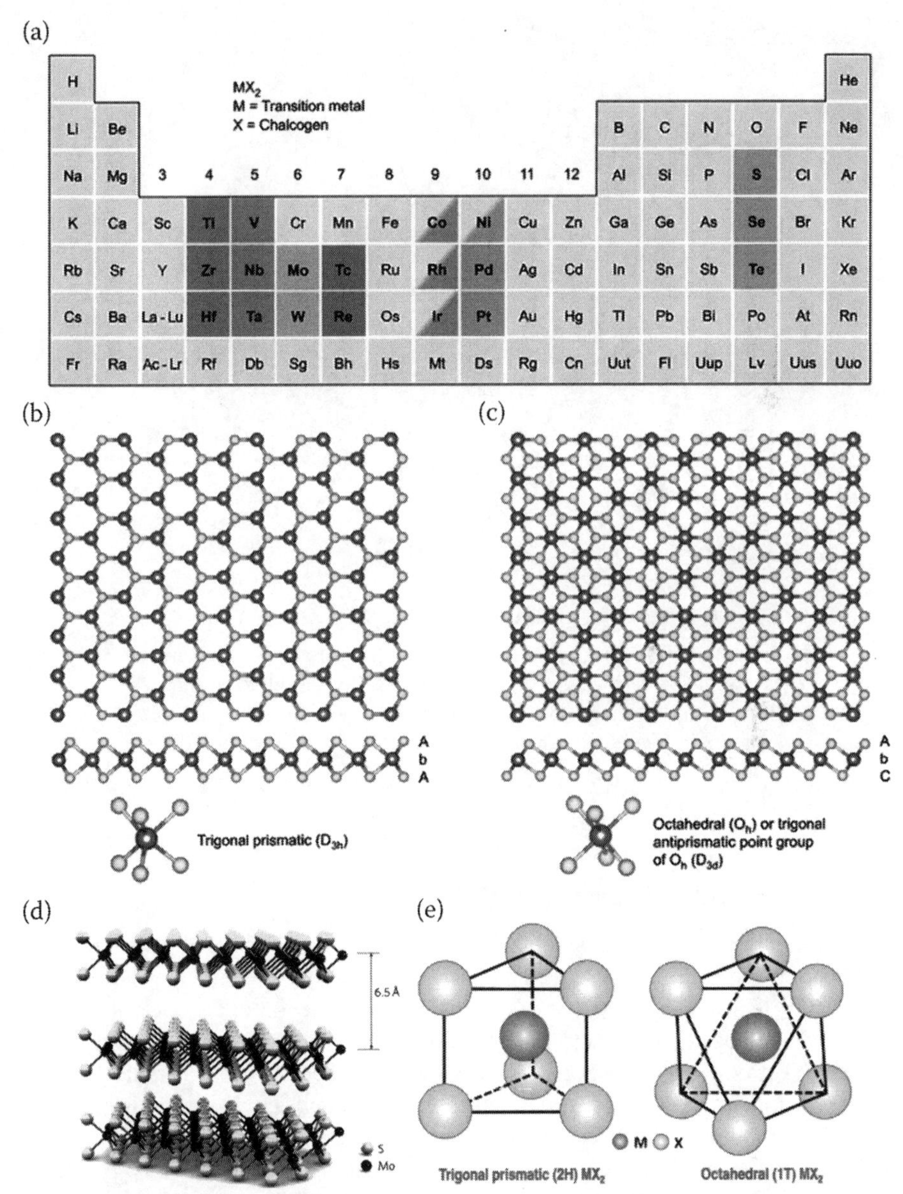

FIGURE 2.17 a) ~40 layered TMDCs compounds periodic table; b, c) trigonal prismatic of TMDCs monolayer (2H) and octahedral (1T) coordination; d, e) 3D-TMDCs crystal structures. Reprinted with permission from Progress in Solid State Chemistry. Copyright, 2020, Elsevier (Tareen et al. 2021).

heteroepitaxial growth methods are not sufficient for ideal growth of thin films on substrates. Thus, it has been found experimentally that major quantum confinement effects arise from less than a 10 nm thickness of single-crystal films and that further reduces the thickness and increases the bandgap.

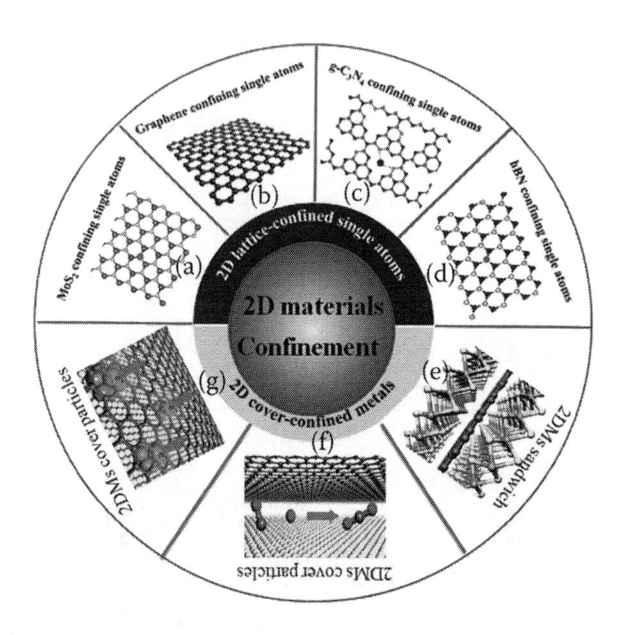

FIGURE 2.18 Confinement electrocatalytic scheme of 2D materials: a) MoS_2; b) graphene; c) g-C_3N_4; d) h-BN; e) graphene or MoS_2; f) space in 2D crystal and metal surface; g) electron transfer from metal to graphene layer. Reprinted with permission from Progress in Solid State Chemistry. Copyright, 2020, Elsevier (Tareen et al. 2021).

2.5 ANISOTROPIC OPTICAL PROPERTIES

Anisotropy is a property of a substance in which various physical properties are shown in different directions. It is a property of the atomic structure that can affect a material's electric, magnetic, optical, or mechanical reaction to an external change. The materials based on anisotropic properties have become more common in advanced electronics, with use in a wide range of fields. An optical polarizer is a classic example of anisotropic material. For liquids and gases, it is reasonable to assume that the optical properties are isotropic, i.e., the same in all directions. This will also be the case for glasses and amorphous materials, which have no preferred physical axes. However, it will not, in general, be a valid assumption for crystals, which have well-defined axes arising from their structure. In this section, we shall discuss the effects of anisotropy on optical materials, beginning with the natural anisotropy found in crystals and then moving on to induced anisotropy caused by strain and external fields. Layered Van der Waals materials are a form of naturally anisotropic material that is particularly fascinating. The anisotropy of VdWs materials is due to the substance's dispersive composition. The inherent anisotropy is useful in a variety of purposes. For example, the interplay between electronic and structural properties in some layered materials is very intense (Gjerding et al. 2017), opening up exciting possibilities for sub-wavelength radiative and imaging emission control (Jacob, Alekseyev, and Narimanov 2006; Lu, Kan, Fullerton, and Liu 2014). Due to the lack of a third dimension, the 2D atomic layers are clearly anisotropic, but they may also show in-plane anisotropy. Due to their strong symmetric crystal structure, the most commonly studied 2D materials, graphene (Novoselov et al. 2004), hBN (Ci et al. 2010), and the family of TMDCs (Mak et al. 2010), have in-plane isotropic properties. The value of MoS_2 and other (non-graphitic) 2D diamond-like-metal carbons have six-fold rotational symmetry and two mirror planes, while graphene and H-Beta have a six-type symmetry and two mirror values. Inhibition of some type of anisotropic reaction occurs when there are significant numbers of crystal symmetries present. Phosphorene is the prototypical in-plane anisotropic 2D material (Li et al. 2014). Phosphorene is made by mechanically exfoliating BP to the monolayer boundary, and it has a highly anisotropic puckered shape that varies in the zigzag and armchair directions. This powerful anisotropy has sparked a slew of theoretical and experimental phosphorene research, revealing the impact of structural anisotropy on optical, optoelectronic, electromechanical, thermal, and excitonic properties (Fei and Yang 2014).

TMDCs in the twisted 1T'-phase, such as WTe_2 (Ma et al. 2016; Torun et al. 2016; Wang et al. 2020; Zhang, Zhang, et al. 2019; Zhang, Wang, et al. 2019), titanium trichalcogenides (most prominently TiS_3 (Island, Buscema, et al. 2014; Island, Barawi, et al. 2015; Jin, Li, and Yang 2015; Kang, Sahin, and Peeters 2015; Silva-Guillén et al. 2017), ReS_2 and $ReSe_2$ (Liu et al. 2015; Yang et al. 2017; Zhang, Wang, et al. 2016), GaTe, and pentagonal structures such as $PdSe_2$ are also prominent examples of in-plane anisotropic 2D materials. Optical and magnetic properties of such materials are anisotropic, which have interesting applications in optical devices.

Since atoms in a solid log into a crystalline is well-defined entry points, we can't assume that optical properties along different crystalline axes are identical, e.g., atom separation cannot be uniform in all directions, resulting in varying vibrational frequencies and, as a result, a difference in refractive index in the related directions. Alternatively, the molecules trapped in the lattice can absorb those polarizations of light preferentially. The phenomenon of birefringence, which can be found in translucent anisotropic crystals, is one of the most obvious manifestations of optical anisotropy, which generalizes the relationship between polarization and an applied electric field to explain the properties of a birefringent crystal. To apply P to how the electric field is applied in some direction relative to the crystalline axes, we must write a tensor equation.

$$P = \varepsilon_o \chi \tag{2.4}$$

where χ represents the susceptibility tensor. It can be:

$$P = \varepsilon_o \sum_j \chi_{ij} \mathcal{E}_j \tag{2.5}$$

This can be conveniently expressed in terms of matrices as:

$$\begin{pmatrix} P_z \\ P_y \\ P_z \end{pmatrix} = \varepsilon_o \begin{pmatrix} \chi_{11} & \chi_{12} & \chi_{13} \\ \chi_{21} & \chi_{22} & \chi_{23} \\ \chi_{31} & \chi_{32} & \chi_{33} \end{pmatrix} \begin{pmatrix} \mathcal{E}_x \\ \mathcal{E}_y \\ \mathcal{E}_z \end{pmatrix} \tag{2.6}$$

Here, Cartesian coordinates x, y, and z may correspond to the crystal's principal axes; we will simplify the shape. The off-diagonal components are 0 in this case, and the susceptibility tensor takes the form

$$\chi = \begin{pmatrix} \chi_{11} & 0 & 0 \\ 0 & \chi_{22} & 0 \\ 0 & 0 & \chi_{33} \end{pmatrix} \tag{2.7}$$

The x, y, and z axes are indistinguishable in cubic crystals, meaning they have $X_{11} = X_{22} = X_{33}$ and their optical properties are isotopic. Uniaxial crystals have hexagonal, tetragonal, or trigonal (rhombohedral) symmetry. The z-axis is commonly assumed to be the single optic axis of these crystals. The optic axis in hexagonal crystals, for example, is determined by the orientation normal to the hexagon's plane. In the x and y axes, the optical properties are same, but not in the z-direction. This indicates that $X_{11} = X_{22} = X_{33}$. Biaxial crystals are crystals with orthorhombic, monolithic, or triclinic symmetry. They have an optic axis, and the susceptibility tensor's three diagonal components are all distinct. Mica is a biaxial crystal that is widely used.

2.6 CHALLENGES TO REALIZE AN EXPERIMENTAL 2D MATERIAL SPR-BASED SENSOR

In recent, the use of 2D materials has increased to enhance the performance of the SPR sensors and photodetector devices. To fabricate the 2D material-based SPR sensor, it is necessary to synthesize the 2D materials and prepare to transfer them over the sensing device. There are different 2D materials available such as transition-metal dichalcogenides, BP semiconductor material, and semimetal graphene, etc. These 2D materials have excellent optoelectronic and electronic properties, such as tunable photon absorption, ultrafast charge transport, and different synthesis processes and transfer processes. Continuous pursuit and challenge are the targets of studying sensitivity, broadband spectrum, ultra-high optical response speed, and other excellent outputs of sensing devices. Due to the unique properties of 2D materials, such as high mobility, no dangling bonds, stable tuning, and much more, 2D material-based sensing instruments such as SPR sensors and field-effect transistors (FETs)–based sensors have become a research hotspot. Then, in the operational wavelength range of visible light to the terahertz band, 2D material-based sensing devices are a common challenge to fabricate systematically. There are mainly two challenges to fabricate 2D material-based SPR sensors: (a) synthesis processes of 2D materials and (b) transfer to the device and characterization process of 2D materials.

2.6.1 Synthesis and Properties of 2D Materials

It is crucial to prepare fine and high-quality 2D materials that play an essential role in the high performance of optical sensors. Several synthesis techniques rely on micromechanical abrasion, chemical vapor deposition, liquid-phase abrasion, and epitaxial growth methods for the synthesis of 2D materials.

2D materials with high quality can be obtained by micromechanical exfoliation using a physical process. Fig. 2.19a illustrates the basic diagram and physical process of the micromechanical exfoliation. Using this process, a single-layer graphene is also processed (Novoselov et al. 2004), while other 2D materials such as MoS_2 (Miremadi and Morrison 1987) and WS_2 (Thripuranthaka and Late 2014) with a single-layer/multi-layer are also generated similarly. Though single or multi-layer 2D materials can be obtained better through micromechanical exfoliation, the sample size is limited, and the thickness of the sample cannot be ensured. The possibilities for future industry growth will be sharply reduced as a result of this limitation of micromechanical exfoliation.

Liquid-phase exfoliation is another 2D material synthesis method. The process is effective because it allows for vast numbers of samples to be taken by exfoliation in a liquid form, as illustrated in Fig. 2.19b (Guo et al. 2015). It disperses bulk materials into particular solvents or surfactants. The energy of ultrasonic waves peels the single-layer or multi-layer 2D materials directly from the bulk material surface, preserving the 2D materials' full appearance and performance. Initially, Coleman et al. (2011) used the N-ethyl pyrrolidone (NMP) organic solvent to remove 2D material from its surfaces with the help of sonification with ultrasonic waves. WS_2 was successfully peeled by Pan, Liu, Xie, and Ye (2016) with the use of the ultrasonic bath method. In this process, the nanosheets of WS_2 dissolved in water rapidly with low peeling efficiency. It can be done on a broad scale by identifying further stripping solutions and developing new processes to properly handle large volumes of 2D materials.

2.6.2 Transfer Process of 2D Materials

The critical process to realize the integrated structure of 2D material and optical elements is wet and dry transfer techniques. In this section is a detailed discussion of their benefits as well as complexities and appropriate applications. In the wet transfer method, polymethyl methacrylate (PMMA) is used to collect 2D material from the copper sheet. First of all, PMMA is spun uniformly over the 2D material sheet and put into the solidification process. After solidified PMMA, a composite structure of PMMA/2D materials/Cu sheet is ready to be placed in a $FeCl_3$ solution. The $FeCl_3$ solution reacts with Cu and is displaced from the composite structure, and 2D materials are assisted by the PMMA materials floating on top of the $FeCl_3$ solution. The lightweight PMMA-2D materials films are then soaked and washed with deionized (DI) water carefully. Secondly, the composite structure of PMMA-2D is transferred to the targeted devices and then dried and shaped, as shown in Fig. 2.20a. In the last step, acetone vapor or acetone is used to remove the PMMA layer from the device-2D material-PMMA composite structure. The wet transfer technique is basic, easy to use,

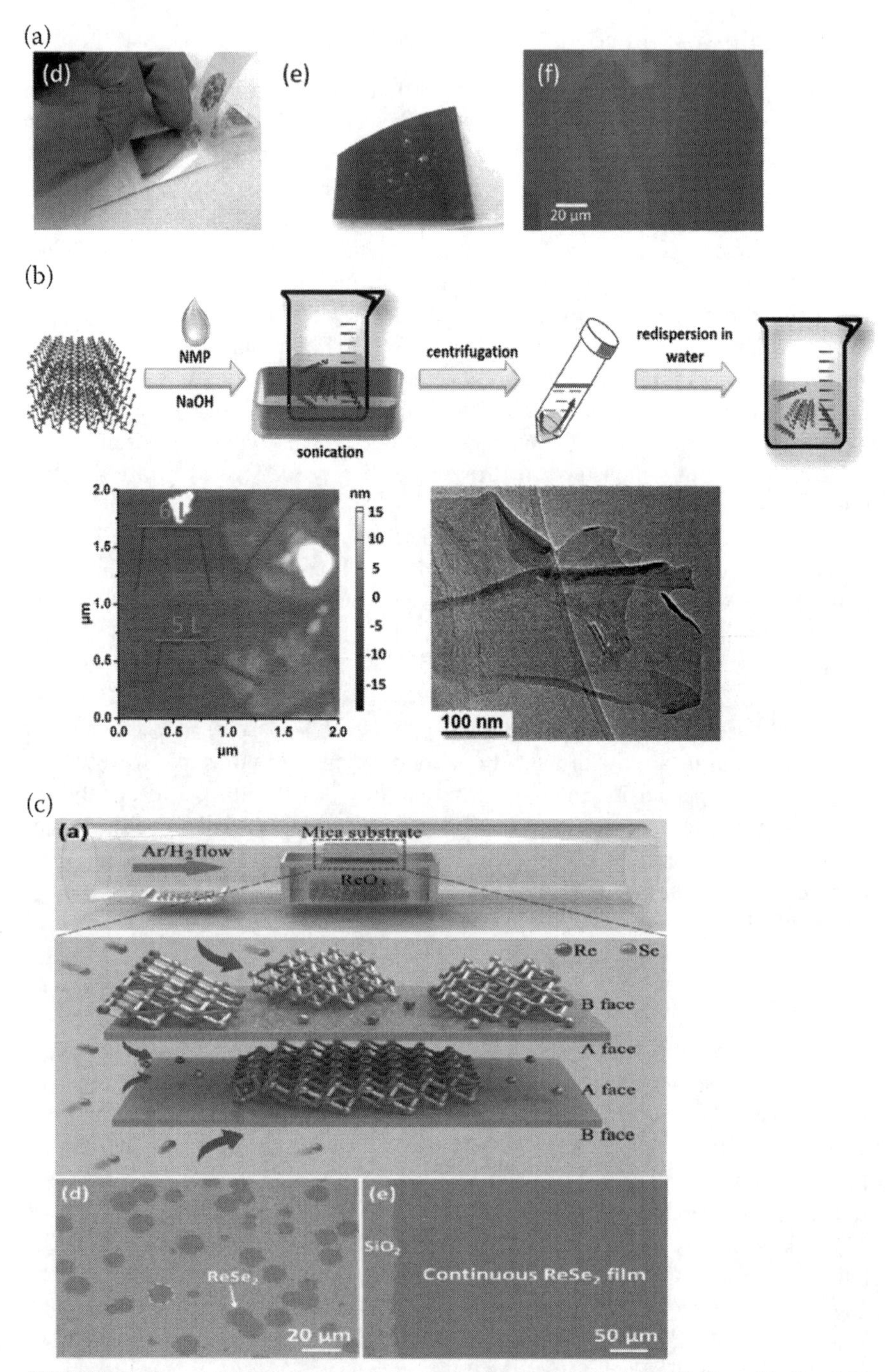

FIGURE 2.19 a) Schematic description of micromechanical exfoliation, b) for phosphorene, c) CVD growth of $ReSe_2$ and those optical images. Reprinted with permission from Nano Express journal. Copyright, 2021, IOP Science (Zhang et al. 2021).

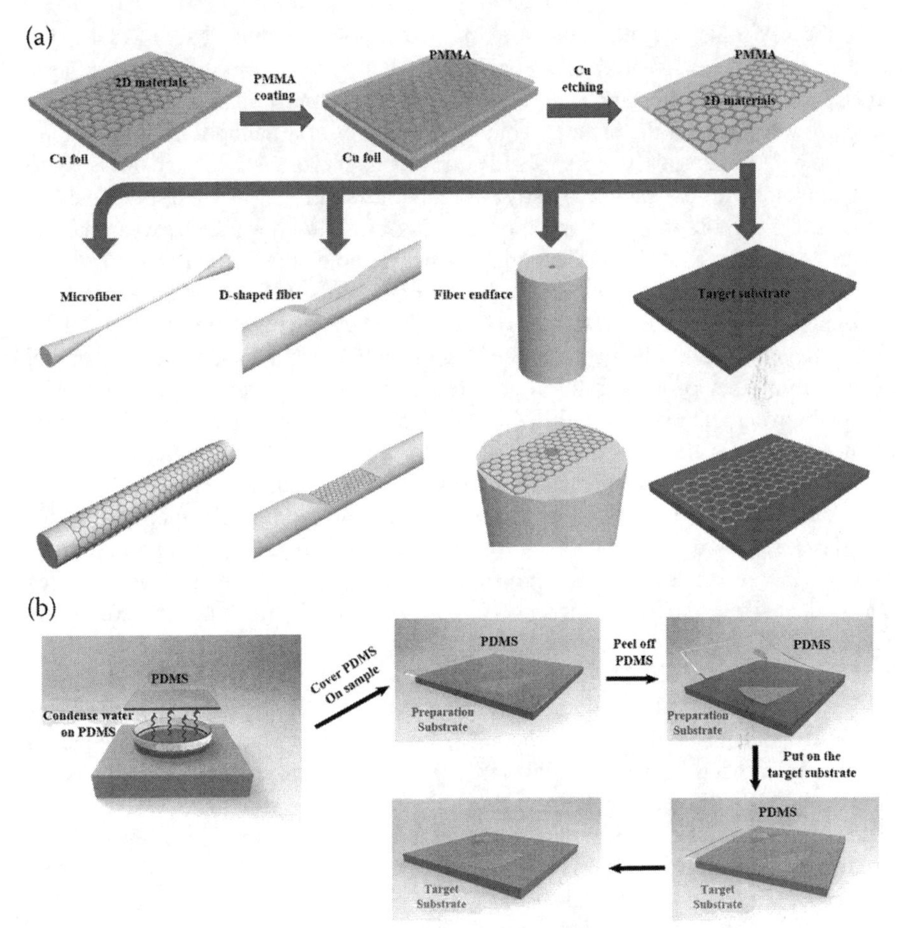

FIGURE 2.20 Schematic of fabrication of 2D materials-based optoelectronic structures. a) Wet transfer technology. b) Capillary-force-assisted dry transfer procedure. Reprinted with permission from Advanced Science journal. Copyright, 2020, Wiley (Tan et al. 2020).

has a good performance rate, and is adapted to many current process requirements of 2D optoelectronic equipment, including waveguide/optical fiber structures with curved surfaces. On the other hand, though, it still has the following challenges:

i. Acetone is commonly used for dissolving PMMA, restricting the use of versatile polymer wood plaque substrates, which are partly soluble.
ii. It is difficult to remove and disinfect PMMA residue thoroughly. As the 2D material is mixed in with foreign contaminants and water molecules, the interface between the target and the substrate and the resulting product is polluted, thereby affecting the device's overall performance.
iii. The surface tension created by solution volatilization during the transfer process quickly causes the supporting layer to curve, stretch, and fold, resulting in the folding and tearing of the 2D material film during the transfer.

In the dry transfer method (Tien et al. 2016), polydimethylsiloxane (PDMS) is used to pick the 2D materials prepared by the mechanical exfoliation synthesis process. After that, the composite structure of glass-2D material-PDMS and sample structure are fixed on the sample table with the help of a nanopositioning system. Then, the 2D materials were explicitly moved to the target spot and lowered to make contact with the sample. Lastly, when the glass-PDMS is lifted, the adhesion between the sample substrate and the 2D material is stronger than between the 2D materials and the PDMS, so the sample structure holds the 2D material film. The exfoliation method is complex due to the poor adhesion of PDMS in 2D exfoliating materials, and its success rate is not high. For the collection process, Ma et al. (2017) have created an immediate glue clean-stamping capillary-assisted technique, using a thin layer of evaporative liquid (e.g., water) for increasing the adherence energy between 2D crystals and PDMS. The thin liquid coating is condensed from its steam process onto the PDMS surface to ensure that 2D products have low corrosion levels and that their chemical and electric properties are largely maintained. Fig. 2.20b illustrates the fabrication steps in detail. This approach is appropriate for the on-chip conversion of materials like BP, TMDCs, and graphene. To date, 2D materials have been transferred in comparatively mature technologies into optical fiber and chips, but some researchers are still working to increase the accuracy and success of the conversion process (Lin et al. 2014; Ma et al. 2017; Zhang, Huang, et al. 2017).

Single-layer planning and identification are also challenging for single-layer characterization. Mechanical exfoliation from bulk crystals is usually the easiest process for moving 2D sheets of layered vdWs solids such as graphene (Novoselov et al. 2004, 2005), metal chalcogenides (Late et al. 2012), and GeH (Bianco et al. 2013) to any substrate. This method provides a 10 μm size of nanoflake, which has a single layer and too few layers. The visualization of single layers without interference approaches on other substrates was also investigated. Certain TMDCs (MoS_2, WS_2) have a single-layer bandgap and thus only display indirect photolumines, making them well suited for use in the study of their photo-induced and photo-chemically generated emissions through fluorescence microscopy. As the layer thickness increases, the distance between these materials' direct and indirect band strengths decreases, though these materials' fluorescence decreases. Hence, accurate characterization of these 2D materials is essential. It is possible with advanced characterization techniques such as fluorescence quenching microscopy (Kim, Kim, and Huang 2010; Kim, Cote, Kim, and Huang 2010; Tan et al. 2013), x-ray diffraction (Bizeto, Shiguihara, and Constantino 2009; Ebina, Sasaki, and Watanabe 2002), Raman spectroscopy (Late et al. 2012; Dresselhaus, Jorio, and Saito 2010), transmission electron microscopy (Aksit, Toledo, and Robinson 2012; Fukuda et al. 2008; Schliehe et al. 2010), atomic force microscopy (Fukuda et al. 2008; Osada et al. 2011), and scanning tunneling microscopy (Li and Andrei 2007; Miller et al. 2009). In addition, it would be essential to incorporate feedback loops between synthesis and system production to refining the increase of high-quality single-layer materials for electronic applications. Some characterization images are shown in Fig. 2.21 after synthesis and transfer to the device.

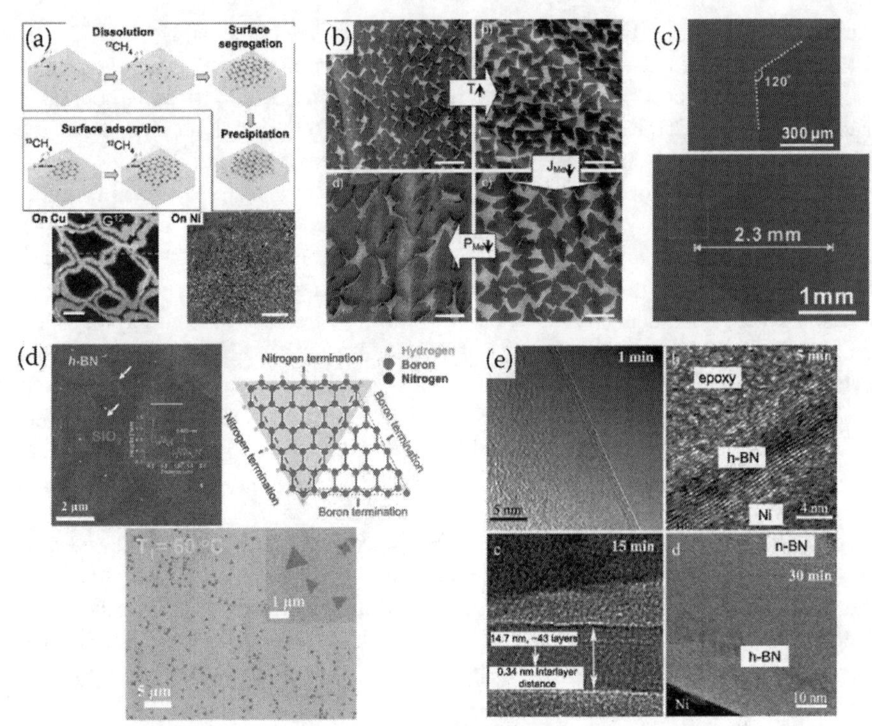

FIGURE 2.21 Characterization images of the 2D materials after synthesis and transfer process over the different devices. Reprinted with permission from ACS Nano journal. Copyright, 2013, American Chemical Society (Butler et al. 2013).

2.7 ASSEMBLE A UNIFORM LAYER OR MULTILAYER OF 2D MATERIALS ON METAL SURFACES

The outstanding achievement of graphene has enormous success, athere has been an equally remarkable boom in the production of other 2D materials capable of forming extraordinary properties of atomic sheets. The 2D library, which includes over 150 exotic layered materials that are simple to work with materials with a thickness of less than a nanometer (Butler et al. 2013) is growing every year. TMDCs are a type of graphene that is transparent, thin, and flexible. Zhang (2015) was the first graphene descendant to spark intense research activity. Many 2D TMDCs, unlike graphene, are semiconductors in nature and have an enormous opportunity for being turned into more efficient ultra-small, low-power transistors than current silicon-based transistors (Zhang et al. 2011).

TMDCs have a tunable bandgap, have a lot of photoluminescence (PL), and a lot of exciton binding capacity, which makes them a good choice for light-emitting diodes and photo-transistors (Bao et al. 2013; Frey et al. 1998). For example, MoS_2 has been extensively studied for electronic and optoelectronics applications due to its direct bandgap (1.8 eV), strong mobility (700 cm^2 V^{-1} s^{-1}), high current on/off ratio of 10^7–10^8, broad optical absorption (10^7 m^{-1} in the visible range), and a giant PL resulting from the direct bandgap (1.8 eV) in a monolayer (Fuhrer and Hone 2013). 2D TMDCs have important features that help in fabrication of sensing capacitive energy storage (e.g., supercapacitors and batteries) application-based devices (Choudhary, Patel, et al. 2016). 2D TMDCs-based sensors benefit from a high surface-to-volume ratio, which improves sensitivity, selectivity, and power consumption.

TMDCs-based sensors do not have physical gates for selectively reacting to the targeted gas molecules or biomolecules, unlike digital sensors (Lee et al. 2014). Carbon, chemical, and biosensors may all benefit from MoS_2-based FET devices. Another advantage of the weakly bonded 2D TMDCs atomic layers is that they can be easily separated and stacked with other TMDCs to create a wide variety of vdWs heterostructures (Choudhary, Park, et al. 2016). By taking advantage of such opportunities, band alignment, for example, is a novel property in these vdWs heterostructures. Several new techniques, such as tunnelling transports and strong interlayer coupling, have been developed. Tunneling transistors, for example, are electronic/opto-electronic instruments. Photodetectors, LEDs, as well as versatile electronics, can be used as barristers.

Fig. 2.22 shows the various devices that were built by exploiting the unique physical, chemical, and optoelectronic properties of 2D TMDCs (Lin et al. 2015). The current TMDC, monolayers of high quality are generated using a cutting-edge mechanical exfoliation process, but it is not scalable. Because its scalability and degree of morphological control are both impressive, chemical vapor deposition

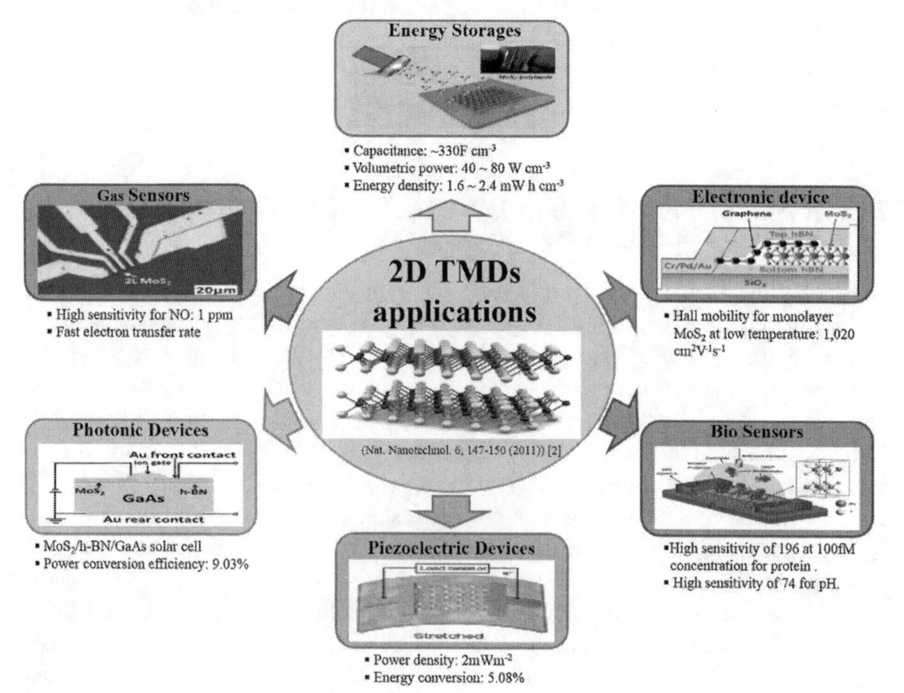

FIGURE 2.22 2D TMDCs are used in mechanical, opto-electronic, and energy applications (TMDs). Reprinted with permission from Materials Today Journal. Copyright, 2017, Elsevier (Choi et al. 2017).

(CVD) is one possible method (Shi, Li, and Li 2015). Recent advancements have resulted in the consistency of TMDC layers and has significantly improved the use of the CVD process. Metal-organic CVD (MOCVD) and atomic layer deposition are two other emerging candidates for the growth of wafer-scale and high-quality TMDC films (Olofinjana et al. 2011). These methods generate TMDCs that are not just consistent across the whole substrate, however, and perform similarly to the regular exfoliation process's atomic layers. The field of 2D materials is constantly evolving, and the quest for alternative 2D materials to graphene isn't limited to TMDCs. In today's fast-paced world, evolving domains of new 2D materials like phosphorene and silicene, as well as existing 2D materials, are good rivals (Oughaddou et al. 2015). Several theoretical studies have been conducted to address the new 2D materials' fundamental properties; but, due to stability problems, perspectives from an experimental standpoint are also in their infancy.

The most recent developments in large-scale and defect-free 2D TMDC fabrication are highlighted in this study. Furthermore, concentrations on recent advances in mechanical areas for TMDCs with logical designs with new optoelectronic and electrochemical properties and new architectures for electronics, sensors, and energy storage devices will all have potential applications. In addition, the latest breakthroughs in the newest 2D material families, such as phosphorene and silicene, are discussed.

2.7.1 Physical Properties and Crystal Structure

TMDCs are layered materials with intriguing properties and applications in emerging technologies in each unit (MX_2). Trigonal prismatic (hexagonal, H) and octahedral structures can be found in 2D TMDCs (tetragonal, T), or twisted phase (T0) based on the arrangement of the atoms, as shown in Fig. 2.23, except in a few situations; 2:3 quintuple layers (M_2X_3) (Zhang et al. 2011) and 1:1 metal chalcogenides are two examples (MX_2). Each metal atom in an H-phase material sends out two tetrahedrons with six divisions in the directions +z and z, with symmetry of hexagons visible in the top view (Fig. 2.23). As a result, the chalcogen–metal–chalcogen structure along the z-axis is called a single layer, and the vdWs interactions are small within mechanical exfoliation from bulk TMDCs is allowed by each layer (chalcogen–chalcogen) to produce flakes with a single layer. In the top view, the T-phase has an upward trigonal chalcogen layer and an underlying rotated, 180° structure (called trigonal antiprism), resulting in a hexagonal arrangement of chalcogen atoms. The T-phase (Keum et al. 2015) involves more atomic distortion (or dimerization in one direction) of metal atoms, resulting in a change in chalcogen atom atomic displacement along the z-axis.

Graphene's exceptional electron mobility (15,000 $cm^2 V^{-1} s^{-1}$, at room temperature) is limited by the absence of a bandgap, which prevents it from being used in FETs as an active component. Nanoribbons are being used to try to open the graphene bandgap, chemical doping, and AB-stacked bilayer graphene have yielded mixed results, together with bandgap openings ranging from 200 meV in most cases (Lee et al. 2015). Recent developments in 2D layers with a single link has been extended to the elements of groups III and V. The synthesis of an ultrahigh vacuum method was recently used to create atomically thin boron layers ("borophene") using pure boron evaporation at high temperatures (450–700°C) on Ag(1 1 1). Scanning tunneling

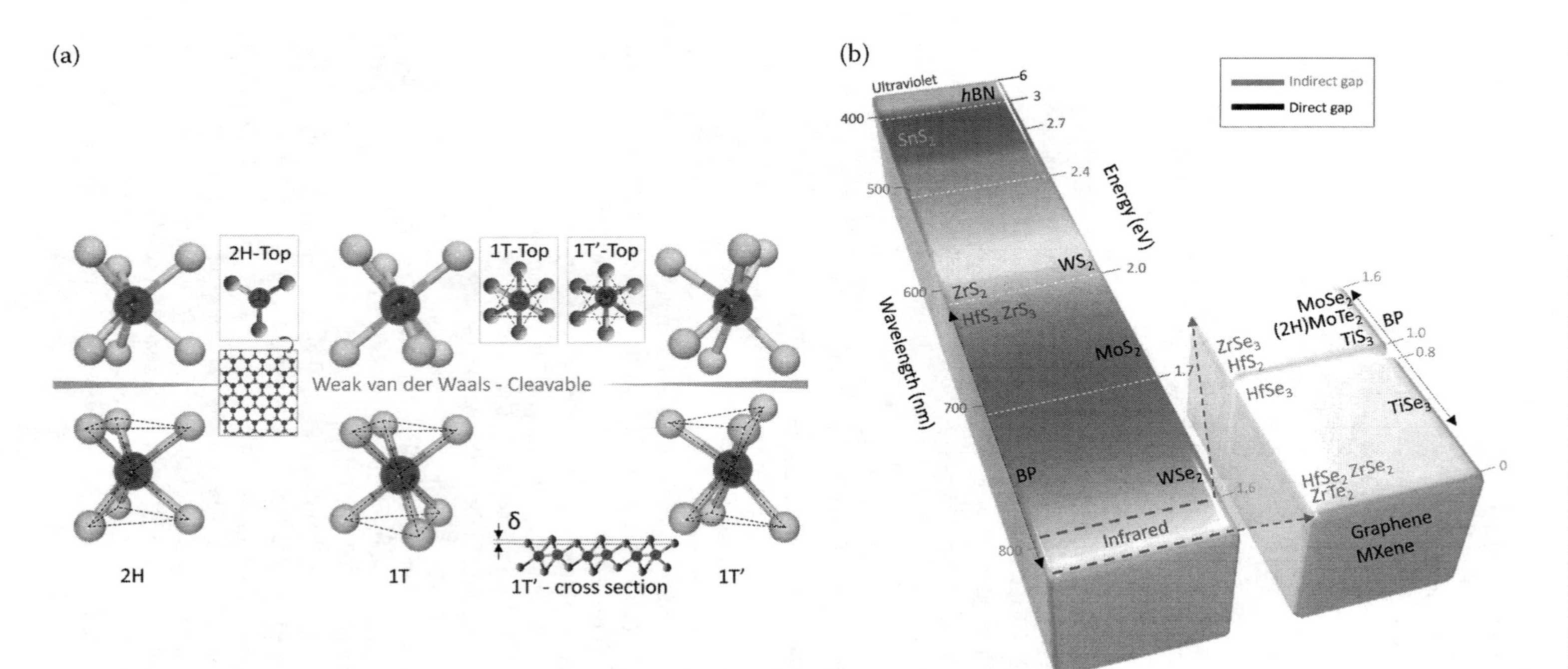

FIGURE 2.23 TMDCs with layers have typical structures. Reprinted with permission from Materials Today Journal. Copyright, 2017, Elsevier (Choi et al. 2017).

spectroscopy confirms anisotropic metallicity; most commonly bulk boron allotropes are semiconductors (Mannix et al. 2015). Chemical vapor transport in the presence of a transport agent red phosphor or pressurization (>1.2 GPa, 200°C) is used to make bulk BP. Because of its high mobility (1000 $cm^2 V^{-1} s^{-1}$) and ambipolarity, BP film is a good candidate for electronic device applications (Liu et al. 2016). MXenes and graphene also exhibit metallic activity and is found at the bottom of the schematic.

2.7.2 Recent Advancements in the Synthesis of 2D TMDCs

Various techniques, both top-down and bottom-up, such as chemical exfoliation, mechanical exfoliation, and chemical vapor deposition, all methods for removing dead skin cells, have been used to control large-scale and uniform atomic layers of various 2D TMDCs and must be synthesized (CVD). Because of its high quality, the exfoliation method has been used in the majority of data and theory. However, the critical flake size and uniformity of the film limitations have slowed its progress beyond the foundational research. The CVD process, on the other hand, has been investigated for the development of wide-area 2D TMDCs in a scalable and reliable manner. Nonetheless, when compared to those who have been exfoliated, CVD-grown TMDCs have a lower consistency. The 2D material-forming process is influenced by substrate lattice parameters and temperatures, as well as the flux of atomic gases (Zhang et al. 2015). We will concentrate our conversation about the expansion of 2D TMDCs using MOCVD, the approaches, as well as their advantages and disadvantages. MOCVD is a common CVD except that the source materials are metal organic or organic compound precursors (Cheon, Gozum, and Girolami 1997). Varying the atomic composition of atoms at the atomic scale can be used to engineer film consistency, the desired thin, high crystallinity film.

Fig. 2.24 depicts a representative MOCVD process diagram, illustrating the different steps participating in 2D materials synthesis. During the MOCVD process, MOCVD has the following benefits in the growth of 2D TMDCs: (i) it allows for 2D TMDCs to evolve at a large scale and in a consistent manner; (ii) it allows for precise

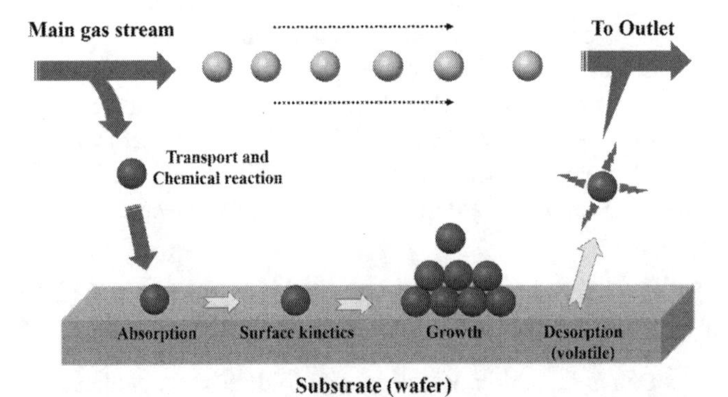

FIGURE 2.24 Deposition process on the substrate and surface processes in MOCVD. Reprinted with permission from Materials Today Journal. Copyright, 2017, Elsevier (Choi et al. 2017).

command of both precursors of metals and chalcogens, allowing for precise control of the morphology and structure of 2D TMDCs. Kang et al. (2015) used molybdenum hexacarbonyl ($Mo(CO)_6$), tungsten hexacarbonyl ($W(CO)_6$), ethylene disulfide ($(C_2H_5)_2S$), and H_2 gas-phase precursors containing an Ar gas carrier to synthesize a monolayer and a few layers on wafer-scale films of MoS_2 and WS_2 on SiO_2 substrates.

On four infused silica substrates, large-scale MoS_2 and WS_2 films (Fig. 2.25a), as well as around 8,000 MoS_2 FET devices are fabricated using a standard photolithography method (Fig. 2.25b). At room temperature, the MoS_2-FETs had high electron mobility (Fig. 2.25c). Fig. 2.25(d) illustrates the time evolution of the monolayer over the entire substratum (t_0). Eichfeld et al. (2015) recently reported they were the first to use $W(CO)_6$ and dimethylselenium ($(CH_3)_2Se$) precursors to develope large areas of mono layer and few layer WSe_2 via MOCVD.

2.8 SURFACE MODIFICATION FOR SPECIFIC AGENT DETECTION

The sensor with specific agent detection is very crucial in industrial and biomedical applications. However, surface modification plays an important role in the sensor design with the detection of such specific agents. Moreover, the detection of these specific agents further depends on the properties of the surface material used. Furthermore, research for various applications based on the surface material, such as metal, semiconductor, oxide layer, and 2D material has been reported and the detailed classification of surface material is shown in Fig. 2.26 (Oliverio et al. 2017).

The shown material consists of properties such as mechanical, thermal, electrical, electronic, and optical that characterize the specific agent detection. The mechanical properties of surface material, for example, characterize pressure detection. Since surface modification with 2D material has a significant effect on optical sensors in the detection of specific agents, the details about the properties of such materials and their applications are discussed in a separate chapter of this book. As a result, the sensor design is classified into three groups based on the specific agent detection: physical, chemical, and biological, with the biological group subdivided into sample detection and interaction groups, as shown in Fig. 2.27. Furthermore, the importance of surface modification is significant for the detection of specific chemical and biological agents because the target sample molecule must interact with the sensor surface molecules rather than the other way around using a solid-liquid interface, a solution-phase involving nanoparticles, or surface chemistry for the target molecules. Therefore, the performance matrix of such chemical and biological sensor greatly depends on the surface material and chemistry used in the sensor surface for its detection (Sonawane and Nimse 2016).

2.8.1 Surface Material and Modification Methods for the Detection of Biological Agents

Surface modification is critical in detecting the biological sample because it needs a reaction between the functional group of target biomolecules and modified surfaces for their immobilization. The most popular biological agent and their functional group of most are listed in Table 2.2. In biological agent detection, surface

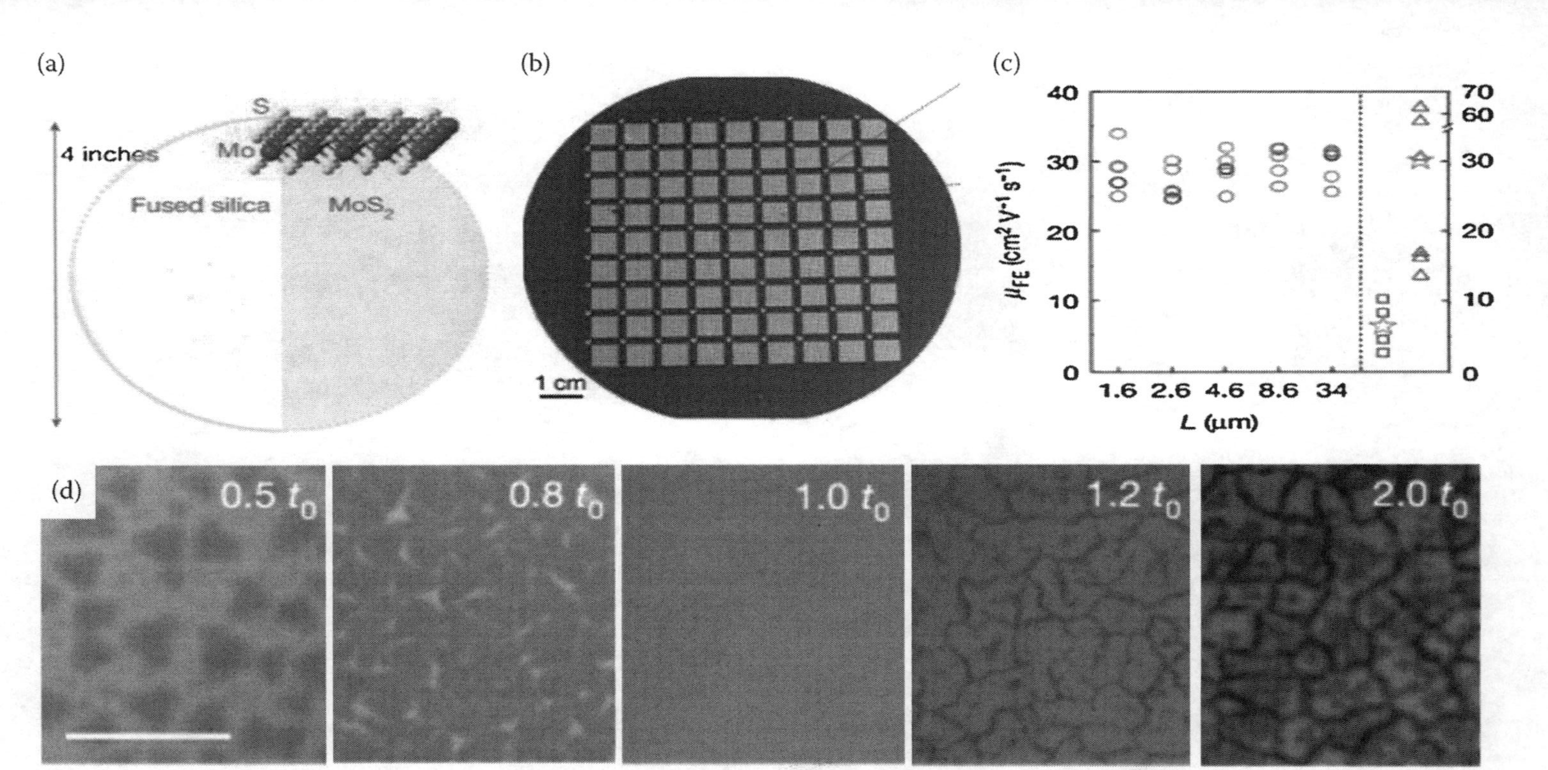

FIGURE 2.25 Using all-gas phase precursors, large-scale MOCVD growth of continuous, a) MoS_2 monolayers on fused silica, b) mass production of 8,000 FET devices is possible due to the scalable development, c) field-effect mobility as determined by five separate length scale FET units, d) optical images of MoS_2 films at various growth times. Reprinted with permission from Materials Today Journal. Copyright, 2017, Elsevier (Choi et al. 2017).

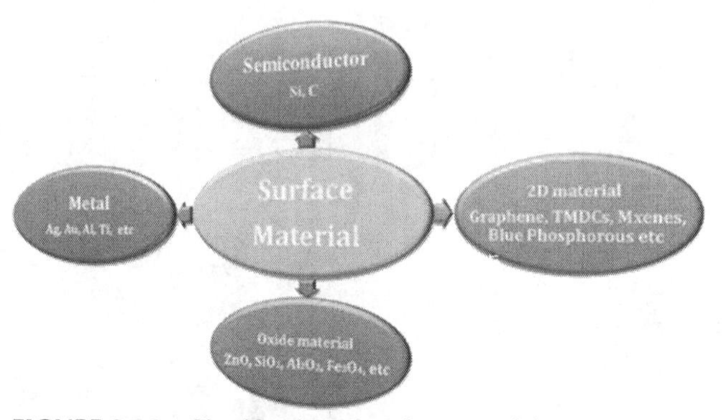

FIGURE 2.26 Classification of surface materials.

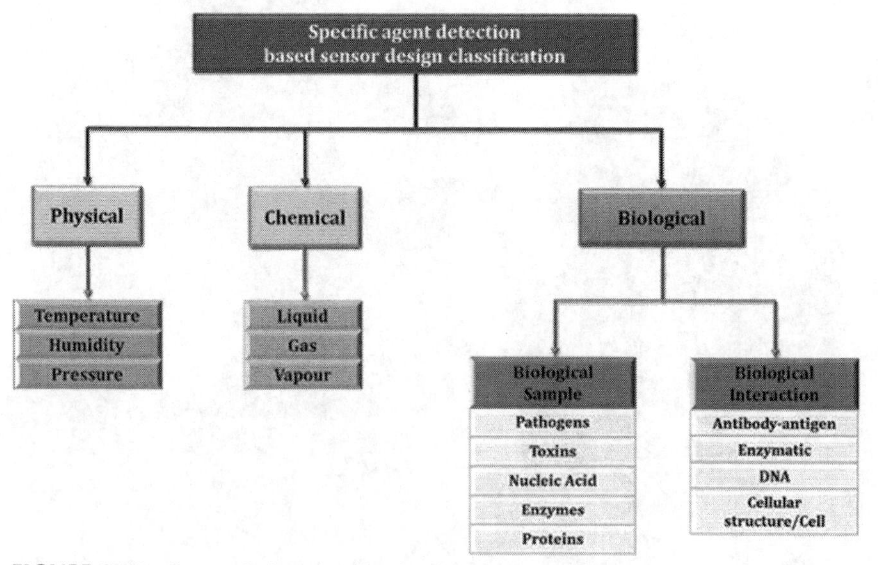

FIGURE 2.27 Sensor design classification based on specific agent detection.

TABLE 2.2
Most Popular Biological Agent and Their Functional Groups

Biological Agent	Functional Group
DNA oligomers	Aldehyde and amine
Proteins	Carboxylic acid, amine, and sulfhydryl
Carbohydrates	Hydroxyl
Glucosamine	Amine

modification mainly depends on its functional group, and the detection efficiency is significantly affected due to the immobilization of the biological agent with the specifically targeted sensor surface.

Surface materials commonly used on sensor substrates include silicon, glass (silica), carbon, gold, and silver, which are discussed in this section, and their modification techniques are summarized in Table 2.2 with their reference. However, other materials such as black phosphorous, TMDCs, graphene, and Maxene are discussed in the respective chapters of this book.

The surface material, silicon, is found mainly in an oxidized state that's stable and is most commonly used as a semiconductor in electronic devices. It is label-free, so it is widely used in real-time detection and electrochemical protein detection. Physical adsorption is a viable option for the immobilization of antibodies because no regulation is required for humidity or temperature regulation. The porous shape of the silicon surface is more effective for immobilization because it has low internal fluorescence, high spot homogeneity, low non-specificity, and low wetting power. Macroporous (pores > 50 nm), mesoporous (10–50 nm), and microporous (pores < 10 nm) are the three forms of porous silicon (larger than 50 nm) where the macroporous silicon surface is particularly well suited to antibody immobilization (Lee et al. 2013).

However, silica (SiO_2) surface material is used to make glass, abundant in nature, and thermally stable because of the large amount of silicon and oxygen. In silica, silicon is protected from the oxide layer from chemical reactions and degradation. The glass substrate is inexpensive, easy to handle, and has good mechanical stability, making it a promising candidate for proteins, DNA, and another biomolecule immobilization. Also, the glass surface serves as a microarray site for identifying various pathogenic DNA and biomarker proteins in the diagnostic area (MacBeath and Schreiber 2000).

Nanoparticle structures like nanowires and nanotubes have numerous applications in biomedical and diagnostic fields. Prior to 1991, the only known types of carbon with potential applications were graphite, diamond, and C60 fullerene; however, carbon nanostructures such as carbon nanotubes (CNT), multiwall faces (MWCNT), carbon dots, carbon fibers, and PDMS have become popular materials in the development of biosensors due to their unique thermal, magnetic, and electrical properties. Chemical vapor deposition, laser ablation, and arc discharge are all popular synthesis methods of these carbon nanostructures. These carbon nanostructures are the latest trend in biosensors due to their diverse properties for biomolecule conjugation because they have a wide surface area, which allows for the most room for the attachment of biomolecules. However, nucleic acid, carbohydrate, and protein immobilization on CNT surfaces allow it into several biological applications. The surface functionalization of carbon-based materials is the most significant step in biomolecule conjugation. The most common techniques for carbon surface modification are covalent and non-covalent modifications (Cui 2007).

Furthermore, inorganic nanoparticles, such as gold, silver, and oxide nanoparticles, have a wide range of applications in colorimetric, drug delivery, cell signaling, and fluorescence detection. But gold nanoparticles (AuNP) have excellent functionality with DNAs, proteins, or Raman probes, so their various applications are reported in gene regulation, molecular diagnostics, protein detection, and cell imaging. Similarly, silver nanoparticles (AgNP) can also be utilized for enzymes and protein immobilization. Since the AgNP surface has the adsorption capability of proteins, it can serve as a biomolecule host matrix. Also, it is an appropriate substrate for isolated enzymes or entire cell immobilization (Sonawane and Nimse 2016).

2.8.2 Surface Material and Modification Methods for the Detection of Chemical Agents

Woo, Na, and Lee (2016) reported a comprehensive review on the physiochemical modified oxide nanowires for specific gas detection with several strategies such as noble metal doping/loading, transition metal oxide doping/decoration, core-shell based structures formation, n-type to p-type nanowires transformation, and preparation of p-type metal oxide nanowires by physicochemical route. However, the comprehensive review of the noble metal catalysts loaded nanowires and metal oxide catalysts doped and decorated nanowires are summarized in Tables 2.2 and 2.3 (Woo, Na, and Lee 2016). Zhang, Zhang, et al. (2019) presented inorganic heterojunction

TABLE 2.3
Surface Modification Techniques for Several Materials

Surface Materials	Surface Modification Techniques	Target Biomolecules Detection	Reference
Silicon	Electrochemical modification	Chemical and bacteria	(Roychaudhuri 2015)
	Covalent modification	Detection of Hela cell	(Guan et al. 2014)
Glass (Silica)	Physisorption Modification	DNA	(Lemeshko et al. 2001)
	Covalent modification		
	A. Aldehyde	DNA	(Fixe et al. 2004)
	B. Epoxy	Phosphorylated DNAs	(Funk et al. 2012)
	C. Carboxydrates	Protein	(Benters et al. 2002)
	D. Diazotization	Protein, DNA	(Moon 1996)
	E. Supramolecules	Proteins	(Oh et al. 2005)
Carbon	Physisorption Modification	Proteins and Enzymes	(Chen et al. 2003)
	Chemical modification		
	A. Carboxyl modification	Protein	(Marshall, Popa-Nita, and Shapter 2006)
	B. Amine modification	DNA, Protein	(Tam 2009)
	C. Epoxy modification	Protein, DNA	(Eitan et al. 2003)
Gold	Covalent modification		
	A. Thiol modification	*E. coli* antibodies	(di Pasqua et al. 2009)
	B. Hydrazide modification	Proteins	(Zhi, Powell, and Turnbull 2006)
Silver	Electrostatic modification	Serum albumins	(Shen 2003)
	Covalent modification	DNA	(J. S. Lee et al. 2007)

sensitization, organic molecules-based surface modification, 1D or 2D materials-based hybrid structures formation, and oxygen vacancy modification to improve gas-sensing performance. Wang et al. (2017) demonstrated the chemical oxidation polymerization of aniline for forming the polyaniline-TiO_2 based nanocomposites sequence on the TiO_2 surface to increase the CO adsorption and result in the transfer of more electrons from CO to polyaniline. The hydrogen band from polyaniline to TiO_2 promotes the photocatalytic oxidation of CO on the composite film sensor that shows the feasibility of creating a sensitive material for gas using hybrid organic-inorganic nanocomposites.

Nulhakim et al. (2017) used radio-frequency magnetron sputtering to develop extremely thin films of Ga-doped ZnO polycrystalline for hydrogen gas sensing that established the relation between the preferred c-axis-oriented thin film microstructure with its sensing efficiency. It is observed that the sample sensitivity to hydrogen increased marginally with the decrease in microcrystalline size at the operating temperature below 330°C. Furthermore, the sensitivity is enhanced greatly with the increase in preferred orientation distribution and concludes that c-axis orientation has a significant impact on the sensitivity.

Yang et al. (2016) utilized hydrothermal techniques to fabricate the MoO_3-based ultralong nanobelts with 200 m length and 200–400 nm width, which shows a good sensing efficiency for trimethylamine (TMS) at 240°C temperature. It also reveals that the sensing response is better in TMA than ammonia, ethanol, methanol, acetone, and toluene during the selectivity test. Liu et al. consider the ultrasonic spray pyrolysis techniques to grow thin-film SnO_2 for the gas sensor. However, it is observed that the gas sensitivity of thin-film SnO_2 is enhanced with the Cu doping at a 0–1 ratio of Cu/Sn.

The synthesis of an Au/NiO-based core-shell structure with uniform dispersion using hydrothermal techniques is presented in Rai et al. (2014). Since Au alone performed exceptionally well for hydrogen sulfide (H_2S) gas sensing, it was found that the Au-NiO core-shell structure has a sensitivity four orders higher than the NiO sphere for H_2S gas. Furthermore, H_2S gas absorbed by the sulfurization layer of the Au particle reduces the potential energy of reaction that makes it easier to transfer an electron from the Au particle to the NiO shell, and increases resistance value. Similarly, Kim et al. (2014) also present the synthesis of a nanostructured NiO hollow sphere, and its surface is decorated by an In_2O_3 layer. Also, the result revealed that the In_2O_3-based decorated nanostructure NiO hollow sphere had nearly five times more sensitivity than NiO spheres for ethanol gas. However, the presence of an n-type semiconductor in In_2O_3 decreased the hole accumulation on the NiO shell surface, significantly reducing the recovery and response time of ethanol gas.

So, this section mainly presented the summary of different types of surface material and its modification and found that the detection of chemical and biological agents needs interaction with surface-attached molecules for its detection.

REFERENCES

Aksit, Mahmut, David P. Toledo, and Richard D. Robinson. 2012. "Scalable Nanomanufacturing of Millimetre-Length 2D Na x CoO_2 Nanosheets." *Journal of Materials Chemistry* 22 (13): 5936–44.

Bao, Wenzhong, Xinghan Cai, Dohun Kim, Karthik Sridhara, and Michael S. Fuhrer. 2013. "High Mobility Ambipolar MoS_2 Field-Effect Transistors: Substrate and Dielectric Effects." *Applied Physics Letters* 102 (4): 42104.

Beiranvand, Razieh. 2021. "Theoretical Investigation of Electronic and Optical Properties of 2D Transition Metal Dichalcogenides MoX_2 (X = S, Se, Te) from First-Principles." *Physica E: Low-Dimensional Systems and Nanostructures* 126: 114416.

Benters, Rüdiger, Christof M. Niemeyer, Denja Drutschmann, Dietmar Blohm, and Dieter Wöhrle. 2002. "DNA Microarrays with PAMAM Dendritic Linker Systems." *Nucleic Acids Research* 30 (2): 10. 10.1093/nar/30.2.e10.

Bianco, Elisabeth, Sheneve Butler, Shishi Jiang, Oscar D. Restrepo, Wolfgang Windl, and Joshua E. Goldberger. 2013. "Stability and Exfoliation of Germanane: A Germanium Graphane Analogue." *ACS Nano* 7 (5): 4414–21.

Bizeto, Marcos A., Ana L. Shiguihara, and Vera R. L. Constantino. 2009. "Layered Niobate Nanosheets: Building Blocks for Advanced Materials Assembly." *Journal of Materials Chemistry* 19 (17): 2512–25.

Bouzidi, Saidi, Mohamed Barhoumi, and Moncef Said. 2021. "Optical Properties of Two-Dimensional AlOCl, BaFCl, and BiOCl Monolayers Using the Density Functional Theory." *Optik* 236: 166678.

Butler, Sheneve Z., Shawna M. Hollen, Linyou Cao, Yi Cui, Jay A. Gupta, Humberto R. Gutiérrez, Tony F. Heinz, et al. 2013. "Progress, Challenges, and Opportunities in Two-Dimensional Materials beyond Graphene." *ACS Nano* 7 (4): 2898–926.

Chang, Pu, Shuo Zhang, Lixiu Guan, Hui Zhang, Guifeng Chen, and Junguang Tao. 2020. "Defect-Mediated Strong Exciton-Phonon Coupling between Flower-Like WS2 Film with Vicinity Layers." *Journal of Luminescence* 226: 117483.

Chaves, A., J. G. Azadani, H. Alsalman, D. R. da Costa, R. Frisenda, A. J. Chaves, S. H. Song, et al. 2020. "Bandgap Engineering of Two-Dimensional Semiconductor Materials." *NPJ 2D Materials and Applications* 4, 29, 1–21.

Chen, Jian, Lilan Zheng, Weiling Yin, Mi Zhang, Yinmei Lu, Zaoli Zhang, Peter J. Klar, Mingkai Li, and Yunbin He. 2020. "Two-Dimensional SnO Ultrathin Epitaxial Films: Pulsed Laser Deposition Growth and Quantum Confinement Effects." *Physica B: Condensed Matter* 599: 412467.

Chen, Robert J., Sarunya Bangsaruntip, Katerina A. Drouvalakis, Nadine Wong Shi Kam, Moonsub Shim, Yiming Li, Woong Kim, Paul J. Utz, and Hongjie Dai. 2003. "Noncovalent Functionalization of Carbon Nanotubes for Highly Specific Electronic Biosensors." *Proceedings of the National Academy of Sciences of the United States of America* 100 (9): 4984–89. 10.1073/pnas.0837064100.

Cheon, Jinwoo, John E. Gozum, and Gregory S. Girolami. 1997. "Chemical Vapor Deposition of MoS2 and TiS2 Films From the Metal-Organic Precursors Mo(S-t-Bu)4 and Ti(S-t-Bu)4." *Chemistry of Materials* 9 (8): 1847–53.

Choi, Wonbong, Nitin Choudhary, Gang Hee Han, Juhong Park, Deji Akinwande, and Young Hee Lee. 2017. "Recent Development of Two-Dimensional Transition Metal Dichalcogenides and Their Applications." *Materials Today* 20 (3): 116–30.

Choudhary, Nitin, Juhong Park, Jun Yeon Hwang, Hee-Suk Chung, Kenneth H. Dumas, Saiful I. Khondaker, Wonbong Choi, and Yeonwoong Jung. 2016. "Centimeter Scale Patterned Growth of Vertically Stacked Few Layer Only 2D MoS2/WS2 van der Waals Heterostructure." *Scientific Reports* 6 (1): 1–7.

Choudhary, Nitin, Patel Mumukshu D., Juhong Park, Ben Sirota, and Wonbong Choi. 2016. "Synthesis of Large Scale MoS2 for Electronics and Energy Applications." *Journal of Materials Research* 31 (7): 824.

Ci, Lijie, Li Song, Chuanhong Jin, Deep Jariwala, Dangxin Wu, Yongjie Li, Anchal Srivastava, et al. 2010. "Atomic Layers of Hybridized Boron Nitride and Graphene Domains." *Nature Materials* 9 (5): 430–5.

Coleman, Jonathan N., Mustafa Lotya, Arlene O'Neill, Shane D. Bergin, Paul J. King, Umar Khan, Karen Young, et al. 2011. "Two-Dimensional Nanosheets Produced by Liquid Exfoliation of Layered Materials." *Science* 331 (6017): 568–71.

Cui, Daxiang. 2007. "Advances and Prospects on Biomolecules Functionalized Carbon Nanotubes." *Journal of Nanoscience and Nanotechnology*, 7(4–5): 1298–314. 10.1166/jnn.2007.654.

De Sousa, D. J. P., L. V. De Castro, Diego Rabelo da Costa, J. Milton Pereira Jr, and Tony Low. 2017. "Multilayered Black Phosphorus: From a Tight-Binding to a Continuum Description." *Physical Review B* 96 (15): 155427.

di Pasqua, Anthony J., Richard E. Mishler, Yan Li Ship, James C. Dabrowiak, and Tewodros Asefa. 2009. "Preparation of Antibody-Conjugated Gold Nanoparticles." *Materials Letters* 63(21): 1876–79. 10.1016/j.matlet.2009.05.070.

Dresselhaus, M. S., A. Jorio, and R. Saito. 2010. "Characterizing Graphene, Graphite, and Carbon Nanotubes by Raman Spectroscopy." *Annual Review of Condensed Matter Physics* 1 (1): 89–108.

Dryfe, Robert A. W. 2019. "2D Transition Metal Chalcogenides and van der Waals Heterostructures: Fundamental Aspects of Their Electrochemistry." *Current Opinion in Electrochemistry* 13: 119–24.

Ebina, Y., T. Sasaki, and M. Watanabe. 2002. "Study on Exfoliation of Layered Perovskite-Type Niobates." *Solid State Ionics* 151 (1–4): 177–82.

Eichfeld, Sarah M., Lorraine Hossain, Yu-Chuan Lin, Aleksander F. Piasecki, Benjamin Kupp, A. Glen Birdwell, Robert A. Burke, et al. 2015. "Highly Scalable, Atomically Thin WSe2 Grown via Metal-Organic Chemical Vapor Deposition." *ACS Nano* 9 (2): 2080–7.

Eitan, Ami, Kuiyang Jiang, Dukes Doug, Rodney Andrews, and Linda S. Schadler. 2003. "Surface Modification of Multiwalled Carbon Nanotubes: Toward the Tailoring of the Interface in Polymer Composites." *Chemistry of Materials* 15 (16): 3198–201.10.1021/cm020975d.

Esaki, Leo, and Ray Tsu. 1970. "Superlattice and Negative Differential Conductivity in Semiconductors." *IBM Journal of Research and Development* 14 (1): 61–5.

Fei, Ruixiang, and Li Yang. 2014. "Strain-Engineering the Anisotropic Electrical Conductance of Few-Layer Black Phosphorus." *Nano Letters* 14 (5): 2884–9.

Fixe, F., M. Dufva, P. Telleman, and C. B. Christensen. 2004. "Functionalization of Poly (Methyl Methacrylate) (PMMA) as a Substrate for DNA Microarrays." *Nucleic Acids Research* 32(1): e9–e9. 10.1093/nar/gng157.

Frey, G. L., S. Elani, M. Homyonfer, Y. Feldman, and R. Tenne. 1998. "Optical-Absorption Spectra of Inorganic Fullerenelike M S 2 (M = Mo, W)." *Physical Review B* 57 (11): 6666.

Fuhrer, Michael S., and James Hone. 2013. "Measurement of Mobility in Dual-Gated MoS2 Transistors." *Nature Nanotechnology* 8 (3): 146.

Fukuda, Katsutoshi, Kosho Akatsuka, Yasuo Ebina, Renzhi Ma, Kazunori Takada, Izumi Nakai, and Takayoshi Sasaki. 2008. "Exfoliated Nanosheet Crystallite of Cesium Tungstate with 2D Pyrochlore Structure: Synthesis, Characterization, and Photochromic Properties." *ACS Nano* 2 (8): 1689–95.

Funk, Christian, Paul M. Dietrich, Thomas Gross, Hyegeun Min, Wolfgang E. S. Unger, and Wilfried Weigel. 2012. "Epoxy-Functionalized Surfaces for Microarray Applications: Surface Chemical Analysis and Fluorescence Labeling of Surface Species." In *Surface and Interface Analysis*, 44: 890–94. John Wiley & Sons, Ltd. 10.1002/sia.3856.

Gjerding, Morten Niklas, René Petersen, Thomas Garm Pedersen, Niels Asger Mortensen, and Kristian Sommer Thygesen. 2017. "Layered van Der Waals Crystals with Hyperbolic Light Dispersion." *Nature Communications* 8 (1): 1–8.

Golovynskyi, Sergii, Matteo Bosi, Luca Seravalli, and Baikui Li. 2021. "MoS2 Two-Dimensional Quantum Dots with Weak Lateral Quantum Confinement: Intense Exciton and Trion Photoluminescence." *Surfaces and Interfaces* 23: 100909.

Guan, Bin, Astrid Magenau, Simone Ciampi, Katharina Gaus, Peter J. Reece, and J. Justin Gooding. 2014. "Antibody Modified Porous Silicon Microparticles for the Selective Capture of Cells." *Bioconjugate Chemistry* 25(7): 1282–89. 10.1021/bc500144u.

Guo, Zhinan, Han Zhang, Shunbin Lu, Zhiteng Wang, Siying Tang, Jundong Shao, Zhengbo Sun, et al. 2015. "From Black Phosphorus to Phosphorene: Basic Solvent Exfoliation, Evolution of Raman Scattering, and Applications to Ultrafast Photonics." *Advanced Functional Materials* 25 (45): 6996–7002.

Irshad, Rabia, Kamran Tahir, Baoshan Li, Zunaira Sher, Jawad Ali, and Sadia Nazir. 2018. "A Revival of 2D Materials, Phosphorene: Its Application as Sensors." *Journal of Industrial and Engineering Chemistry* 64: 60–9.

Island, Joshua O., Michele Buscema, Mariam Barawi, José M. Clamagirand, José R. Ares, Carlos Sánchez, Isabel J. Ferrer, Gary A. Steele, Herre S. J. van der Zant, and Andres Castellanos-Gomez. 2014. "Ultrahigh Photoresponse of Few-Layer TiS3 Nanoribbon Transistors." *Advanced Optical Materials* 2 (7): 641–5.

Island, Joshua O., Mariam Barawi, Robert Biele, Adrián Almazán, José M. Clamagirand, José R. Ares, Carlos Sánchez, et al. 2015. "TiS3 Transistors with Tailored Morphology and Electrical Properties." *Advanced Materials* 27 (16): 2595–601.

Jacob, Zubin, Leonid V. Alekseyev, and Evgenii Narimanov. 2006. "Optical Hyperlens: Far-Field Imaging Beyond the Diffraction Limit." *Optics Express* 14 (18): 8247–56.

Jayakumar, Ashtami, Anju Surendranath, and P. V. Mohanan. 2018. "2D Materials for Next Generation Healthcare Applications." *International Journal of Pharmaceutics* 551 (1–2): 309–21.

Jin, Yingdi, Xingxing Li, and Jinlong Yang. 2015. "Single Layer of MX 3 (M = Ti, Zr; X = S, Se, Te): A New Platform for Nano-Electronics and Optics." *Physical Chemistry Chemical Physics* 17 (28): 18665–9.

Kang, Jun, Hasan Sahin, and François M. Peeters. 2015. "Mechanical Properties of Monolayer Sulphides: A Comparative Study between MoS2, HfS2 and TiS3." *Physical Chemistry Chemical Physics* 17 (41): 27742–9.

Kang, Kibum, Saien Xie, Lujie Huang, Yimo Han, Pinshane Y. Huang, Kin Fai Mak, Cheol-Joo Kim, David Muller, and Jiwoong Park. 2015. "High-Mobility Three-Atom-Thick Semiconducting Films with Wafer-Scale Homogeneity." *Nature* 520 (7549): 656–60.

Keum, Dong Hoon, Suyeon Cho, Jung Ho Kim, Duk-Hyun Choe, Ha-Jun Sung, Min Kan, Haeyong Kang, et al. 2015. "Bandgap Opening in Few-Layered Monoclinic MoTe2." *Nature Physics* 11 (6): 482–6.

Kim, Jaemyung, Franklin Kim, and Jiaxing Huang. 2010. "Seeing Graphene-Based Sheets." *Materials Today* 13 (3): 28–38.

Kim, Jaemyung, Laura J. Cote, Franklin Kim, and Jiaxing Huang. 2010. "Visualizing Graphene Based Sheets by Fluorescence Quenching Microscopy." *Journal of the American Chemical Society* 132 (1): 260–7.

Kim, Hyo Joong, Hyun Mook Jeong, Tae Hyung Kim, Jae Ho Chung, Yun Chan Kang, and Jong Heun Lee. 2014. "Enhanced Ethanol Sensing Characteristics of In2O3-Decorated NiO Hollow Nanostructures via Modulation of Hole Accumulation Layers." *ACS Applied Materials and Interfaces* 6(20): 18197–204. 10.1021/am5051923.

Kroemer, Herbert. 2001. "Nobel Lecture: Quasielectric Fields and Band Offsets: Teaching Electrons New Tricks." *Reviews of Modern Physics* 73 (3): 783.

Kumbhakar, Partha, Chinmayee Chowde Gowda, Preeti Lata Mahapatra, Madhubanti Mukherjee, Kirtiman Deo Malviya, Mohamed Chaker, Amreesh Chandra, et al. 2021. "Emerging 2D Metal Oxides and Their Applications." *Materials Today* 45: 142–68.

Lashgari, Hamed, Arash Boochani, Ashkan Shekaari, Shahram Solaymani, Elmira Sartipi, and Rohollah Taghavi Mendi. 2016. "Electronic and Optical Properties of 2D Graphene-Like ZnS: DFT Calculations." *Applied Surface Science* 369: 76–81.

Late, Dattatray J., Anha Bhat, and Chandra Sekhar Rout. 2019. "Fundamentals and Properties of 2D Materials in General and Sensing Applications." In*Fundamentals and Sensing Applications of 2D Materials* (Woodhead Publishing Series in Electronic and Optical Materials): 5–24. Woodhead Publishing. ISBN 9780081025772.

Late, Dattatray J., Bin Liu, H. S. S. Ramakrishna Matte, C. N. R. Rao, and Vinayak P. Dravid. 2012. "Rapid Characterization of Ultrathin Layers of Chalcogenides on SiO2/Si Substrates." *Advanced Functional Materials* 22 (9): 1894–905.

Lee, Joonhyung, Piyush Dak, Yeonsung Lee, Heekyeong Park, Woong Choi, Muhammad A. Alam, and Sunkook Kim. 2014. "Two-Dimensional Layered MoS2 Biosensors Enable Highly Sensitive Detection of Biomolecules." *Scientific Reports* 4 (1): 1–7.

Lee, Si Young, Dinh Loc Duong, Quoc An Vu, Youngjo Jin, Philip Kim, and Young Hee Lee. 2015. "Chemically Modulated Band Gap in Bilayer Graphene Memory Transistors with High on/off Ratio." *ACS Nano* 9 (9): 9034–42.

Lee, Sang Wook, Soyoun Kim, Johan Malm, Ok Chan Jeong, Hans Lilja, and Thomas Laurell. 2013. "Improved Porous Silicon Microarray Based Prostate Specific Antigen Immunoassay by Optimized Surface Density of the Capture Antibody." *Analytica Chimica Acta* 796: 108–14. 10.1016/j.aca.2013.06.041

Lee, Jae Seung, Abigail K. R. Lytton-Jean, Sarah J. Hurst, and Chad A. Mirkin. 2007. "Silver Nanoparticle - Oligonucleotide Conjugates Based on DNA with Triple Cyclic Disulfide Moieties." *Nano Letters* 7(7): 2112–15. 10.1021/nl071108g.

Lemeshko, S. v., T. Powdrill, Y. Y. Belosludtsev, and M. Hogan. 2001. "Oligonucleotides Form a Duplex with Non-Helical Properties on a Positively Charged Surface." *Nucleic Acids Research* 29(14): 3051–3058. 10.1093/nar/29.14.3051.

Li, Guohong, and Eva Y. Andrei. 2007. "Observation of Landau Levels of Dirac Fermions in Graphite." *Nature Physics* 3 (9): 623–7.

Li, Likai, Jonghwan Kim, Chenhao Jin, Guo Jun Ye, Diana Y. Qiu, H. Felipe, Zhiwen Shi, et al. 2017. "Direct Observation of the Layer-Dependent Electronic Structure in Phosphorene." *Nature Nanotechnology* 12 (1): 21.

Li, Likai, Yijun Yu, Guo Jun Ye, Qingqin Ge, Xuedong Ou, Hua Wu, Donglai Feng, Xian Hui Chen, and Yuanbo Zhang. 2014. "Black Phosphorus Field-Effect Transistors." *Nature Nanotechnology* 9 (5): 372.

Lin, Wei-Hsiang, Ting-Hui Chen, Jan-Kai Chang, Jieh-I. Taur, Yuan-Yen Lo, Wei-Li Lee, Chia-Seng Chang, Wei-Bin Su, and Chih-I. Wu. 2014. "A Direct and Polymer-Free Method for Transferring Graphene Grown by Chemical Vapor Deposition to Any Substrate." *ACS Nano* 8 (2): 1784–91.

Lin, Shisheng, Xiaoqiang Li, Peng Wang, Zhijuan Xu, Shengjiao Zhang, Huikai Zhong, Zhiqian Wu, Wenli Xu, and Hongsheng Chen. 2015. "Interface Designed MoS 2/GaAs Heterostructure Solar Cell with Sandwich Stacked Hexagonal Boron Nitride." *Scientific Reports* 5 (1): 1–9.

Liu, Erfu, Yajun Fu, Yaojia Wang, Yanqing Feng, Huimei Liu, Xiangang Wan, Wei Zhou, et al. 2015. "Integrated Digital Inverters Based on Two-Dimensional Anisotropic ReS2 Field-Effect Transistors." *Nature Communications* 6 (1): 1–7.

Liu, Han, Adam T. Neal, Zhen Zhu, Zhe Luo, Xianfan Xu, David Tománek, and Peide D. Ye. 2014. "Phosphorene: An Unexplored 2D Semiconductor with a High Hole Mobility." *ACS Nano* 8 (4): 4033–41.

Liu, Jun-Jie, Chu-Ya Niu, Yao Wu, Dan Tan, Yang Wang, Ming-Da Ye, Yang Liu, et al. 2016. "CryoEM Structure of Yeast Cytoplasmic Exosome Complex." *Cell Research* 26 (7): 822–37.

Low, Tony, A. S. Rodin, A. Carvalho, Yongjin Jiang, Han Wang, Fengnian Xia, and A. H. Castro Neto. 2014. "Tunable Optical Properties of Multilayer Black Phosphorus Thin Films." *Physical Review B* 90 (7): 75434.

Lu, Dylan, Jimmy J. Kan, Eric E. Fullerton, and Zhaowei Liu. 2014. "Enhancing Spontaneous Emission Rates of Molecules Using Nanopatterned Multilayer Hyperbolic Metamaterials." *Nature Nanotechnology* 9 (1): 48–53.

Ma, Jinlong, Yani Chen, Zheng Han, and Wu Li. 2016. "Strong Anisotropic Thermal Conductivity of Monolayer WTe2." *2D Materials* 3 (4): 45010.

Ma, Xuezhi, Qiushi Liu, Da Xu, Yangzhi Zhu, Sanggon Kim, Yongtao Cui, Lanlan Zhong, and Ming Liu. 2017. "Capillary-Force-Assisted Clean-Stamp Transfer of Two-Dimensional Materials." *Nano Letters* 17 (11): 6961–7.

MacBeath, G., and S. L. Schreiber. 2000. "Printing Proteins as Microarrays for High-Throughput Function Determination." Science 289(5485): 1760–63. 10.1126/science.289.5485.1760.

Mak, Kin Fai, Changgu Lee, James Hone, Jie Shan, and Tony F. Heinz. 2010. "Atomically Thin MoS_2: A New Direct-Gap Semiconductor." *Physical Review Letters* 105 (13): 136805.

Mannix, Andrew J., Xiang-Feng Zhou, Brian Kiraly, Joshua D. Wood, Diego Alducin, Benjamin D. Myers, Xiaolong Liu, et al. 2015. "Synthesis of Borophenes: Anisotropic, Two-Dimensional Boron Polymorphs." *Science* 350 (6267): 1513–6.

Marshall, Matthew W. , Simina Popa-Nita, and Joseph G. Shapter. 2006. "Measurement of Functionalised Carbon Nanotube Carboxylic Acid Groups Using a Simple Chemical Process." *Carbon* 44 (7): 1137–41. 10.1016/j.carbon.2005.11.010.

Miller, David L., Kevin D. Kubista, Gregory M. Rutter, Ming Ruan, Walt A. de Heer, Phillip N. First, and Joseph A. Stroscio. 2009. "Observing the Quantization of Zero Mass Carriers in Graphene." *Science* 324 (5929): 924–7.

Miremadi, Bijan K., and S. Roy Morrison. 1987. "High Activity Catalyst from Exfoliated MoS2." *Journal of Catalysis* 103 (2): 334–45.

Moon, Joong Ho, Ji Won Shin, Sang Youl Kim, and Joon Won Park. 1996. "Formation of Uniform Aminosilane Thin Layers: An Imine Formation to Measure Relative Surface Density of the Amine Group." Langmuir 12(20): 4621–24. 10.1021/la9604339.

Novoselov, Kostya S., Andre K. Geim, Sergei V. Morozov, Dingde Jiang, Yanshui Zhang, Sergey V. Dubonos, Irina V. Grigorieva, and Alexandr A. Firsov. 2004. "Electric Field Effect in Atomically Thin Carbon Films." *Science* 306 (5696): 666–9.

Novoselov, Kostya S., D. Jiang, F. Schedin, T. J. Booth, V. V. Khotkevich, S. V. Morozov, and Andre K. Geim. 2005. "Two-Dimensional Atomic Crystals." *Proceedings of the National Academy of Sciences* 102 (30): 10451–3.

Nulhakim, Lukman, Hisao Makino, Seiichi Kishimoto, Junichi Nomoto, and Tetsuya Yamamoto. 2017. "Enhancement of the Hydrogen Gas Sensitivity by Large Distribution of C-Axis Preferred Orientation in Highly Ga-Doped ZnO Polycrystalline Thin Films." *Materials Science in Semiconductor Processing* 68: 322–326. 10.1016/j.mssp.2017.06.045.

Oh, Sang Wook, Jung Dae Moon, Hyo Jin Lim, Sang Yeol Park, Taisun Kim, Jaebum Park, Moon Hi Han, Michael Snyder, and Eui Yul Choi. 2005. "Calixarene Derivative as a Tool for Highly Sensitive Detection and Oriented Immobilization of Proteins in a Microarray Format through Noncovalent Molecular Interaction." *The FASEB Journal* 19(10): 1335–37. 10.1096/fj.04-2098fje.

Olofinjana, Bolutife, Gabriel Egharevba, Bidini Taleatu, Olumide Akinwunmi, Ezekiel Oladele Ajayi, et al. 2011. "MOCVD of Molybdenum Sulphide Thin Film via Single Solid Source Precursor Bis-(Morpholinodithioato-s, s')-Mo." *Journal of Modern Physics* 2 (05): 341.

Oliverio, Manuela, Sara Perotto, Gabriele C. Messina, Laura Lovato, and Francesco de Angelis. 2017. "Chemical Functionalization of Plasmonic Surface Biosensors: A Tutorial Review on Issues, Strategies, and Costs." *ACS Applied Materials and Interfaces* 9 (35): 29394–411. 10.1021/acsami.7b01583

Osada, Minoru, Genki Takanashi, Bao-Wen Li, Kosho Akatsuka, Yasuo Ebina, Kanta Ono, Hiroshi Funakubo, Kazunori Takada, and Takayoshi Sasaki. 2011. "Controlled Polarizability of One-Nanometer-Thick Oxide Nanosheets for Tailored, High-$κ$ Nanodielectrics." *Advanced Functional Materials* 21 (18): 3482–7.

Oughaddou, Hamid, Hanna Enriquez, Mohammed Rachid Tchalala, Handan Yildirim, Andrew J. Mayne, Azzedine Bendounan, Gérald Dujardin, Mustapha Ait Ali, and Abdelkader Kara. 2015. "Silicene, a Promising New 2D Material." *Progress in Surface Science* 90 (1): 46–83.

Pan, Long, Yi-Tao Liu, Xu-Ming Xie, and Xiong-Ying Ye. 2016. "Facile and Green Production of Impurity-Free Aqueous Solutions of WS2 Nanosheets by Direct Exfoliation in Water." *Small* 12 (48): 6703–13.

Rai, Prabhakar, Ji Wook Yoon, Hyun Mook Jeong, Su Jin Hwang, Chang Hoon Kwak, and Jong Heun Lee. 2014. "Design of Highly Sensitive and Selective Au@NiO Yolk-Shell Nanoreactors for Gas Sensor Applications." *Nanoscale* 6 (14): 8292–8299. 10.1039/c4nr01906g.

Roychaudhuri, C. 2015. "A Review on Porous Silicon Based Electrochemical Biosensors: Beyond Surface Area Enhancement Factor." *Sensors and Actuators, B: Chemical* 210: 310–323. 10.1016/j.snb.2014.12.089.

Shen, Xingcan. 2003. "Hysteresis Effects of the Interaction between Serum Albumins and Silver Nanoparticles." *Science in China Series B* 46(4): 387. 10.1360/02yb0062.

Sonawane, Mukesh Digambar, and Satish Balasaheb Nimse. 2016. "Surface Modification Chemistries of Materials Used in Diagnostic Platforms with Biomolecules." *Journal of Chemistry*, 2016. 10.1155/2016/9241378

Schliehe, Constanze, Beatriz H. Juarez, Marie Pelletier, Sebastian Jander, Denis Greshnykh, Mona Nagel, Andreas Meyer, et al. 2010. "Ultrathin PbS Sheets by Two-Dimensional Oriented Attachment." *Science* 329 (5991): 550–3.

Shaikh, Gaushiya A., Dhara Raval, Bindiya Babariya, Sanjeev K. Gupta, and P. N. Gajjar. 2020. "An Ab-Initio Study of Blue Phosphorene Monolayer: Electronic, Vibrational and Optical Properties." *Materials Today: Proceedings* 1–4, ISSN 2214-7853.

Shi, Yumeng, Henan Li, and Lain-Jong Li. 2015. "Recent Advances in Controlled Synthesis of Two-Dimensional Transition Metal Dichalcogenides via Vapour Deposition Techniques." *Chemical Society Reviews* 44 (9): 2744–56.

Silva-Guillén, J. A., Enric Canadell, Pablo Ordejón, Francisco Guinea, and Rafael Roldán. 2017. "Anisotropic Features in the Electronic Structure of the Two-Dimensional Transition Metal Trichalcogenide TiS_3: Electron Doping and Plasmons." *2D Materials* 4 (2): 25085.

Sun, J., X. Li, C. A. Ullrich, and J. Yang. 2020. "Excitons in Bent Black Phosphorus Nanoribbons: Multiple Excitonic Funnels." *Materials Today Advances* 7: 100096.

Tam, Phuong Dinh, Nguyen van Hieu, Chien Nguyen Duc, Le Anh Tuan, Mai Anh Tuan. 2009. "DNA Sensor Development Based on Multi-Wall Carbon Nanotubes for Label-Free Influenza Virus (Type A) Detection." *Journal of Immunological Methods* 350 (1–2): 118–24. 10.1016/j.jim.2009.08.002.

Tan, Alvin T. L., Jaemyung Kim, Jing-Kai Huang, Lain-Jong Li, and Jiaxing Huang. 2013. "Seeing Two-Dimensional Sheets on Arbitrary Substrates by Fluorescence Quenching Microscopy." *Small* 9 (19): 3253–8.

Tan, Teng, Xiantao Jiang, Cong Wang, Baicheng Yao, and Han Zhang. 2020. "2D Material Optoelectronics for Information Functional Device Applications: Status and Challenges." *Advanced Science* 7 (11): 2000058.

Tao, Xiaoming. 2005. "Wearable Photonics Based on Integrative Polymeric Photonic Fibres." In Woodhead Publishing Series in Textiles, Wearable Electronics and Photonics, 136–54. Woodhead Publishing. ISBN 9781855736054.

Tareen, Ayesha Khan, Karim Khan, Muhammad Aslam, Xinke Liu, and Han Zhang. 2021. "Confinement in Two-Dimensional Materials: Major Advances and Challenges in the Emerging Renewable Energy Conversion and Other Applications." *Progress in Solid State Chemistry* 61: 100294.

Thripuranthaka, M., and Dattatray J. Late. 2014. "Temperature Dependent Phonon Shifts in Single-Layer WS_2." *ACS Applied Materials & Interfaces* 6 (2): 1158–63.

Tien, Dung Hoang, Jun-Young Park, Ki Buem Kim, Naesung Lee, Taekjib Choi, Philip Kim, Takashi Taniguchi, Kenji Watanabe, and Yongho Seo. 2016. "Study of Graphene-Based 2D-Heterostructure Device Fabricated by All-Dry Transfer Process." *ACS Applied Materials & Interfaces* 8 (5): 3072–8.

Torun, E., H. Sahin, S. Cahangirov, Angel Rubio, and F. M. Peeters. 2016. "Anisotropic Electronic, Mechanical, and Optical Properties of Monolayer WTe2." *Journal of Applied Physics* 119 (7): 74307.

Wang, Gang, Alexey Chernikov, Mikhail M. Glazov, Tony F. Heinz, Xavier Marie, Thierry Amand, and Bernhard Urbaszek. 2018. "Colloquium: Excitons in Atomically Thin Transition Metal Dichalcogenides." *Reviews of Modern Physics* 90 (2): 21001.

Wang, Chong, Shenyang Huang, Qiaoxia Xing, Yuangang Xie, Chaoyu Song, Fanjie Wang, and Hugen Yan. 2020. "Van der Waals Thin Films of WTe2 for Natural Hyperbolic Plasmonic Surfaces." *Nature Communications* 11 (1): 1–9.

Woo, Hyung Sik, Chan Woong Na, and Jong Heun Lee. 2016. "Design of Highly Selective Gas Sensors via Physicochemical Modification of Oxide Nanowires: Overview." Sensors 16(9), 1531. 10.3390/s16091531.

Wang, Zhongming, Xiaoying Peng, Chuyun Huang, Xun Chen, Wenxin Dai, and Xianzhi Fu. 2017. "CO Gas Sensitivity and Its Oxidation over TiO2 Modified by PANI under UV Irradiation at Room Temperature." *Applied Catalysis B: Environmental* 219: 379–90. 10.1016/j.apcatb.2017.07.080.

Yang, He, Henri Jussila, Anton Autere, Hannu-Pekka Komsa, Guojun Ye, Xianhui Chen, Tawfique Hasan, and Zhipei Sun. 2017. "Optical Waveplates Based on Birefringence of Anisotropic Two-Dimensional Layered Materials." *ACS Photonics* 4 (12): 3023–30.

Yang, Shuang, Yueli Liu, Wen Chen, Wei Jin, Jing Zhou, Han Zhang, and Galina S. Zakharova. 2016. "High Sensitivity and Good Selectivity of Ultralong MoO3 Nanobelts for Trimethylamine Gas." *Sensors and Actuators, B: Chemical* 226: 478–85. 10.1016/j.snb.2015.12.005.

You, J. W., S. R. Bongu, Q. Bao, and N. C. Panoiu. 2018. "Nonlinear Optical Properties and Applications of 2D Materials: Theoretical and Experimental Aspects." *Nanophotonics* 8 (1): 63–97.

Zhang, Hua. 2015. "Ultrathin Two-Dimensional Nanomaterials." *ACS Nano* 9 (10): 9451–69.

Zhang, Guowei, Andrey Chaves, Shenyang Huang, Fanjie Wang, Qiaoxia Xing, Tony Low, and Hugen Yan. 2018. "Determination of Layer-Dependent Exciton Binding Energies in Few-Layer Black Phosphorus." *Science Advances* 4 (3): eaap9977.

Zhang, Rui, and Rebecca Cheung. 2016. "Mechanical Properties and Applications of Two-Dimensional Materials." *Two-Dimensional Materials-Synthesis, Characterization and Potential Applications*. Rijeka, Croatia: InTech: 219–46.

Zhang, Guohui, Aleix G. Güell, Paul M. Kirkman, Robert A. Lazenby, Thomas S. Miller, and Patrick R. Unwin. 2016. "Versatile Polymer-Free Graphene Transfer Method and Applications." *ACS Applied Materials & Interfaces* 8 (12): 8008–16.

Zhang, Guowei, Shenyang Huang, Andrey Chaves, Chaoyu Song, V Ongun Özçelik, Tony Low, and Hugen Yan. 2017. "Infrared Fingerprints of Few-Layer Black Phosphorus." *Nature Communications* 8 (1): 1–9.

Zhang, Xin-Quan, Chin-Hao Lin, Yu-Wen Tseng, Kuan-Hua Huang, and Yi-Hsien Lee. 2015. "Synthesis of Lateral Heterostructures of Semiconducting Atomic Layers." *Nano Letters* 15 (1): 410–5.

Zhang, Jun, Zeping Peng, Ajay Soni, Yanyuan Zhao, Yi Xiong, Bo Peng, Jianbo Wang, Mildred S Dresselhaus, and Qihua Xiong. 2011. "Raman Spectroscopy of Few-Quintuple Layer Topological Insulator Bi2Se3 Nanoplatelets." *Nano Letters* 11 (6): 2407–14.

Zhang, Yuanbo, Tsung-Ta Tang, Caglar Girit, Zhao Hao, Michael C. Martin, Alex Zettl, Michael F. Crommie, Y. Ron Shen, and Feng Wang. 2009. "Direct Observation of a Widely Tunable Bandgap in Bilayer Graphene." *Nature* 459 (7248): 820–3.

Zhang, Yue-Jiao, Rui-Ning Wang, Guo-Yi Dong, Shu-Fang Wang, Guang-Sheng Fu, and Jiang-Long Wang. 2019. "Mechanical Properties of 1 T-, 1 T′-, and 1 H-MX2 Monolayers and Their 1 H/1 T′-MX2 (M= Mo, W and X= S, Se, Te) Heterostructures." *AIP Advances* 9 (12): 125208.

Zhang, Enze, Peng Wang, Zhe Li, Haifeng Wang, Chaoyu Song, Ce Huang, Zhi-Gang Chen, et al. 2016. "Tunable Ambipolar Polarization-Sensitive Photodetectors Based on High-Anisotropy ReSe2 Nanosheets." *ACS Nano* 10 (8): 8067–77.

Zhang, Qiankun, Rongjie Zhang, Jiancui Chen, Wanfu Shen, Chunhua An, Xiaodong Hu, Mingli Dong, Jing Liu, and Lianqing Zhu. 2019. "Remarkable Electronic and Optical Anisotropy of Layered 1T'-WTe2 2D Materials." *Beilstein Journal of Nanotechnology* 10 (1): 1745–53.

Zhang, Kaixuan, Libo Zhang, Li Han, Lin Wang, Zhiqingzi Chen, Huaizhong Xing, and Xiaoshuang Chen. 2021. "Recent Progress and Challenges Based on Two-Dimensional Material Photodetectors." *Nano Express* 2 (1): 12001.

Zhi, Zheng Liang, Andrew K. Powell, and Jeremy E. Turnbull. 2006. "Fabrication of Carbohydrate Microarrays on Gold Surfaces: Direct Attachment of Nonderivatized Oligosaccharides to Hydrazide-Modified Self-Assembled Monolayers." *Analytical Chemistry* 78(14): 4786–93. 10.1021/ac060084f.

3 Theoretical Design and Analysis of Surface Plasmon Resonance Chemical Sensors Assisted by 2D Materials

3.1 INTRODUCTION

Surface plasmon resonance (SPR) has now been a mature technology for various real-time in situ monitoring systems. SPR has been fascinated by the interaction of photons and electrons at metallic and dielectric interfaces. These phenomena have been utilized to sense various biochemical, physical, and chemical parameters quickly and accurately. In this chapter, theoretical discussion and analysis are presented based on the Kretschmann configuration utilizing gold (Au) and silver (Ag) metals with 2D material for the purpose of improving the sensitivity of SPR-sensing chips. In this chapter, six-layer SPR structural sensors (Kretschmann geometry) assisted by gold and black phosphorus, sandwich graphene layer is presented which holds a nice absorption capability. Theoretical analysis is then extended by introducing a silver layer in between black phosphorus and graphene layers. Since gold is chemically stable and not prone to oxidizing as compared to silver; hence, one more layer of different material, such as 2D materials, is introduced after a coating of silver to avoid an oxidation problem (Singh and Raghuwanshi 2019; Singh et al. 2020b; Srivastava and Jha 2019). Further improvement is possible by involving a WSe_2 layer instead of graphene and deploying different types of prisms to improve sensitivity, detection accuracy, and FWHM parameter. In fact, all these performance parameters are depending on metal/graphene/WSe_2 layer thickness and number of layers of 2D materials. Recently, an SPR sensor has been analyzed with a combination of silica and blue phosphorene/MOS_2-based heterostructure. Again the performance of the heterostructure depends on thickness and number of layers of silica as well. Here, blue phosphorene MOS_2, graphene, WSe_2, etc. are new 2D materials for various chemical and biosensing applications. A MOS_2/Blue phosphorene-based SPR sensor reveals a high sensitivity, as any other form of configuration (Sharma and Pandey 2018). The newly emerging graphene material has an application in various optical and electronics devices, leading to the exploration of other promising 2D materials with a prediction of their excellent integration with Si-technology. Almost all 2D materials are having excellent optical, physical, and chemical properties due to their excellent

DOI: 10.1201/9781003190738-3

quantum confinement effect. Surface waves (SWs) arise due to direct contact of two materials, with their opposite sign of relative permittivity coefficients. Also, the field is evanescently decaying outside of the interface or contact points. The dispersion relation to the SPP waves for the case of two different materials, have dielectric constants ε_d and ε_m given by

$$k_{SPP} = \frac{\omega}{c}\sqrt{\frac{\varepsilon_d \varepsilon_m}{\varepsilon_d + \varepsilon_m}} \tag{3.1}$$

In fact, an SPP wave travels along the axis of an interface; also they are in TM mode, while the TE mode travels along with the interface. To the case of metal-dielectric interface, where metal has a dielectric constant, $\varepsilon_m = \varepsilon'_m + i\varepsilon''_m$ then from Eq. (3.1)

$$k'_{SPP} = k_0\sqrt{\frac{\varepsilon_d \varepsilon'_m}{\varepsilon_d + \varepsilon'_m}} \Rightarrow \lambda_{SPP} = \lambda_0\sqrt{\frac{\varepsilon_d + \varepsilon'_m}{\varepsilon_d \varepsilon'_m}} \tag{3.2}$$

where SPP momentum k_{SPP} would have a real part k'_{SPP} and imaginary part k''_{SPP} as follows:

$$k''_{SPP} = k_0 \frac{\varepsilon''_m}{2(\varepsilon'_m)^2}\left(\frac{\varepsilon_d \varepsilon'_m}{\varepsilon_d + \varepsilon'_m}\right)^{3/2} \tag{3.3}$$

where SPP propagation length is given by

$$\delta_{SPP} = \frac{\varepsilon_0}{2\pi}\frac{(\varepsilon'_m)^2}{\varepsilon'_m}\left(\frac{\varepsilon_d + \varepsilon'_m}{\varepsilon_d \varepsilon'_m}\right)^{3/2} \tag{3.4}$$

It is recommended to have a real, large part of metal and dielectric permittivity constant and a small imaginary part for a long SPP propagation length. After some mathematical manipulation, the field penetration depth in metal (δ_m) and dielectric part (δ_d) may be given as

$$\delta_m = \frac{1}{k_0}\left|\frac{\varepsilon_d + \varepsilon'_m}{\varepsilon_d \varepsilon'_m}\right|^{1/2} \ \& \ \delta_d = \frac{1}{k_0}\left|\frac{\varepsilon_d + \varepsilon'_m}{\varepsilon_d^2}\right|^{1/2} \tag{3.5}$$

In fact, the behavior of 2D materials is quite different from their bulk form, as discussed above. Recently, a SPR biosensor coated by graphene has been analyzed for DNA hybridization (Shushama, Rana, Inum, and Hossain 2017). In this study, a four-layer model has been analyzed, as shown in Fig. 3.1, where graphene is acting as a bio-recognition element (BRE). In this work, a highly sensitive refractive index (RI) sensor is proposed, which can successfully differentiate the bonding between thymine-adenine and cytosine-guanine. Numerical analysis reveals the successful variation of SPR wavelength and transmitted power irrespective of DNA strands.

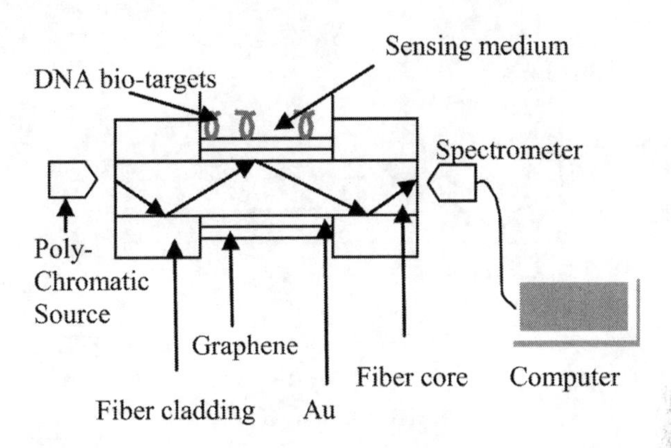

FIGURE 3.1 Schematic representation of proposed experimental setup for DNA hybridization. Reprinted with permission from Optics Communications Journal. Copyright, 2017, Elsevier. (Shushama, Rana, Inum, and Hossain 2017).

The proposed sensor in this work is capable enough for online monitoring and remote sensing applications. Mishra, Mishra, and Verma (2016) have already fabricated and demonstrated the Au-Graphene/MOS_2-based SPR fiber-optic sensing chip along with improved performance parameters over a prism-based SPR sensor chip; hence, the proposed work has a great impact on the future trends of a DNA hybridization biosensing chip.

Fig. 3.2(a,b) shows the proposed sensing probe for a metal/graphene/MoS_2 structure along with their SPR spectra with ambiguity in measurement of full-width half maximum (FWHM), where MoS_2 is used as a biorecognition layer. Mishra, Mishra, and Verma (2016) have studied the effect of different metals like Cu, Au, and Ag on sensitivity of the proposed biosensor. In Fig. 3.2, the light is launched axially on a fiber axis with variation in incident angles θ and $\theta + d\theta$; then the power is received at the fiber end, $dP \propto P(\theta)d\theta$, where $P(\theta)$ is the angular dependency of power related by (Mishra, Mishra, and Verma 2016)

$$P(\theta) = \frac{n_{core}^2 sin\theta cos\theta}{(1 - n_{core}^2\, cos\theta)^2} \tag{3.6}$$

where n_{core} stands for fiber core RI. The normalized transmitted power through a sensing probe for the p-polarized mode case may be given as (Mishra, Mishra, and Verma 2016)

$$P_{trans} = \frac{\int_{\theta cr}^{\pi/2} R_p{}^{Nref(\theta)} P(\theta) d\theta}{\int_{\theta cr}^{\pi/2} P(\theta) d\theta} \tag{3.7}$$

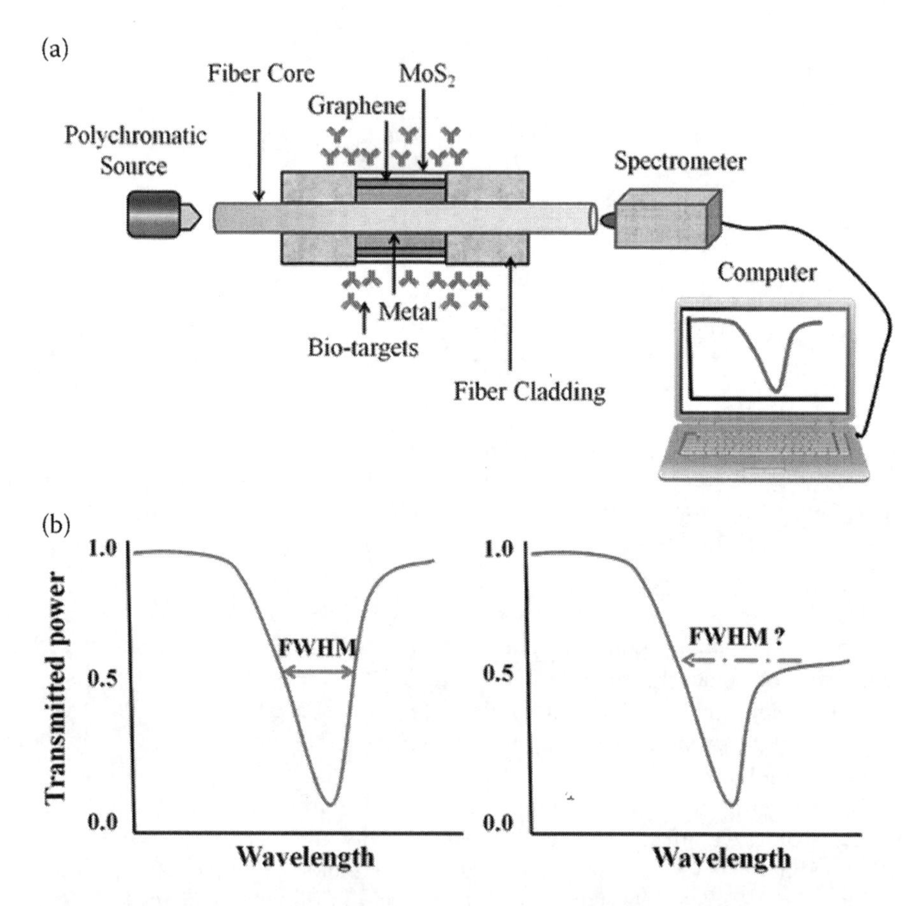

FIGURE 3.2 a) Schematic representation of the proposed sensing probe with a configuration of metal/Graphene/MoS_2 and b) SPR dip along with FWHM definition. Reprinted with permission from The Journal of Physical Chemistry C. Copyright, 2016, ACS publication (Mishra, Mishra, and Verma 2016).

where $Nref(\theta) = \frac{L}{D\tan\theta}$ is the total number of reflections of rays inside the sensing probe having an incident angle θ, L is the length of sensing region, D is the fiber core diameter, and θ_{cr} is the critical angle of the fiber. In that case, the reflectivity of p-polarized light is given by $R_p = |r_p|^2$

$$r_p = \frac{(M_{11} + M_{11}q_N)q_1 - (M_{21} + M_{22}q_N)}{(M_{11} + M_{22}q_N)q_1 + (M_{21} + M_{21}q_N)} \quad (3.8)$$

where $M = \prod_{k=2}^{N-1} M_k$, and $M_k = \begin{vmatrix} \cos\beta_k & -i\sin\beta_k/q_k \\ -i\sin\beta_k/q_k & \cos\beta_k \end{vmatrix}$, $q_k = \frac{(\varepsilon_k - n_c^2\sin^2\theta)^{1/2}}{\varepsilon_k}$, $\beta_k = \frac{2\pi d_k}{\lambda}(\varepsilon_k - n_c^2\sin^2\theta)^{1/2}$ respectively.

3.2 EXCITATION OF SURFACE PLASMONS IN DIFFERENT MATERIALS

Thin-film technology has the potential to meet the demand of sensing applications due to its capability to determine the thickness of ultrathin film (d) and their RI ($N = n + ik$). Field emission scanning electron microscopy (FE-SEM), transmission electron microscopy (TEM), and x-ray reflectivity are some useful techniques to determine the thickness of material with a nanometer scale; however, they are not meant to estimate the optical constant of material. Ellipsometry has proved to be a very sensitive method to determine the RI and thickness of material, simultaneously (Hu et al. 2017). However, it has its limitations in terms of their angstrom thin-film thickness measurement capabilities and fails to estimate the geometrical features due to interconnected thin films.

3.2.1 Excitation of Surface Plasmons by Light

Graphene-assisted surface plasmon polaritons (GSPPs) have emerged recently due to their strong localization, low power, and tunable resonance wavelength characteristics. Due to the mismatch of wave vectors, GSPPs cannot be directly excited by an incident wave in free space. There are several schemes proposed to overcome this momentum mismatch problem through grating coupling, prism coupling (Otto and Kretschmann-Raether schemes), dipolar sources, patterned graphene, electron beams, etc. SPR-assisted ellipsometry has matured enough for detection of biologicals, chemicals, and gases.

3.2.2 Otto Configuration

The binding energy and absorption efficiency of a material is a crucial role into the optimum utilization of conductivity and surface area of any developed sensor. The recently developed 2D nanomaterials, such as BP, transition metal dichalcogenides (TMDCs), etc., have a great potential to meet the demand of present sensor technology. A typical Otto configuration of power coupling of incident light to metal surface is shown in Fig. 3.2. In this proposed configuration, incident light is coupled through a prism to surface plasmon wave via a metal layer; an air gap is maintained between the prism and metal layer. However, there is a flaw in this geometry due to the air gap, which restricts the field decay exponentially through the air-to-metal interface. To circumvent this effect, Otto's configuration is further refined to develop a more advanced technique called the ellipsometric technique. SPR-assisted ellipsometry has matured enough for detection of biologicals, chemicals, and gases. Fig. 3.3 reveals one of the SPR-assisted ellipsometry setup as suggested in Hu et al. (2017). The Otto configuration is also involved in ellipsometry in terms of a prism, air gap, and thin films of metal and substrate region, which assist to excite the SPR.

In Fig. 3.4, a xenon lamp is emitting unpolarized light, which is passing through the polarizer and then incident on prism and sample assembly. A micrometer size spot is created on the center of the lens assembly, which is then reflected to a modulator, secondary polarizer, and finally on the detector side. The convex lens is perfectly matched at the tip point with a rectangular prism by using index matching

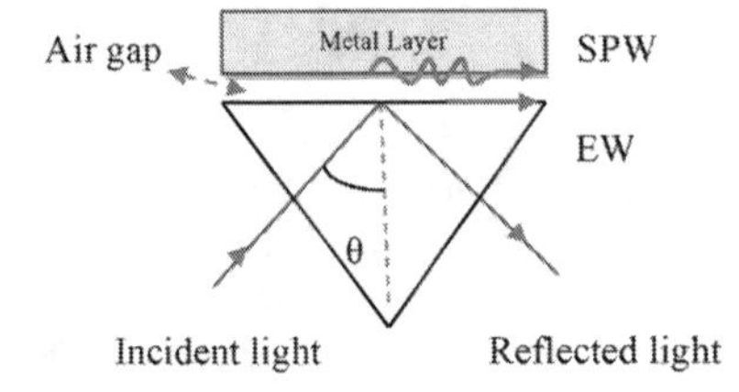

FIGURE 3.3 A typical Otto configuration where SPW stands for surface plasmon wave and EW for evanescent wave. Reprinted with permission from Micromachines. Copyright, 2020, MDPI (Singh et al. 2020b).

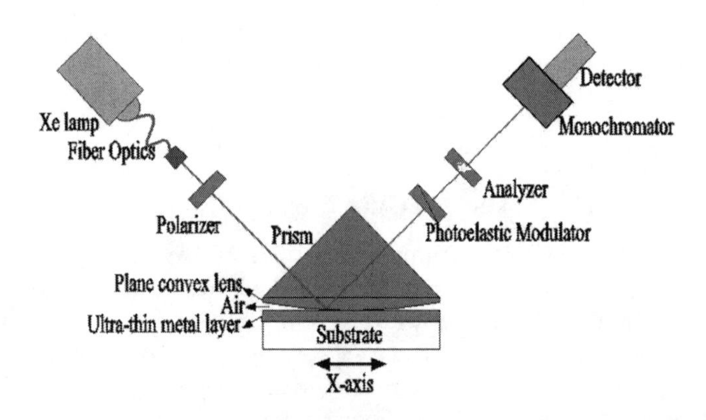

FIGURE 3.4 Proposed experiment of SPR-assisted ellipsometry setup for the measurement of optical constants of thin film. Reprinted with permission from Optics Express. Copyright, 2017, Optical Society of America (Hu et al. 2017).

oil. The air gap thickness, d_{air}, between the curved surface and flat surface of the sample at the tip point (contact point) changes with having distance r from their tip point along with the plane surface

$$d\_{air} = R - \sqrt{R^2 - r^2} \tag{3.9}$$

to eliminate the effect thickness perturbation of air gap, a micrometer spot beam is considered. The proposed arrangement is allowed for angular scans, provided that their angle of incidence on the prism is greater than their critical angle of the structure. The composite structure can be considered a multi-layer substrate stack. The characteristics or eigenvalue equation for the TE mode (s-polarization) of each layer can be written as follows:

$$M_i = \begin{bmatrix} cos\,(k_0 nzcos\theta) & -\,i/psin\,(k_0 nzcos\theta) \\ -\,ipsin\,(k_0 nzsin\theta) & cos\,(k_0 nzcos\theta) \end{bmatrix} \tag{3.10}$$

where $p = \sqrt{\varepsilon/\mu}\,cos\theta$, n, k_0, ε, z, μ are RI, wave number, permittivity, thickness, and permeability of the respective layer and θ is the angle of incidence for the layer. For the case of the TM mode (p-polarization), the above relation holds, but

$p = \sqrt{\mu/\varepsilon}\,cos\theta$. The transfer matrix of the multilayer structure can be described as usual

$$M = M_1 M_2\, M_3 \ldots \ldots .. M_j = \begin{bmatrix} M_{11} & M_{12} \\ M_{21} & M_{22} \end{bmatrix} \tag{3.11}$$

where $j = 1, 2, 3, \ldots$ denotes the layer number. After some trivial manipulation, the TE mode reflectance can be described as

$$r_s = \frac{(m_{11} + m_{12}P_l)P_1 - (m_{21} + m_{22}P_l)}{(m_{11} + m_{12}P_l)P_1 - (m_{21} + m_{22}P_l)} \tag{3.12}$$

where $P_1 = \sqrt{\frac{\varepsilon_l}{\varepsilon_1}}\,cos\theta_1$ and $P_l = \sqrt{\frac{\varepsilon_l}{\varepsilon_l}}\,cos\theta_l$. For the case of the TM mode, Eq. (3.12) holds, except $q_1 = \sqrt{\frac{\mu_1}{\varepsilon_1}}\,cos\theta_1$ and $q_l = \sqrt{\frac{\mu_l}{\varepsilon_l}}\,cos\theta_l$ hence,

$$r_p = \frac{(m_{11} + m_{12}P_l)q_1 - (m_{21} + m_{22}q_l)}{(m_{11} + m_{12}P_l)q_1 - (m_{21} + m_{22}q_l)} \tag{3.13}$$

which will lead to complex reflectance $\rho = \frac{r_p}{r_s} = \tan(\Psi)e^{i\Delta}$. Based on the ellipsometric parameters, amplitude ratio, Ψ, and phase shift, Δ, can be obtained. It is indeed much easier to estimate the optical constants of ultrathin metal film by the proposed ellipsometry setup subject to minimizing their mean square error (MSE) between the theoretical and experimental value. The MSE is calculated by

$$MSE = \frac{1}{N}\sum_{i=1}^{N}\left[\left(\frac{\Psi_i^{mod} - \Psi_i^{exp}}{\Psi_i^{exp}}\right)^2 + \left(\frac{\Delta_i^{mod} - \Delta_i^{exp}}{\Delta_i^{exp}}\right)^2\right] \tag{3.14}$$

where N is the number of measurements and *mod* and *exp* denote the theoretical and experimental data of the proposed ellipsometry setup. The dielectric permittivity of ultrathin metal film can be expressed by the Drude mode, $(\omega) = \varepsilon_\infty - \frac{\omega_p^2}{\omega^2 + i\omega}$, where meaning of the various terms is as usual (Singh et al. 2020b).

3.2.3 Kretschmann–Raether Configuration

A localized surface plasmon resonance sensor (LSPR) is less sensitive around 10,000 times as compared to an SPR sensor (Singh et al. 2020b). However, the sensitivity of LSPR can be significantly improved due to its short decay length characteristics. Kretschmann et al. (Otto 1968) modified the Otto configuration significantly (Hu et al. 2017), as revealed in Fig. 3.4. It is apparent that in Kretschmann geometry as compared to Otto configuration, a metal layer is directly deposited on the prism by the glue gel for the purpose of matching RI on either side of the interface.

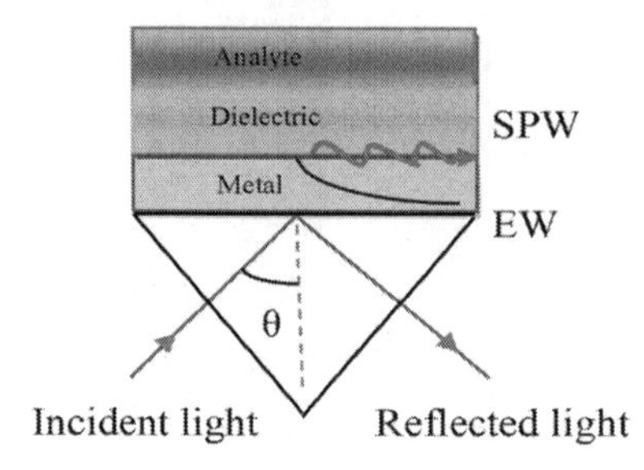

FIGURE 3.5 A typical Kretschmann configuration where SPW stands for surface plasmon wave and EW for evanescent wave. Reprinted with permission from Micromachines. Copyright, 2020, MDPI (Singh et al. 2020b).

In Fig. 3.5, it is apparent that the air gap is completely removed but the Otto configuration has an air gap between the prism and the metal layer. Hence, this configuration improves the sensitivity substantially as compared to the Otto configuration. In fact, an evanescent wave generates both sides of the interface while decaying exponentially on either side of it depending on the dielectric constant of the metal and dielectric. There is a sensing medium (analyte) on which a bimolecular interaction takes place. However, the method of excitation plays a great role in the generation of SPR. There are several excitation ways available, such as wavelength interrogation (optical fiber based), grating coupled, and angular interrogation (prism based) to mention a few. The Kretschmann configuration utilizes the prism-based technique to excite the SPR phenomena. Fig. 3.6(a,b,c) shows the summary of various methods of excitation of SPR. In general, rigorous coupled-wave analysis (RCWA) is used to analyze the grating-assisted SPR configuration. Pechprasarn et al. (2016) have numerically proved that the modified grating assisted Otto configuration, as shown in Fig. 3.6d, supports many different types of modes with enhanced field confinement properties within a sensing chip and eventually improving sensitivity on the sensor surface.

Liedberg, Nylander, and Lunström (1983) demonstrated the gas-sensing application of SPR for the first time. Since then, the SPR technology has been widely researched worldwide and adopted as well. Even the Kretschmann configuration was supported to enhance sensitivity of the SPR sensor subject to the choice of 2D nanomaterials. In the recent past, 2D nanomaterials, like black phosphorus, graphene, TMDCs, i.e., WSe_2, WS_2, $MoSe_2$, and MoS_2, have played a vital role in the enhancement of sensitivity for SPR.

3.3 SPR PHENOMENON FOR CHEMICAL SENSING FOR AU/BP/AG/GRAPHENE CONFIGURATION

SPR biosensors are optical sensors used to test the interaction between biomolecules with sensor surfaces using a surface plasmon polariton (SPP). Deposition of a thin metallic film on a single side of the prism in the traditional SPR biosensor configuration is used to separate the prism and the sensing medium. The metallic film is usually made of noble metals such as silver (Akimov, Koh, and Ostrikov 2009; Akimov, Koh, Sian, and Ren 2010; Chu, Ewe, Koh, and Li 2008) and gold (Hutter and Fendler 2004; Maier 2007; Reather 1988), which enable SPP

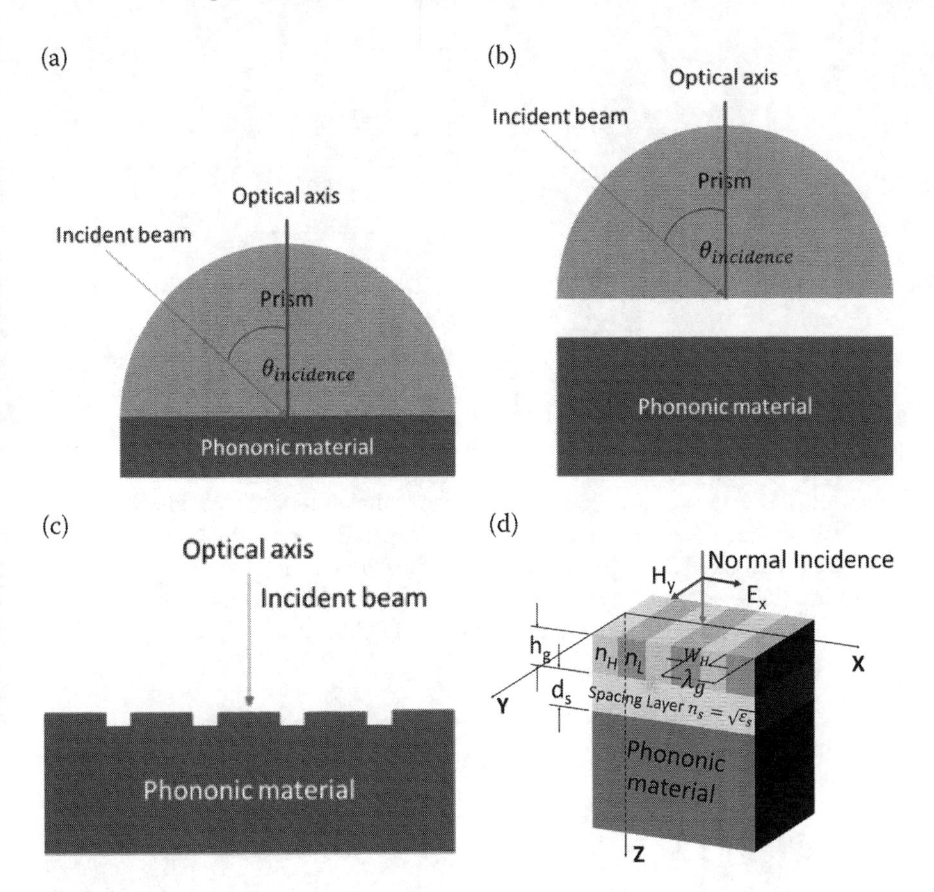

FIGURE 3.6 Various ways of excitation of SPR: a) prism-assisted Kretschmann configuration, b) Otto configuration assisted by prism, c) SPR excitation based on grating assisted, and d) improved grating assisted Otto configuration. Reprinted with permission from Optics Express. Copyright, 2016, Optical Society of America (Pechprasarn et al. 2016).

propagation at visible light frequencies. However, gold is commonly favored because it is resistant to oxidation and degradation in various conditions. On the other hand, biomolecule absorption on the gold layer is not significant. Hence, the traditional SPR biosensor's sensitivity is limited as a result of this limitation. In traditional SPR biosensors, the strength of angular interrogation is reduced, consisting of the prism, single metal, and sensing medium. As a result, several researchers have been working to improve the performance parameters of biosensors with the help of a bimetallic configuration. There are two types of bimetallic arrangements considered in SPR sensors, as shown in Fig. 3.7.

The bimetallic structure has advantages over the conventional structure, such as enhanced propagation length, generating long-range surface plasmon resonance (LRSPR), improving sensitivity, etc. Ong et al. (2006) demonstrate the silver/gold-based bimetallic design experimentally for biosensing. In contrast, for most of the industrial sensors that use only single gold films, the double layer configuration

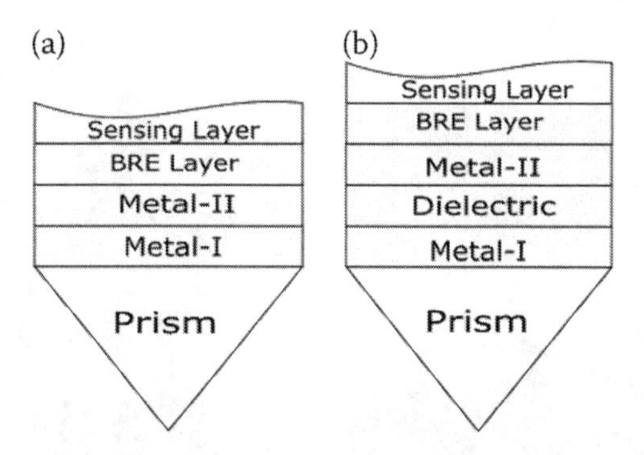

FIGURE 3.7 Different configurations of the bimetallic structure of SPR sensors.

enhanced the evanescent field at the analyte–metal interface and allows for detecting biomolecular interactions that exist deeper inside the analyte. This work shows that the silver-layer single and bimetallic SPR systems increase the intensity of electric fields at the analyte–metal interface by 5.7 and 3.0 times, respectively, compared to a single gold layer, as shown in Fig. 3.8.

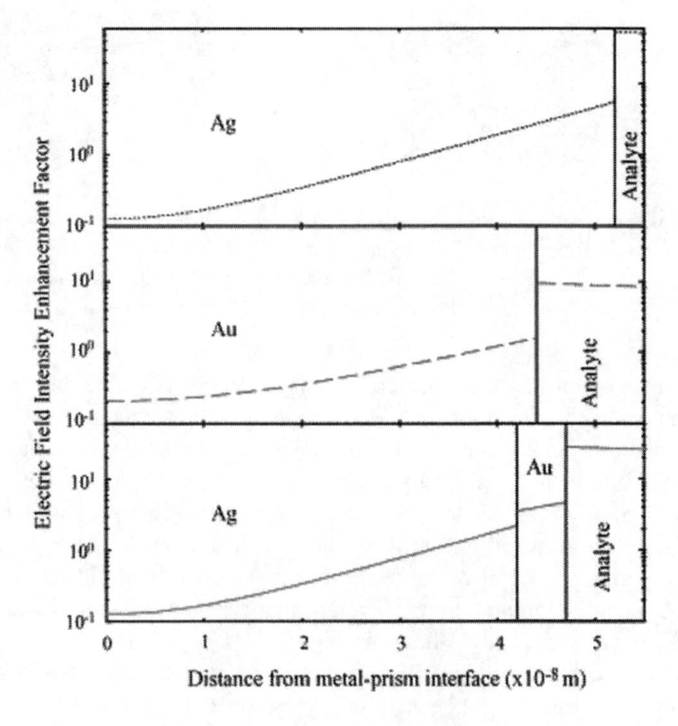

FIGURE 3.8 Comparison of electric field variation in conventional and bimetallic structures. Reprinted with permission from Sensors and Actuators B: Chemical Journal. Copyright, 2006, Elsevier (Ong et al. 2006).

Nowadays, 2D materials, e.g., TMD, graphene, and BP, are used as dielectric materials in bimetallic configurations. To evaluate the sensitivity of graphene and BP with Au/Ag-based optical sensors, it is essential to understand BP's and graphene's optical properties. A recent work (Nair et al. 2008) on light propagation through membranes of suspended graphene demonstrated that graphene opacity is a fundamental constant

$$n = 3.0 + i\frac{C_1}{3}\lambda \tag{3.15}$$

where the wavelength of a vacuum is denoted by λ, and C_1 is calculated by opacity measurement (Nair et al. 2008) and is approximately equal to the 5.446 μm^{-1}. In most cases, a single-layer graphene is also used as a BRE layer or top layer in SPR sensors. Liu et al. (2019) proposed Ag/Au-based bimetallic structures with graphene and barium titanate biosensors. At a wavelength of 633 nm, an Ag–Au bimetallic film with sensitivity and a figure of merit (FOM) of 294°/RIU and 42.13 RIU^{-1} can be acquired by optimizing parameters. They also compared the results with conventional SPR sensors with only Ag and graphene and found the sensitivity enhancement, as shown in Fig. 3.9.

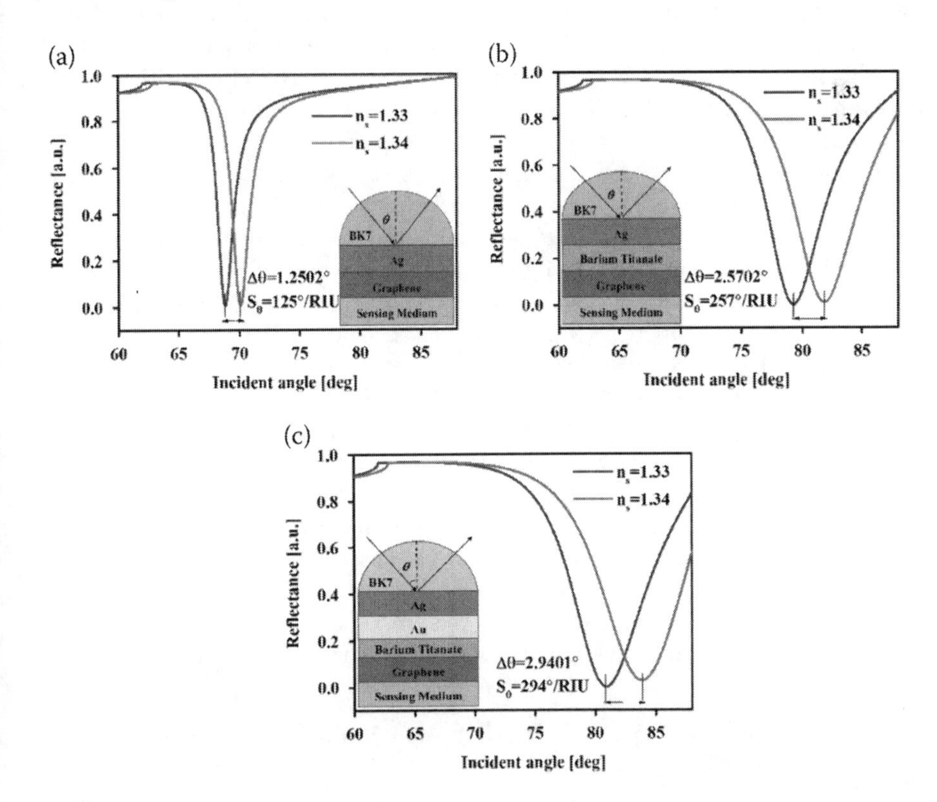

FIGURE 3.9 Variation of $\Delta\theta_{SPR}$ in different configurations. Reprinted with permission from Journal of the Optical Society of America B. Copyright, 2019, OSA (Liu et al. 2019).

Singh and Raghuwanshi (2019) demonstrate an SPR sensor using a multilayered structure. The modified Kretschmann configuration consists of two Au layers: black phosphorus and graphene. The operating wavelength is 633 nm, and their highest sensitivity is 218°/RIU for the proposed sensor. Pal and Jha (2021) theoretically demonstrate an SPR sensor using a multilayered structure. The modified Kretschmann configuration consists of two Ag layers: barium titanate and graphene. The proposed sensor operating wavelength is 633 nm, and the highest sensitivity is 280°/RIU. It is suitable for biomedical and biochemical sensing applications. Chabot, Miron, Grandbois, and Charette (2012) analyzed an LRSPR sensor to detect the toxicity in the living cells. The proposed sensor has a BK7 prism, Teflon, silver, and a gold layer. Due to the bimetallic multilayered structure, the sensor shows more sensitivity. Use of the Teflon ensures the surface plasmons generated more at the prism and metal interface. Wu et al. (2016) demonstrated a multilayer SPR biosensor using cytop material, graphene, and a single layer of metal of Ag/Al/Cu. Multiple combinations of cytop, graphene, and metal are analyzed and the maximum sensitivity of the sensor is 1,589 RIU^{-1}. The proposed biosensor is suitable for biological detection, medical diagnosis, and chemical examination. Cu and graphene combination-based sensors offered the highest sensitivity. Su et al. (2020) numerically validate the SPR biosensor for biological and chemical detection. As shown in Fig. 3.10a, the proposed sensor has black phosphorus and few layers of 2D material, such as graphene. Here, 2S2G is used as a coupling prism attached with Ag as an SPR active material. The optimized sensitivity of 148.2°/RIU is achieved using this sensor. The effect of BP layers, graphene layers, and Si is shown in Fig. 3.10c and Fig. 3.10d, respectively.

Zekriti (2021) has theoretically analyzed the modified Kretschmann structure. The proposed multilayer design consists of a silver metal layer and the graphene layer. Here, the design of the biosensor is analyzed with a BK7 prism and a LaSFN9 prism. The designed biosensor can measure with high precision. Maharana et al. (2014) investigated multilayered biosensors for near-infrared biosensing applications. It consists of a chalcogenide (2S2G) prism for coupling, one layer of gold, and graphene. Air is used between the prism and the metal to improve the performance of the proposed biosensor. The maximum sensitivity of the sensor is 43.18°/RIU. Verma, Prakash, and Tripathi (2015) proposed a biosensor design using chromium, graphene, and an Ag layer. BK7 is used as a prism of coupling, and the air is used as a dielectric material between the prism and chromium. The designed biosensor is investigated using the angle interrogation technique. The maximum sensitivity of 52.94 (°/RIU) is achieved. Jha and Sharma (2009) proposed an Ag–Au-based bimetallic layer-based biosensor for the infrared region. Chalcogenide glass prism is used as the coupling prism. The nanoparticle layer of Ag and Au is deposited over the prism. The proposed sensor can detect environmental analysis, breath analysis, homeland security, and minerals detection.

3.4 SPR PHENOMENON FOR CHEMICAL SENSING USING THE WSE_2 LAYER

A large number of materials were considered as 2D nanomaterial in previous SPR-based sensors and have gathered a lot of attention, given their special optical,

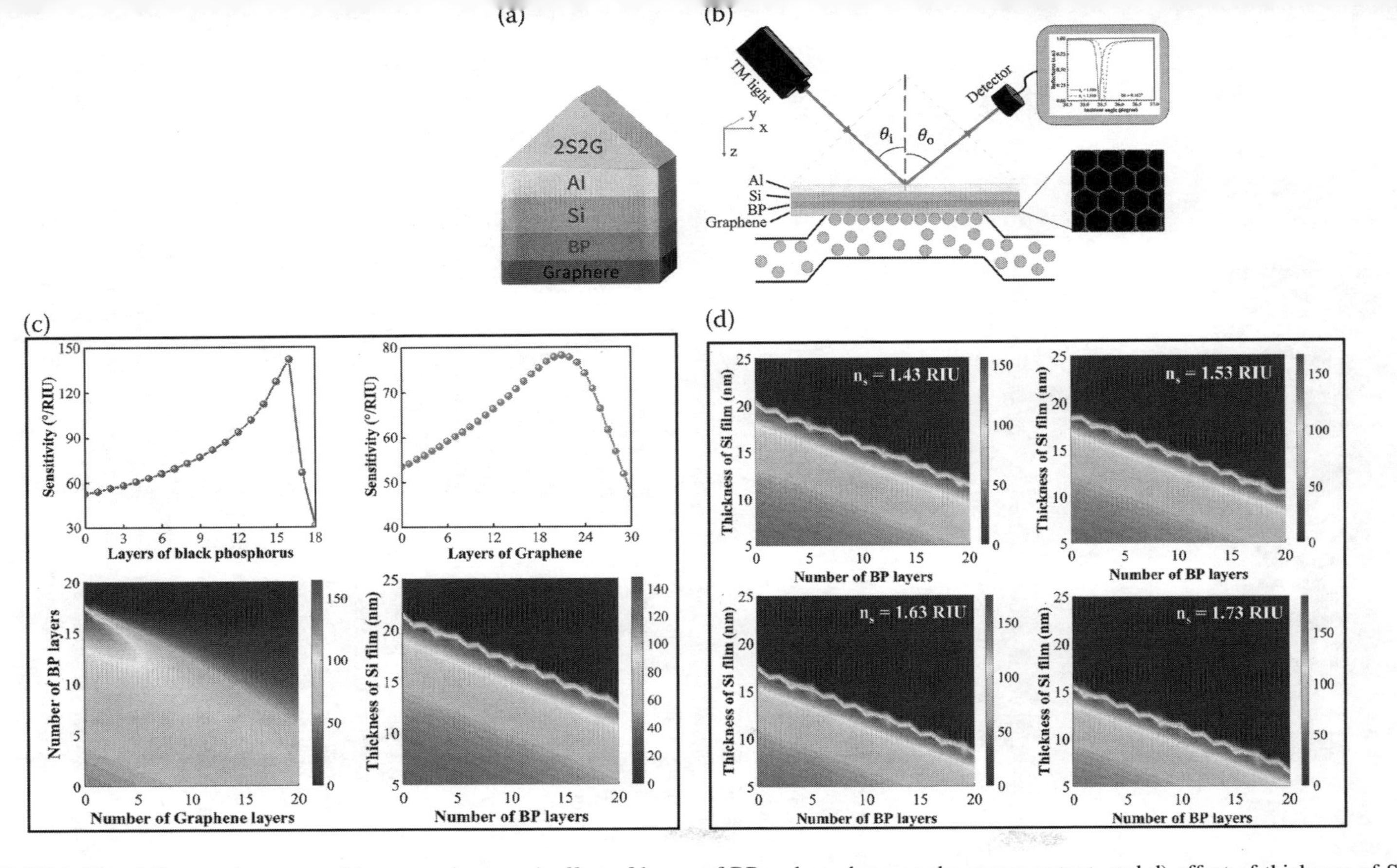

FIGURE 3.10 a) Proposed structure, b) proposed setup, c) effect of layers of BP and graphene on the sensor output, and d) effect of thickness of Si layers w.r.t. the number of BP layers on the performance of the sensor. Reprinted with permission from Nanophotonics journal. Copyright, 2020, De Gruyter (Su et al. 2020).

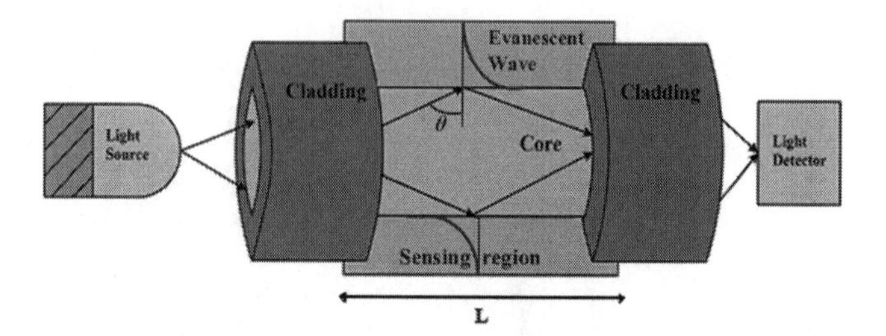

FIGURE 3.11 Schematic diagram of EW-based fiber-optic sensor. Reprinted with permission from Optik Journal Copyright, 2019, Elsevier (Sharma, Gupta, and Sharma 2019).

catalytic, and electronic features. The working principle of fiber-optic evanescent wave sensors was investigated by Sharma, Gupta, and Sharma (2019). The optical fiber usually consists of a core with RI, n_{co}, and a clad with RI and n_{cl} (satisfying condition $n_{co} > n_{cl}$), shown in Fig. 3.11. The fiber is used for the electromagnetic light (EM) based on the concept of attenuated total reflection (ATR). The lighting simulates the interface of the core with a greater angle than the critical angle so that it can fully reflect the waves in the core. However, some vibrations will occur in the cladding to avoid breaking equations of Maxwell and related boundary conditions. The evanescent waves are known to these EM vibrations. An evanescent wave (EW) is an optical fiber-clad exponentially decaying field (EM), corresponding to radioactive light losses in the optical fiber. The field can penetrate exponentially into the medium that is placed in fiber contact. By modifying their cover portion of the fiber, it can manipulate the interaction between the EW and the surroundings. This results in variation in output energy depending on the sensing medium's optical property.

Because of its outstanding optical and electrical properties, the 2D TMDCs were commonly used in photodetectors and transistors. The TMDCs family consists of over 40 members, usually MX_2, wherein M indicates the metal of transition from group IV to group VII, as well as Nb, W, Ta, and Mo; the chalcogen is also indicated by X, such as Te, S, and Se. A monolayer of the MX_2 consists of three nuclear layers in which two layers of chalcogen contain the transition layer of metal. van der Waals forces stack every sheet. The main subjects of this study were group IV dichalcogenide WX_2 and MoX_2, consisting of diselenide of molybdenum ($MoSe_2$), disulfide (MoS_2), diselenide of tungsten (WSe_2), and tungsten disulfide (WS_2). Rahman, Anower, and Abdulrazak (2019) designed a fiber-optical biosensor based on a performance-enhanced SPR by using phosphorene with (2D) materials such as molybdenum disulfide (MoS_2), molybdenum diselenide ($MoSe_2$), graphene, tungsten disulfide (WS_2), and tungsten diselenide (WSe_2). With 10 layers of black phosphorene and three layers of graphene, the sensitivity is increased to 4,050 nm/RIU. The opportunity is to create a very responsive SPR phosphorene-based fiber biosensor that detects DNA hybridization using 2D materials. The proposed sensor had a very high sensitivity, with 10 phosphorene layers of 3,725 nm/RIU. The performance parameter of FOM increased to 64 RIU^{-1} when just six layers of phosphorene was used (increased by 36%). In a promising way to improve the 2D materials' gas sensing efficiency, Ni et al. (2020) developed a monolayer based on WSe_2 doping with Pt, Au, Ag, and Pd. WSe_2 adsorption of SO_2, NO_2, and CO_2 gas molecules with doped monolayers Pt, Au, Ag, and Pd is investigated in the band structure, charge transfer, and adsorption energy. Chemical doping has been an efficient way to accurately modulate WSe_2 performances with the use of solar, gas sensing, and materials based on photocatalytic, including MoS_2, graphene, and blue phosphorous in target technological applications. It has been examined that the interaction of gas molecules affects NO_2, NO, H_2O, CO, N_2, monolayers, and N_2O of Se vacancies on the surfaces of WSe_2. If S_e is vacant, the sensing capacity of WSe_2 may improve significantly. The CO and NO gas molecules preferring molecular chemisorbing on the single-layer WSe_2 surface are extremely vulnerable to physisorption at the surface of WSe_2 with an oxygen-free atom on the place of Se vacancy. The atoms Se and W are responsible for a conversion in the WSe_2 monolayer after adsorption and the main contribution of atoms Se is lower than 0 eV and atom d orbits of W are higher than 0 eV. Fig. 3.12 shows the sum of density of states (DOS) doped for Ag, Au, and Pd, and Pt-doped WSe_2. On the basis of the Fermi level, it consist of three different large peaks where the first peak comes from −15 eV to −12 eV and contributed an Se atom to the s orbital, the second peak comes from −8 eV to 0 eV and contributed an Se atom and SO_2 to the d orbital, and the third peak comes above from 0 eV and contributed a W atom to the s orbital. Accumulation of heavy charge is indicated by a red color, and the red color comes in the replacement position of the S atom. So at this place the charged transfer is very strong and the darker color shows the variation of the Au and Ag atoms.

The effect of SO_2, NO_2, and CO_2 adsorption on the Au, Ag, Pd, and Pt/WSe_2 monolayers, in Fig. 3.13, is easy to demonstrate how the work functions (Φ) are different for each adsorption device.

It is based on the minimum energy necessary to remove an electron from the Fermi into infinity, and is expressed as follows:

$$\Phi = V(\infty) - E_F \tag{3.16}$$

where Φ is the work function, V (∞) is electrostatic potential, and monolayer's Fermi energy Au, Ag, Pd, and Pt/WSe_2 is E_F. The value of Φ for a monolayer WSe_2-coated gas molecule of varying intensity is between 5.11 and 5.34 eV. The work function is relieved after the adsorption of SO_2, NO_2, and CO_2 in the TM-doped monolayer WSe_2, indicating that electrons are not more restricted to the sheet. Chiu and Lin (2018) proposed a successful change to the monochloroacetic acid (MCA) for the affinity capture area, to enhance the SPR immune sensor sensitivity. The modified carboxylation technique MoS_2 could boost sensitivity and biosensitivity, mainly because of dielectric improvements to the energy connector features of MoS_2–COOH. Moreover, –COOH had a high protein affinity and a covalent bond reaction formation, which increased the efficiency of protein binding. In comparison to the traditional SPR chip, the angle response chip of MoS_2–COOH increased by 3.1-fold with a constant increase in the rate of kinetic connection. The MoS_2–COOH (1 g l^{-1}) chip has highly affinity to interactions with protein because

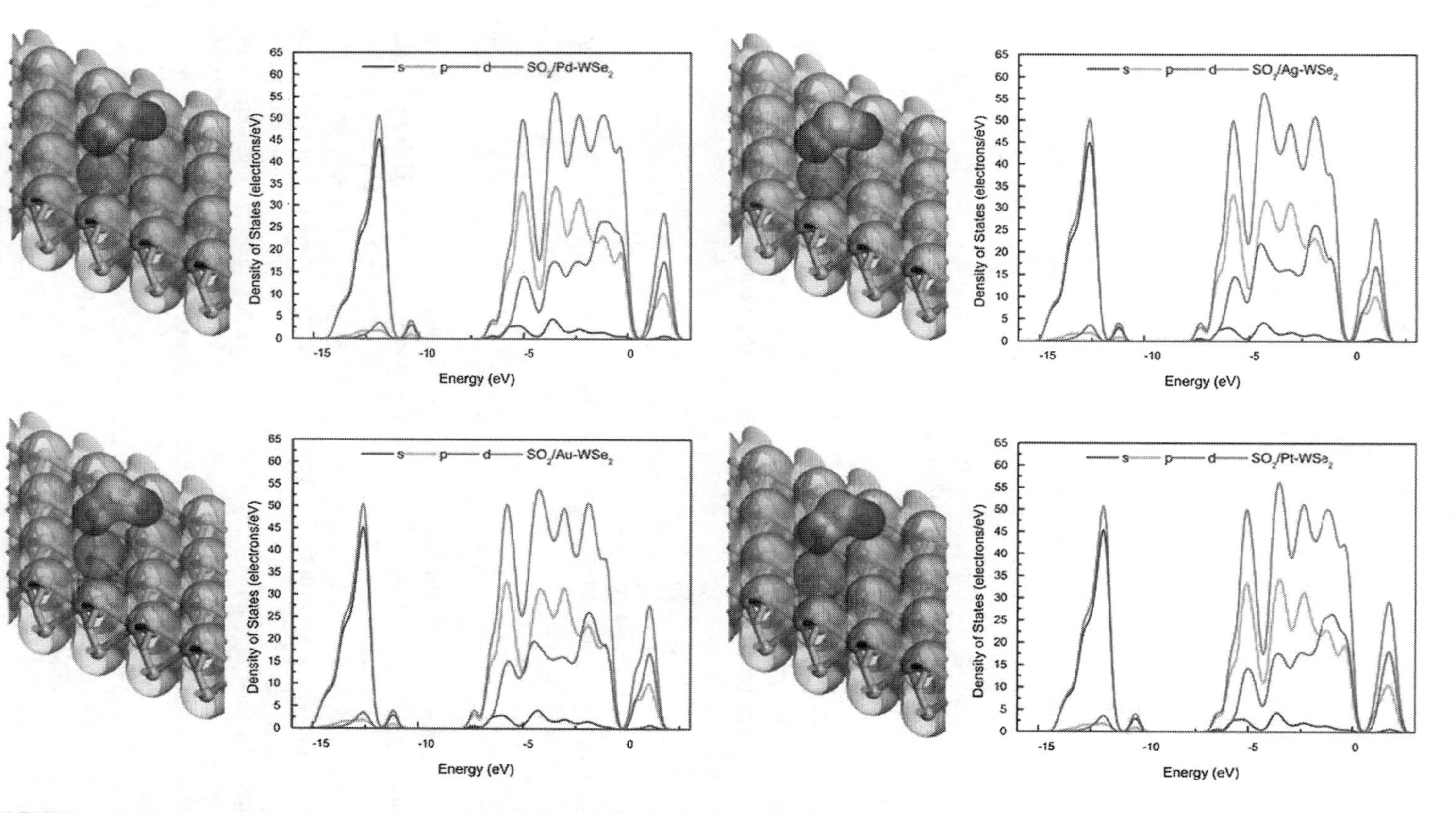

FIGURE 3.12 Structure and density of states (DOS) electrostatic potential of SO_2 gas molecules in the X-WSe_2 monolayer. Reprinted with permission from Applied Surface Science Journal Copyright, 2020, Elsevier (Ni et al. 2020).

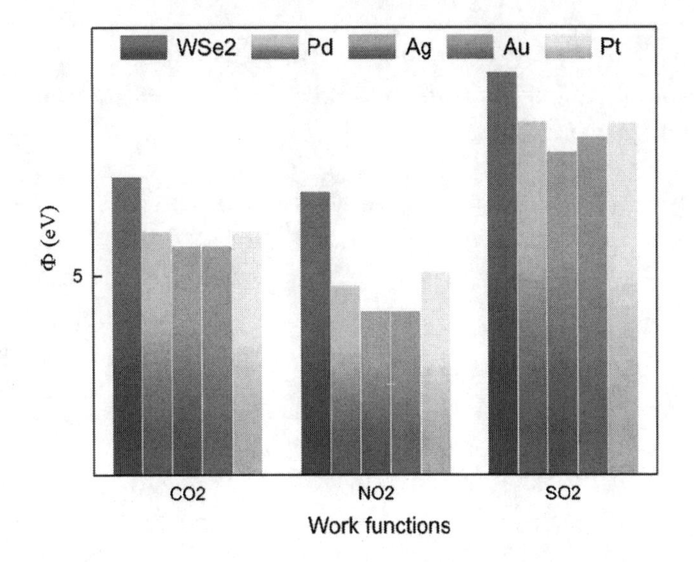

FIGURE 3.13 Work function of CO_2 and NO_2; and SO_2/Pd, Au, and Ag; and Pt/WSe_2. Reprinted with permission from Applied Surface Science Journal Copyright, 2020, Elsevier (Ni et al. 2020).

the value of Ka was 212 times greater than the standard SPR. Guo et al. (2019) proposed ultrasensitive NO_2 sensors at room temperature based on WSe_2-exfoliated nanosheets from the liquid phase. In the case of a dispersant N-methyl-2 Pyrrolidone (NMP), a tungsten diselenide (WSe_2) exfoliation method was formed into a few nanosheets of the layer. The gas sensor based on the as-syntheses WSe_2 nanosheets (50 ppb, 5.06) shows an extremely high response. The properties of the WSe_2 bulk nanosheet and exfoliated gaseous sensors are 50 ppb of NO_2 radiation. When 50 ppb NO_2 for 60 s, which is around three times as large WSe_2, is sustained, the nanosheet NO_2 gas sensor instantly reacts to NO_2 and reaches 5.06. Wu et al. (2020) suggested tungsten chemical vapor deposition diselenide monolayer NO_2 gas sensors and it is made of a monolayer of diselenide (WSe_2) synthesized with chemical vapor deposition, operating at various temperatures and concentrations ranging between 100 ppb and 5 ppm. The WSe_2 sensor showed a fast recovery of 18 s and 38 s when used at a maximum temperature of 250°C, exposed to NO_2 with a value of 100 ppb. The WSe_2 sensor operates between 0.1 and 5 ppm and is therefore suitable for environmental surveillance in the detection range. Table 3.1 gives a comparison of the NO_2 gases and WSe_2, including WS_2, WS_2–graphene aerogel, and MoS_2 at the most effective room temperature; the sensor shown in this chapter is competitive and shows optimized temperatures at the fastest recovery (250°C).

For the trace concentration of NO_2 to be detected at room temperature (25°C), Yang et al. (2021) proposed a highly selective and versatile WSe_2 nanosheet gas sensor. A simple exfoliation method for liquid phase processes is used to develop the high-quality WSe_2-layered nanosheets, and different characteristics and gas-sensing measurements have been fully investigated for their structure and electric

TABLE 3.1
Comparison of Parameters with Other 2D of NO_2 Gas Sensor for a WSe_2 Compound (Wu et al. 2020)

Materials	NO_2 Concentration (ppm)	Recovery Speed (min.)	Sensor Response (%)	Operating Temp. (°C)
WS_2	10	–	84.7	RT
MoS_2	100	No recovery	~14	RT
WS_2/GA	2	5	3	180
WS_2/GA	2	Incomplete	6.5	RT
WSe_2	500	85	4140	RT
WSe_2	500	21.5	–	100
WSe_2	0.05	17.5	5.06	RT
WSe_2	2	–	~10	RT
WSe_2	0.5	–	238.35	RT
WSe_2	2	24.5	18.8	RT
WSe_2	1	4.8	13.45	250
WSe_2	0.1	0.63 (38 s)	6.15	250

properties. An ultra-low 8 ppb limit of detection (LOD) is reported to detect UV NO_2. The molecules of gases, as shown in Fig. 3.14a, will adsorb the material on a spontaneous surface and receive electrons from the valence band of WSe_2, leading to an increase in hole number density when the WSe_2, a typical electron accepter, is exposed to NO_2. The increase in hole concentration provides additional tunneling paths with the WSe_2 hole conductive, leading to a reduction of the sensor resistance. The photo-generated electrons, displayed in Fig. 3.14b, were attracted to the WSe_2 surface by the incorporated electric field when UV lighting was imposed. Therefore, more free electrons will participate in the NO_2 reaction and an improved sensing response is maintained, which indicates the superb gas sensor flexibility.

Biolayer interferometry-systematic evolution of ligands by exponential enrichment (SELEX) has also been proposed by Kaur, Shorie, and Sabherwal (2020) for

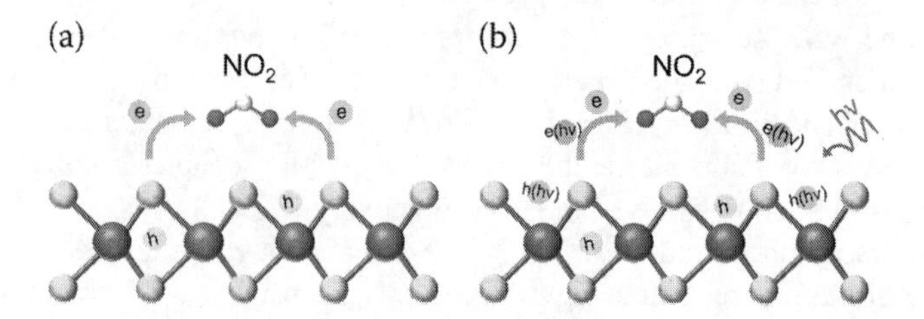

FIGURE 3.14 WSe_2 nanosheet schematic sensing system with NO_2: a) non-UV lighting and b) UV illumination. Reprinted with permission from Sensors and Actuators: B. Chemical Journal Copyright, 2021, Elsevier (Yang et al. 2021).

aptamers and chitosan WSe_2 detection by Shiga toxin antigenic peptide. BLI-SELEX is a one-step technology for quickly generating protein biomarkers aptamers in a microtiter platform, eliminating the necessity to collect high-affinity aptamers in multiple enhancement rounds, as is the case with conventional SELEX. To manufacture voltammetric diagnostics assays by immobilization on chitosan 2D-exfoliated tungsten diselenide (WSe_2) nanosheet platforms, two unique aptamers with picomolar K_d (~44 pM and ~29 pM) are selected against subtypes of Shiga toxins (stx1) and (stx2). These aptamers were nanosensors of modified nanosensing with high sensitivity of approximately 5.0 μA ng^{-1} mL, dynamically responding to the Shiga toxin subtypes from 50 pg mL^{-1}–100 ng mL^{-1} and showing low cross-reactivity in urine, milk, and serum samples of 44.5 pg mL^{-1} and 41.4 pg mL^{-1}, respectively. The new strategy for removing SO_x from the highly effective molecular sensor N-doped titanium dioxide/tungsten diselenide (TiO_2/WSe_2) was investigated in Abbasi and Sardroodi (2017). Density function theory (DFT) was corrected by van der Waals. The adsorbing process for N-doped TiO_2/WSe_2 is considered more high energy efficient than the intrinsic adsorption, indicating a higher sensing capacity than the undoped nanocomposites for N-doped TiO_2/WSe_2. The SO_3 molecule dissolves into SO_2 and a separate atom of oxygen over the TiO_2/WSe_2 nanocomposites. Abbasi and Sardroodi (2018) investigate the adsorption by DFT computations of ozone molecules into TiO_2/WSe_2, for gas-sensor applications. WSe_2 facilitates the interaction between five TiO_2 coordinated sites, and the O_3 molecule. To achieve balanced geometries in the TiO_2/WSe_2 interface for O_3 molecules, the consequences of contact with van der Waals have been taken into account. This study reveals that interactions with nanocomposites of TiO_2 /WSe_2 between TiO_2 and O_3 are greater than between bare TiO_2 and O_3, which indicates WSe_2 helps reinforce ozone molecule interaction with TiO_2 particulates. Adsorption of O_3 causes the O–O molecule to weaken with some extension of the molecules O–O. O_3 adsorption leads. TiO_2 anatase nitrogen doping can play a key role in enhancing TiO_2/WSe_2 abilities of adsorption. The high-performance gas sensor for ethanol has been proposed by Pan, Zhang, and Zhang (2020), based on TiO_2/WSe_2. The nanospheres of TiO_2 (NS_s) and WSe_2 were synthesized through an economic route of hydrothermal. The nanofilm sensor TiO_2/WSe_2 was developed using a self-assembly technology on an epoxy substrate. The research results indicated the high response of a TiO_2/WSe_2 nanofilm having 42.8 with 100 ppm, a very fast response per recovery times, high selectivity and repetitiveness, stability, and a low ethanol gas detector detection limit of 2 s/1 s with 30 ppm.

3.5 BLUE PHOSPHORENE/MOS_2 HETEROSTRUCTURE-BASED SPR CHEMICAL SENSOR USING GOLD METAL

SP would be required materials that are capable of generating SPs after light interaction. SP would be generated when light hits on a metal surface such as aluminum (Al), gold (Au), copper (Cu), sodium (Na), silver (Ag), and indium (In). The condition is known as the SPR conditions when wave vectors of incident light and SPs matched. Singh et al. (2020a) investigated materials with a large surface area and also good conductivity for absorption and binding energy in the development of all sensor systems. This work highlighting the structure of the Kretschmann configuration, the sensing principle and characteristic of SPR, and its application in different fields of SPR, i.e., silver, gold, or copper, or flat film sensors, have been used to detect analyte binding on close to a metal surface for about 30 years. This has often been used to detect many analyte-surface interactions such as small molecular adsorption, link-receptor linkage, monolayer protein adsorption in the self-assembled form, antibody-anti-factor binding, RNA, and DNA hybridization, and DNA-protein interactions. The proposed structure would be depending on a metal layer 10–100 nm thickness, which deposited on the base of the prism with the glue gel technique. There is a significant field penetration in substrates subject to choosing a metal and prism combination. There is not any air gap between the metal and prism for this case. Kretschmann is a very useful technique for improving the SPR sensor sensitivity. In both media, like metal and dielectric media, the associated EW decreases exponentially. The dielectric layer is also linked to a sensing medium and analyte for the detection of biomolecules. SPs are generated depending on the optical arousal method. There are different optical arousal methods, like prisms, gratings, wavelength couplings, etc.

Based on the configuration of Kretschmann, certain nanomaterials, like BP, graphene, TMDCs, e.g., MoS_2, WS_2, $MoSe_2$, and WSe_2, also serve for increasing the sensitivity of the SPR sensors. Table 3.2 shows that the 2D materials have a real dielectric constant, so their ability to absorb light energy is stronger. 2D materials are protected against oxidation by the layering on the metal surface as a protective layer. These benefits help the 2D material-based SPR sensor to be developed.

Rahman, Anower, and Abdulrazak (2019) designed an SPR sensor with 2D phosphorene material, including graphene, WSe_2, WS_2, $MoSe_2$, and MoS_2. When built with

TABLE 3.2

Refractive Index Values for Various 2D Materials and Dielectric Constants (Singh et al. 2020a)

2D materials	Wavelength (nm)	Monolayer Thickness (nm)	Refractive Index ($n_c = n + ik$)	Dielectric Constant ($\varepsilon_n + i\varepsilon_k$)	Ratio ($\varepsilon_n/\varepsilon_k$)
Graphene	633	0.34	3.0 + 1.1487i	7.68 + 6.89i	1.114
BP	633	0.53	3.5 + 0.01i	–	–
MoS_2	633	0.65	5.0805 + 1.1724i	24.4368 + 11.9122i	2.05
$MoSe_2$	633	0.70	4.6226 + 1.0062i	20.3560 + 9.3040i	2.19
WS_2	633	0.80	4.8937 + 0.3123i	23.8511 + 3.0580i	7.80
WSe_2	633	0.70	4.5501 + 0.4332i	20.5156 + 3.9423i	5.20

10 phosphorene layers, the high sensitivity of the proposed sensor is 3,725 nm/RIU. The sensitivity of 2D materials was up to 4,050 nm RIU when integrated into phosphorene. The figure of merit (FOM) was enhanced to 64 RIU^{-1} (increase by 36%) when only six phosphorene layers are used. Liu and Zhou (2017) synthesized gas adsorption by molecular beam epitaxial growth on the monolayer blue phosphorus in the substrate of Au. This particular study shows that O_2, an exothermic process with a small energy block, tends to dissociate and chemisorb on blue phosphorus. The other molecules may be physisorbed on a stable blue phosphorus sheet and make significant changes to the electronic characteristics of the bandgap, carrier-effective mass, and working function makes the material suitable for the sensing of gas. Due to their uniqueness in optical, mechanical, electric, and sensing capability for multifunctional electronic and optoelectronic applications, Singh et al. (2018) investigated 2D metal TMDCs. TMDC monolayers like (MoS_2) and graphene also demonstrate excellent optical and electrical characteristics, having a direct band spacing of 1.8 eV coupled with a high level of mechanical flexibility. This exfoliates MoS_2 from bulk and transfers it mechanically from bulk to a substrate containing a range of micro-fabricated circular holes, which is illustrated in Fig. 3.15. Indication experiments with an atomic force microscope (AFM) with a standard silicone cantilever were carried out to test their mechanical features of the suspended stand-alone

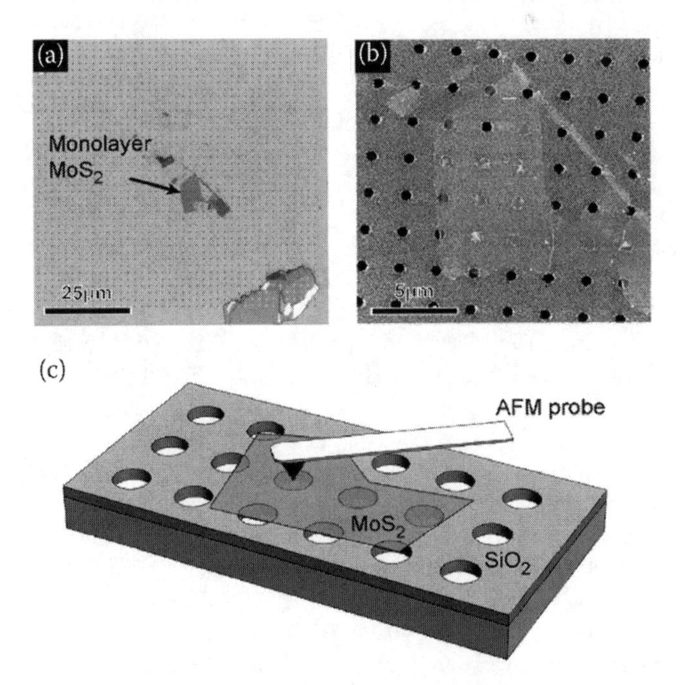

FIGURE 3.15 a) MoS_2 flake optics of one layer transmitted to the pre-patterned SiO_2 substratum with an array of 550 nm circular holes. b) AFM picture with the same MoS_2 monolayer. c) An experimental example of schematic indentation. The AFM tip is positioned over the center and steadily decreased during measurements. Reprinted with permission from Applied Materials Today Journal Copyright, 2018, Elsevier (Singh et al. 2018).

membranes. It was determined that the in-plane rigidity of the MoS_2 monolayer is 180 ± 60 N m^{-1}, which corresponds to the 270 ± 100 GPa Young Modulus.

Rahman, Anower, Abdulrazak, and Rahman (2019) proposed the use of a 2D phosphorene heterostructure as an interactive layer with the analyte for an enhanced biosensor based on fiber optics (SPR). This study shows that the fiber-optic biosensor with a FOM of 61 RIU^{-1} shows an extremely high sensitivity of 3.725 nm as well as a high sensitivity of 10 phosphorene on the Ag-layer. With six layers of phosphorene, FOM reaches a maximum value of 64 RIU^{-1}.

Zhang et al. (2015) proposed the construction of blue phosphorene and MoS_2 bandgap engineering in van der Waals. Both the characteristics of BP and MoS_2 are retained by the heterostructures BP/MoS_2, but the interlayer effect creates relatively narrower bandgaps. By applying external electric fields, the tape structures are further developed. In BP/MoS_2, the electrical field is applied to about 0.6 eV $Å^{-1}$, making an indirect-direct transition. This shows that the symmetry of the integrated electrical dipole fields controls the transition from indirect to direct bandgap. Therefore, the lattice mismatch between BP and MoS_2 is estimated to be around 3%. Then, the bilayer heterostructures BP/MoS_2 were built using the former's grill to stack BP on top of MoS_2. Two starting configurations are considered to achieve equilibrium geometry, as shown in Fig. 3.16(a,d). We refer to them respectively as AB and AB′ stacking. It is evident to note that the BP layer in the x and y directions for the MoS_2 layer can be used to obtain some other possible configurations.

Thus, from a thermodynamic perspective, the formation of heterostructures is exothermic. The lattice constants are optimized in the form of 3.24 Å for two heterostructures. The distance between the layers (dL) for AB and AB′ is estimated at approximately 3.11 and 3.20 Å. The effect of blue P/MoS_2 heterostructure layer has been proposed by Prajapati and Srivastava (2019) on the SPR sensor output parameters, i.e., Structure-3 (BK7 Prism and Au+ Blue P/MoS_2 + Graphene heterostructures). The blue P/TMD heterostructure's thermodynamic stability varies according to these factors: (i) lattice constant and structure, (ii) binding energy, and (iii) formation energy. Blue P and MoS_2 have both 3.26 A and 3.16 A hexagonal lattice structures that are constant and almost the same. The thermodynamic stability of the heterostructure BlueP/MoS_2 is in charge of the formation energy. For stability based on thermodynamic and experimental existence, the value must be negative. A value of −165.6 meV of the formation energy is given by BlueP/TMD heterostructures. This is why the biosensor is the most appropriate to improve their stability by a BlueP/MoS_2 heterostructure. In comparison with its single layer similarity, the multilayered BlueP/MoS_2 heterostructure and 2D graphene have different values of the RI. The difference is due to the optical conductivity. The RI variation of three and five graphene layers is ±3%. Sensors having different structures (structure 1, 2, and 3) are distributed in the transverse-magnetic fields of the sensor in Fig. 3.17(a, b, and c).

The depth of penetration observed from Fig. 3.17 is 45.20 nm, 91.62 nm, and 181.5 nm for the first structure, second structure, and third structure. Thus, in comparison with other structures presented, structure 3 is the proposed sensor with a high depth of penetration and thus is sensitive. The change in parameters for various structures is shown in Table 3.3.

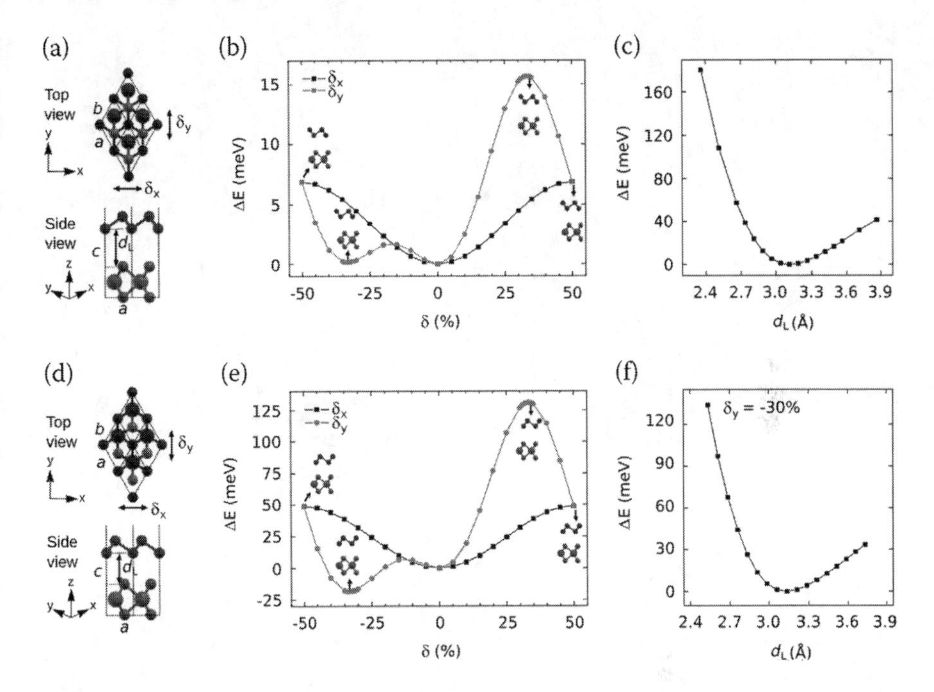

FIGURE 3.16 An overview of both top and side MoS_2 Heterostructure images. The P, Mo, and S stand for blue and red: the changes to total energy with a δx (δy) from b) and e). δx (δy) shows the shifting of the BP layer for MSI along the direction of x (y). This structure shows how b) measures from left to the bottom to the right 33% toward the lower left, and e) measures from top to bottom. Total interlayer distance in f) AB and AB′. Reprinted with permission from Journal of Solid State Chemistry Copyright, 2015, Elsevier (Zhang et al. 2015).

Sharma and Pandey (2018) proposed the 2D (BlueP/MoS_2) heterostructure that is used as an analyte-interacting layer to enhance the sensitivity of an (SPR) sensor. The proposed sensor structure has a substantially higher sensitivity than graphene-based and traditional SPR sensors. The Kretschmann configuration underpins the proposed four-layer SPR sensor scheme. They looked at the sensor's performance by taking into account the number of heterostructure layers and the wavelength as a combined effect. Setareh and Kaatuzian (2021) used an (FK51A) prism with a low RI with a combination of a Blue Phosphorene/MoS_2 heterostructure and barium titanate to theoretically investigated the sensitivity enhancement of an SPR sensor. For this proposed device, a sensitivity of 347.82°/RIU was obtained. The sensitivity of a sensor is improved by adding $BaTiO_3$, and a value of 239.12°/RIU was achieved when the RI of the sensing medium has been modified from 1.33 to 1.335. Fig. 3.18 shows an improvement in sensitivity and minimum reflection of the SPR curve as the number of heterostructure Blue P/MoS_2 layers increases from two to three, with 8 nm $BaTiO_3$.

The value of S and Q is observed as 347.82°/RIU and 60.521 RIU with the use of three layers of heterostructure instead of two layers of BlueP/MoS_2. This arrangement raises not only the sensitivity parameter but also improves the quality factor.

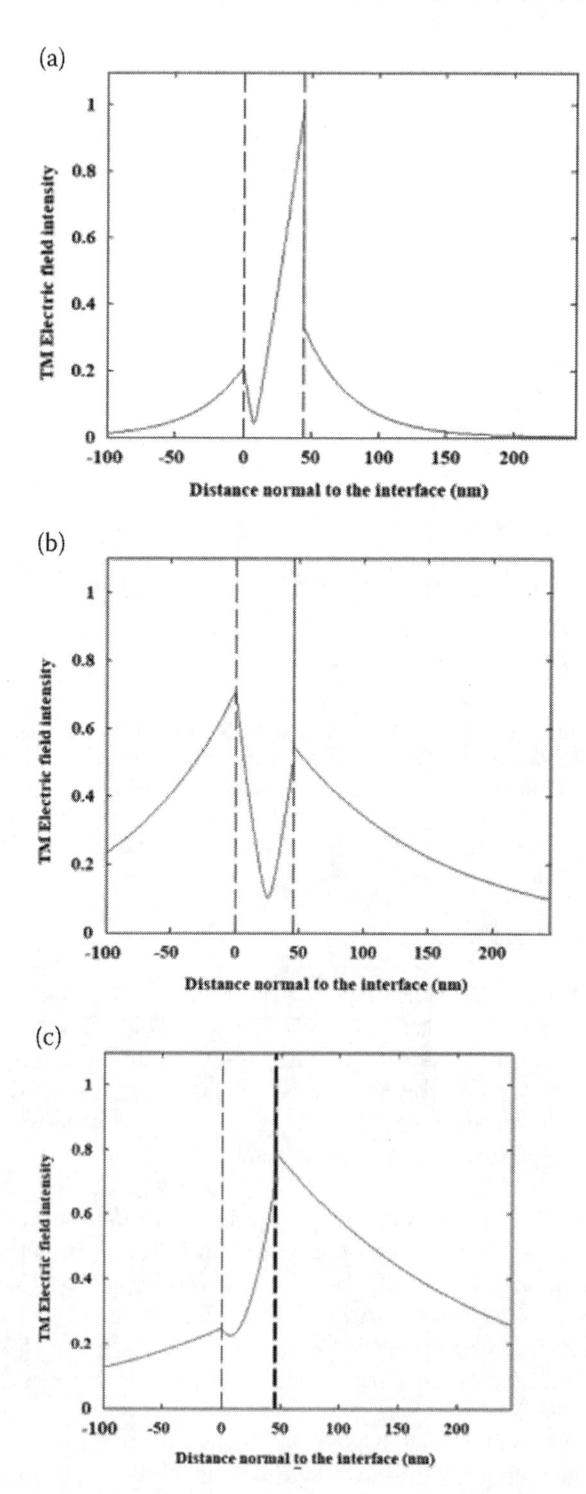

FIGURE 3.17 Plotting for TM field response of different structures: a) first structure, b) second structure, and c) third structure. Reprinted with permission from Superlattices and Microstructures Journal Copyright, 2019, Elsevier (Prajapati and Srivastava 2019).

TABLE 3.3
Performed on the Proposed Model at the Wavelength (λ) is 633 nm (Prajapati and Srivastava 2019)

Structure of Sensors	$\Delta\theta_{SPR}$	R_{min}	Sensitivity (°/RIU)	FWHM (Full Width at Half Maximum)	Detection Accuracy	Quality Parameter (RIU^{-1})
BK7 Prism + Au (Structure 1)	5.7238	0.0080	190.66	9.30	0.61	20.43
BK7 Prism + Au+ Graphene (Structure 2)	5.8613	0.0002	195.33	10.30	0.56	18.96
BK7 Prism + Au+ Blue P/ MoS_2 heterostructure + Graphene (Structure 3)	6.50	0.0028	204	11.10	0.58	18.37

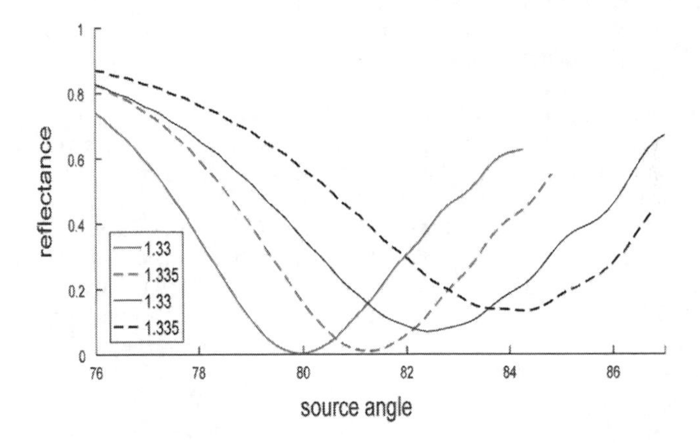

FIGURE 3.18 Absorption curve vs. angle for a sensor with 50 nm Silver/8 nm $BaTiO_3$ thickness, or to Silver/$BaTiO_3$, from 1.33 to 1.335, at a wavelength of 585 nm. Reprinted with permission from Superlattices and Microstructures Journal Copyright, 2021, Elsevier (Setareh and Kaatuzian 2021).

$$Q = \frac{S}{FWHM} = S.\ (D.\ A) \tag{3.17}$$

The quality factor is Q, the sensitivity is S, the full width at half maximum is FWHM, and the detection accuracy is DA. The detection accuracy, minimum reflectance, resonance angle, quality factor, and sensitivity for the sensor proposed were calculated using Fig. 3.18 and Eq. (3.17).

An SPR sensor based on the heterostructure blue phosphorene/MoS_2 proposed by Srivastava and Prajapati (2019) improves performance parameters such as quality factor, detection accuracy, and sensitivity. The proposed structure serves as a layer of interaction with the analyte, increasing the sensitivity of sensors. After the use of the blue P/MoS_2 heterostructure, the sensitivities increased considerably from 150.66°/RIU to 230.66°/RIU were observed when compared to the conventional SPR sensor. Furthermore, between the blue P/MoS_2 and metal layer, an additional silicon nanolayer with a thickness of 2.0 nm is used. As a result, sensitivity and accuracy are significantly improved with a moderate quality factor.

REFERENCES

Abbasi, Amirali, and Jaber Jahanbin Sardroodi. 2017. "A Novel Strategy for SOx Removal by N-Doped TiO_2/WSe_2 Nanocomposite as a Highly Efficient Molecule Sensor Investigated by van Der Waals Corrected DFT." *Computational and Theoretical Chemistry* 1114: 8–19. 10.1016/j.comptc.2017.05.020.

Abbasi, Amirali, and Jaber Jahanbin Sardroodi. 2018. "Investigation of the Adsorption of Ozone Molecules on TiO_2/WSe_2 Nanocomposites by DFT Computations: Applications to Gas Sensor Devices." *Applied Surface Science* 436: 27–41. 10.1016/j.apsusc.2017.12.010.

Akimov, Yu A., Wee Song Koh, and Kostya Ostrikov. 2009. "Enhancement of Optical Absorption in Thin-Film Solar Cells through the Excitation of Higher-Order Nanoparticle Plasmon Modes." *Optics Express* 17 (12): 10195–205.

Akimov, Yu A., W. S. Koh, S. Y. Sian, and S. Ren. 2010. "Nanoparticle-Enhanced Thin Film Solar Cells: Metallic or Dielectric Nanoparticles?" *Applied Physics Letters* 96 (7): 73111.

Chabot, Vincent, Yannick Miron, Michel Grandbois, and Paul G. Charette. 2012. "Long Range Surface Plasmon Resonance for Increased Sensitivity in Living Cell Biosensing through Greater Probing Depth." *Sensors and Actuators B: Chemical* 174: 94–101.

Chiu, Nan Fu, and Ting Li Lin. 2018. "Affinity Capture Surface Carboxyl-Functionalized MoS_2 Sheets to Enhance the Sensitivity of Surface Plasmon Resonance Immunosensors." *Talanta* 185 (January): 174–81. 10.1016/j.talanta.2018.03.073.

Chu, H. S., W. B. Ewe, W. S. Koh, and E. P. Li. 2008. "Remarkable Influence of the Number of Nanowires on Plasmonic Behaviors of the Coupled Metallic Nanowire Chain." *Applied Physics Letters* 92 (10): 103103.

Guo, Rensong, Yutong Han, Chen Su, Xinwei Chen, Min Zeng, Nantao Hu, Yanjie Su, Zhihua Zhou, Hao Wei, and Zhi Yang. 2019. "Ultrasensitive Room Temperature NO_2 Sensors Based on Liquid Phase Exfoliated WSe_2 Nanosheets." *Sensors and Actuators, B: Chemical* 300 (2): 127013. 10.1016/j.snb.2019.127013.

Hu, Guohang, Hongbo He, Anna Sytchkova, Jiaoling Zhao, Jianda Shao, Marialuisa Grilli, and Angela Piegari. 2017. "High-Precision Measurement of Optical Constants of Ultra-Thin Coating Using Surface Plasmon Resonance Spectroscopic Ellipsometry in Otto-Bliokh Configuration." *Optics Express* 25 (12): 13425–34.

Hutter, Eliza, and Janos H. Fendler. 2004. "Exploitation of Localized Surface Plasmon Resonance." *Advanced Materials* 16 (19): 1685–706.

Jha, Rajan, and Anuj K. Sharma. 2009. "Chalcogenide Glass Prism Based SPR Sensor with Ag–Au Bimetallic Nanoparticle Alloy in Infrared Wavelength Region." *Journal of Optics A: Pure and Applied Optics* 11 (4): 45502.

Kaur, Harmanjit, Munish Shorie, and Priyanka Sabherwal. 2020. "Biolayer Interferometry-SELEX for Shiga Toxin Antigenic-Peptide Aptamers & Detection via Chitosan-WSe2 Aptasensor." *Biosensors and Bioelectronics* 167 (July): 112498. 10.1016/j.bios.2020.112498.

Liedberg, Bo, Claes Nylander, and Ingemar Lunström. 1983. "Surface Plasmon Resonance for Gas Detection and Biosensing." *Sensors and Actuators* 4: 299–304.

Liu, Nanshu, and Si Zhou. 2017. "Gas Adsorption on Monolayer Blue Phosphorus: Implications for Environmental Stability and Gas Sensors." *Nanotechnology* 28 (17): 175708. 10.1088/1361-6528/aa6614.

Liu, Lili, Mei Wang, Lipeng Jiao, Ting Wu, Feng Xia, Meijie Liu, Weijin Kong, Lifeng Dong, and Maojin Yun. 2019. "Sensitivity Enhancement of a Graphene–Barium Titanate-Based Surface Plasmon Resonance Biosensor with an Ag–Au Bimetallic Structure in the Visible Region." *Journal of the Optical Society of America B* 36 (4): 1108–16.

Maharana, Pradeep Kumar, Rajan Jha, and Srikanta Palei. 2014. "Sensitivity Enhancement by Air Mediated Graphene Multilayer Based Surface Plasmon Resonance Biosensor for near Infrared." *Sensors and Actuators B: Chemical* 190: 494–501.

Maier, Stefan Alexander. 2007. *Plasmonics: Fundamentals and Applications*. New York City, United States: Springer Science & Business Media.

Mishra, Akhilesh Kumar, Satyendra Kumar Mishra, and Rajneesh Kumar Verma. 2016. "Graphene and beyond Graphene MoS_2: A New Window in Surface-Plasmon-Resonance-Based Fiber Optic Sensing." *The Journal of Physical Chemistry C* 120 (5): 2893–900.

Nair, Rahul Raveendran, Peter Blake, Alexander N. Grigorenko, Konstantin S. Novoselov, Tim J. Booth, Tobias Stauber, Nuno M. R. Peres, and Andre K. Geim. 2008. "Fine Structure Constant Defines Visual Transparency of Graphene." *Science* 320 (5881): 1308.

Ni, Jiaming, Wei Wang, Mildred Quintana, Feifei Jia, and Shaoxian Song. 2020. "Adsorption of Small Gas Molecules on Strained Monolayer WSe_2 Doped with Pd, Ag, Au, and Pt: A Computational Investigation." *Applied Surface Science* 514 (December 2019): 145911. 10.1016/j.apsusc.2020.145911.

Ong, Biow Hiem, Xiaocong Yuan, Swee Chuan Tjin, Jingwen Zhang, and Hui Min Ng. 2006. "Optimised Film Thickness for Maximum Evanescent Field Enhancement of a Bimetallic Film Surface Plasmon Resonance Biosensor." *Sensors and Actuators B: Chemical* 114 (2): 1028–34.

Otto, Andreas. 1968. "Excitation of Nonradiative Surface Plasma Waves in Silver by the Method of Frustrated Total Reflection." *Zeitschrift Für Physik A Hadrons and Nuclei* 216 (4): 398–410.

Pal, Amrindra, and Ankit Jha. 2021. "A Theoretical Analysis on Sensitivity Improvement of an SPR Refractive Index Sensor with Graphene and Barium Titanate Nanosheets." *Optik* 231: 166378.

Pan, Wenjing, Yong Zhang, and Dongzhi Zhang. 2020. "Self-Assembly Fabrication of Titanium Dioxide Nanospheres-Decorated Tungsten Diselenide Hexagonal Nanosheets for Ethanol Gas Sensing Application." *Applied Surface Science* 527 (April): 146781. 10.1016/j.apsusc.2020.146781.

Pechprasarn, Suejit, Supannee Learkthanakhachon, Gaige Zheng, Hong Shen, Dang Yuan Lei, and Michael G. Somekh. 2016. "Grating-Coupled Otto Configuration for Hybridized Surface Phonon Polariton Excitation for Local Refractive Index Sensitivity Enhancement." *Optics Express* 24 (17): 19517–30.

Prajapati, Y. K., and Akash Srivastava. 2019. "Effect of BlueP/MoS_2 Heterostructure and Graphene Layer on the Performance Parameter of SPR Sensor: Theoretical Insight." *Superlattices and Microstructures* 129 (January): 152–62. 10.1016/j.spmi.2019.03.016.

Rahman, M. Saifur, Md. Shamim Anower, and Lway Faisal Abdulrazak. 2019. "Utilization of a Phosphorene-Graphene/TMDC Heterostructure in a Surface Plasmon Resonance-Based Fiber Optic Biosensor." *Photonics and Nanostructures – Fundamentals and Applications* 35 (May): 100711. 10.1016/j.photonics.2019.100711.

Rahman, M. Saifur, Md. Shamim Anower, Lway Faisal Abdulrazak, and Md. Maksudur Rahman. 2019. "Modeling of a Fiber-Optic Surface Plasmon Resonance Biosensor

Employing Phosphorene for Sensing Applications." *Optical Engineering* 58 (03): 1. 10.1117/1.oe.58.3.037103.

Reather, Heinz. 1988. "Surface Plasmons on Smooth and Rough Surfaces and on Gratings." *Springer Tracts in Modern Physics* 111: 1–3.

Setareh, Maryam, and Hassan Kaatuzian. 2021. "Sensitivity Enhancement of a Surface Plasmon Resonance Sensor Using Blue Phosphorene/MoS_2 Hetero-Structure and Barium Titanate." *Superlattices and Microstructures* 153 (August 2020): 106867. 10.1016/j.spmi.2021.106867.

Sharma, Anuj K., Jyoti Gupta, and Ishika Sharma. 2019. "Fiber Optic Evanescent Wave Absorption-Based Sensors: A Detailed Review of Advancements in the Last Decade (2007–18)." *Optik* 183 (December 2018): 1008–25. 10.1016/j.ijleo.2019.02.104.

Sharma, Anuj K., and Ankit Kumar Pandey. 2018. "Blue Phosphorene/MoS_2 Heterostructure Based SPR Sensor with Enhanced Sensitivity." *IEEE Photonics Technology Letters* 30 (7): 595–8. 10.1109/LPT.2018.2803747.

Shushama, Kamrun Nahar, Md Masud Rana, Reefat Inum, and Md Biplob Hossain. 2017. "Graphene Coated Fiber Optic Surface Plasmon Resonance Biosensor for the DNA Hybridization Detection: Simulation Analysis." *Optics Communications* 383: 186–90.

Singh, Arun Kumar, P. Kumar, D. J. Late, Ashok Kumar, S. Patel, and Jai Singh. 2018. "2D Layered Transition Metal Dichalcogenides (MoS_2): Synthesis, Applications and Theoretical Aspects." *Applied Materials Today* 13: 242–70. 10.1016/j.apmt.2018.09.003.

Singh, Yadvendra, and Sanjeev Kumar Raghuwanshi. 2019. "Sensitivity Enhancement of the Surface Plasmon Resonance Gas Sensor with Black Phosphorus." *IEEE Sensors Letters* 3 (12): 1–4.

Singh, Sachin, Pravin Kumar Singh, Ahmad Umar, Pooja Lohia, Hasan Albargi, L. Castañeda, and D. K. Dwivedi. 2020a. "2D Nanomaterial-Based Surface Plasmon Resonance Sensors for Biosensing Applications." *Micromachines* 11 (8): 1–28. 10.3390/mi11080779.

Singh, Sachin, Pravin Kumar Singh, Ahmad Umar, Pooja Lohia, Hasan Albargi, L Castañeda, and D K Dwivedi. 2020b. "2D Nanomaterial-Based Surface Plasmon Resonance Sensors for Biosensing Applications." *Micromachines* 11 (8): 779.

Srivastava, Triranjita, and Rajan Jha. 2018. "Black Phosphorus: A New Platform for Gaseous Sensing Based on Surface Plasmon Resonance." *IEEE Photonics Technology Letters* 30 (4): 319–22.

Srivastava, Akash, and Y. K. Prajapati. 2019. "Performance Analysis of Silicon and Blue Phosphorene/MoS_2 Hetero-Structure Based SPR Sensor." *Photonic Sensors* 9 (3): 284–92. 10.1007/s13320-019-0533-1.

Su, Mingyang, Xueyu Chen, Linwei Tang, Bo Yang, Haijian Zou, Junmin Liu, Ying Li, Shuqing Chen, and Dianyuan Fan. 2020. "Black Phosphorus (BP)–Graphene Guided-Wave Surface Plasmon Resonance (GWSPR) Biosensor." *Nanophotonics* 9(14): 4265–4272.

Verma, Alka, Arun Prakash, and Rajeev Tripathi. 2015. "Sensitivity Enhancement of Surface Plasmon Resonance Biosensor Using Graphene and Air Gap." *Optics Communications* 357: 106–112.

Wu, Leiming, Zhitao Ling, Leyong Jiang, Jun Guo, Xiaoyu Dai, Yuanjiang Xiang, and Dianyuan Fan. 2016. "Long-Range Surface Plasmon with Graphene for Enhancing the Sensitivity and Detection Accuracy of Biosensor." *IEEE Photonics Journal* 8 (2): 1–9.

Wu, Yichuan, Nirav Joshi, Shilong Zhao, Hu Long, Liujiang Zhou, Ge Ma, Bei Peng, Osvaldo N. Oliveira, Alex Zettl, and Liwei Lin. 2020. "NO_2 Gas Sensors Based on CVD Tungsten Diselenide Monolayer." *Applied Surface Science* 529 (2): 2–8. 10.1016/j.apsusc.2020.147110.

Yang, Chen, Jiayue Xie, Chengming Lou, Wei Zheng, Xianghong Liu, and Jun Zhang. 2021. "Flexible NO_2 Sensors Based on WSe_2 Nanosheets with Bifunctional Selectivity and Superior Sensitivity under UV Activation." *Sensors and Actuators, B: Chemical* 333 (2): 129571. 10.1016/j.snb.2021.129571.

Zekriti, M. 2021. "Theoretical Comparison between Graphene-on-Silver and Gold-on-Silver Based Surface Plasmon Resonance Sensor." *Materials Today: Proceedings* 45 (8), 7571–7575.

Zhang, Z. Y., M. S. Si, S. L. Peng, F. Zhang, Y. H. Wang, and D. S. Xue. 2015. "Bandgap Engineering in van Der Waals Heterostructures of Blue Phosphorene and MoS_2: A First Principles Calculation." *Journal of Solid State Chemistry* 231: 64–9. 10.1016/j.jssc.2015.07.043.

4 Black Phosphorus

Narrow Gap and Wide Application in SPR Sensors

4.1 INTRODUCTION

The advent of 2D materials-based devices for chemical sensing have seen a boost in performance due to the exceptional mechanical, surface, electrical, as well as optical properties of these materials. The density of several adsorption sites increases due to the large surface area offered by the 2D materials (Varghese, Lonkar, et al. 2015; Varghese, Varghese, et al. 2015; Yuan and Shi 2013), as well as the sensing channel. It may be possible due to being easily customized with an appropriate bandgap by varying the number of the layers of the 2D materials in the device structure (Wang et al. 2012; Yang et al. 2014). These materials offer numerous different active sites for selective molecular adsorption. These include edge, vacancy, defects, and basal planes (Cui et al. 2014; Jaramillo et al. 2007; Li, Yang, et al. 2016; Narayan-Maiti et al. 2014). Another advantage of 2D materials apart from the above is that these are operated at room temperature, which is unattainable with devices having metal oxide-based semiconductors (Lu, Ocola, and Chen 2009). There are so many 2D materials that have been developed with a variety of extraordinary optical properties (e.g., ultrafast broad optical response varies from ultraviolet to radio waves, large optical nonlinearities, and tunable and strong interaction between light and mater) due to the different band gaps and electronic properties.

A nanoscale planar crystal constituted by stacking several layer(s) of a single atom(s) is termed a two-dimensional (2D) crystal. In 2004, Andre Geim along with Konstantin Novoselov became the first group to successfully demonstrate the stripping off of graphene with the help of an adhesive tape (Corbridge 2016), this provided a phenomenal stimulus in the research, exploration, and potential applications of 2D crystals. Recent developments have brought to light another basic 2D material known as phosphorene (Castellanos-Gomez 2015) with a unique crumpled structure as well as fascinating in-plane anisotropy. This material has properties that make it a potential material suitable for energy band compensation applications. Prof. M. Peruzzini, National Research Council, Italy, called phosphorene a new competitor to graphene, which is the front runner of the 2D material class. He even praised it to the extent of calling the emergence of phosphenes as a black swan that got converted from an ugly duckling. The renaissance of BP is termed from recognition of black phosphorus (BP) along with the corresponding research in the field of materials (Fig. 4.1).

DOI: 10.1201/9781003190738-4

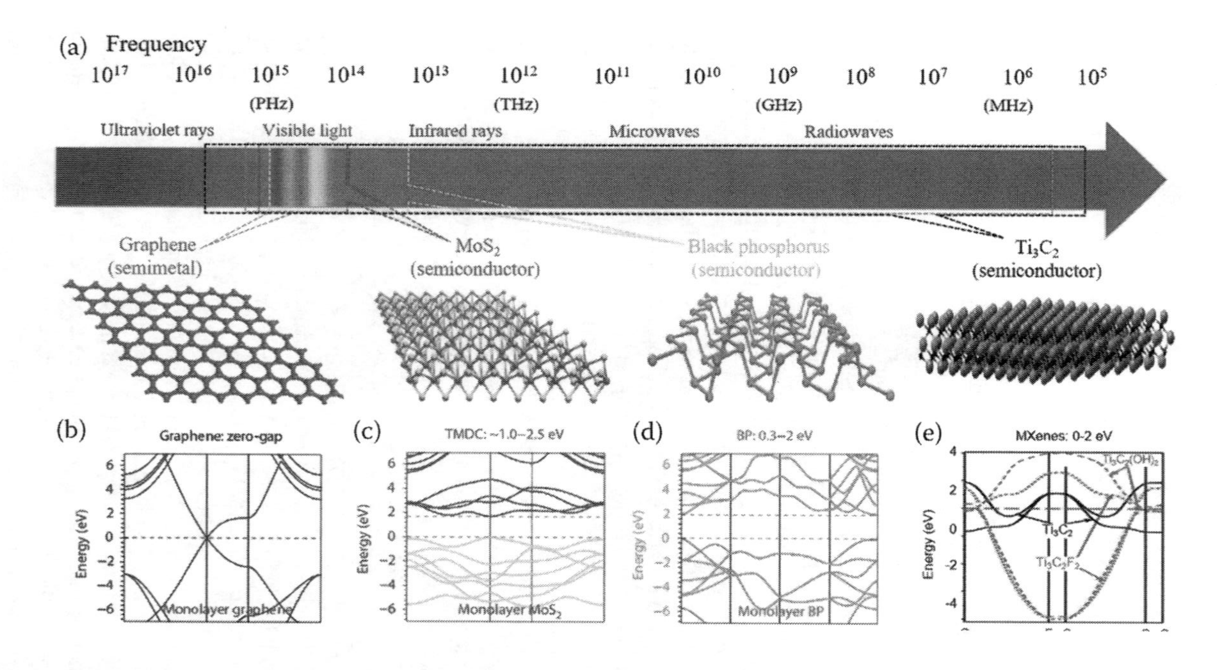

FIGURE 4.1 Broadband spectrum of 2D materials. a) Atomic structures of 2D materials (graphene, MoS_2, BP, and Ti_3C_2) with their respective electromagnetic spectrum, b) band diagram of single-layer graphene, c) band diagram of MoS_2, d) band diagram of BP, and e) band structure of MXene monolayer. Reprinted with permission from Advanced Science journal. Copyright, 2020, Wiley (Tan et al. 2020).

BP is a crystal that resembles graphene, it is thin, layered, and Van der Waals (VdW) forces hold its atomic layers together. VdW forces are weak in nature, thus single or thin nanofilms may be formed by separating BP layers. Under normal pressure, phosphorous has three stable allotropes, namely red phosphorus, white phosphorus, and BP. The crystalline structure of single layer of BP is a folded-honeycomb structure of covalent bonds with the three neighboring atoms (of phosphorus). BP, unlike other similar materials of its class (2D materials), has an specific crystalline structure arrangement. It aligns to the material atoms in two directions (perpendicular to each other), namely the armchair and the z-direction of a font. Thus, the structure of the crystal demonstrates a ridges form and a folding form. Ridges form in the path of the two atoms between covalent bond lengths; the bond angle varies, resulting in major variations in thermal, mechanical, chemical, and physical properties, as well as various crystal directions (Fig. 4.2).

BP offers some unique optical properties when compared to other optical materials; for instance, with molybdenum sulphide and silicon, the former being a direct bandgap whereas the latter is an indirect bandgap. By absorbing optical energy, the material (BP) can change from an insulator to a conductor, whereas in silicon and molybdenum sulphide materials the change is brought about by absorbing the energy changes as well as the change in the momentum (position) simultaneously. This implies that BP can be used in places where there is a need to couple with light, and it offers a wide band range that can absorb light from visible to near-infrared regions (wavelength of 0.6–4.0 μm). Its distinct anisotropic quality distinguishes it from other materials like transition metal, graphene, sulphur compounds, and other 2D materials. BP is seen as an excellent replacement candidate substance in developing new micro-nano and/or photoelectric devices as well as for development of the next sensing devices with enhanced performance. It also has the potential for work related to space applications. BP has also been associated with many other domains (e.g., the electrical, optical, biological, and chemical) in various application prospects.

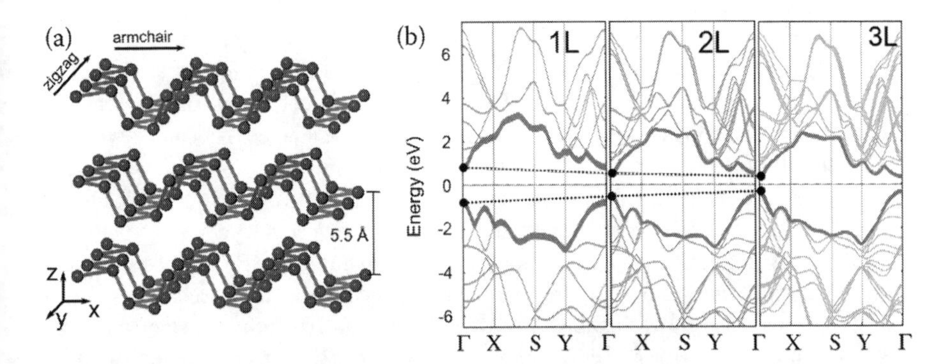

FIGURE 4.2 a) The crystalline structure of BP. b) The tunable bandgap structure of single-layer, double-layer, and triple-layer BP. Reprinted with permission from Journal of Physical Chemistry Letters. Copyright, 2015, American Chemical Society (Castellanos-Gomez 2015).

The next section will go into the properties, synthesis, instruments, and applications of BP.

4.2 HISTORY OF BLACK PHOSPHORUS (BP) AND PHOSPHORENE

BP crystals are not a recent discovery; they were discovered in 1914. Synthesis of BP from white phosphorus (Bridgman 1914) was first demonstrated by Percy Bridgman. Bridgman, in his experiment to study the impact of high pressure on white phosphorus (Xu et al. 2019b), observed that there was a phase transition that led to the conclusion of the new allotrope, which was named "black phosphorus." The x-ray revealed that BP crystals were an orthorhombic formation (Hultgren, Gingrich, and Warren 1935). The 2D-layered structure of BP was the most stable among the allotropes of phosphorus at high temperatures and under high pressure. Properties such as high carrier mobility, tunable narrow/direct band gaps, photothermal property, large specific surface area, biodegradability, biocompatibility, and many interesting in-layer anisotropies have drawn considerable interest for applications such as oxygen evolution, storage and energy conversion, photocatalytic hydrogenation, optoelectronics, electronics, and others.

The structure and fundamental properties of BP, as well as a few preparation methods, are summarised in the following section.

4.2.1 Synthesis and Characterization

Yi et al. (2017) investigated and intensely studied BP as a 2D semiconductor material, using a new and advanced class of 2D materials family. BP's tunable and moderate bandwidth and high carrier mobility are promising for optoelectronic and electronic applications, and the unusual structure with puckered edges creates remarkable mechanical, thermal, electronic, transport, and optical properties that can be used in new device designs. BP crystal syntheses can be classified by various parameters, including bulk crystal growth (Bridgman 1935, 1937; Shirotani 1982), bulk crystal development, liquid exfoliation (Brent et al. 2014; Coleman et al. 2011; Kang et al. 2015), electronic structure (Asahina and Morita 1984; Takao, Asahina, and Morita 1981), etc. The growth of bulk crystals is shown by changes in the glass structural elements dependent upon both temperature and pressure, with the addition of tin-iodide, tin, and gold in less amounts to quality wise superior BP single crystals in conditions of low-pressure at 600°C from red phosphorus (RP). It is found from some studies that can convert red phosphorus into BP with the pressure of 4.5 GPa/8.0 GPa at room temperature with/without shearing tension. BP is a synthesis without the toxic or "dirty" flux at low pressure. With the mechanical layering feature, thin, atomic BP layers can be cleaved with adhesive tape. The parent crystal of BP is used to peel the BP thin flakes, diluted with the aid of adhesive tape by the repeated adherence/separation technique. Direct thin-film growth (Du, Lin, Xu, and Chu 2015; Jiang et al. 2016; Smith, Hagaman, and Ji 2016; Yang et al. 2015) shows wafer-scale, crystalline BP films and chemical vapor deposition (CVD) is available to monolayer graph and TMD wafer-scale, which helps to produce layer-controllable BP on a broad and efficient basis. Liquid exfoliation permits the mass processing of various BP nanostructures, including nanosheets and quantum dots. For 2D materials, such as graphene, hexagonal boron nitride (short BN or h-BN), and TMD, it (liquid exfoliation) serves as a strong guide for large-scale BP nanosheet growth and can also be used to produce BP quantum dots (BPQDs), which show unique quantum preservation and edge effects. Two different techniques are capable of electronic structure, one being a layer-dependent structure for the electronic band, and one anisotropic. More modification of the surface by decomposition risk and encapsulation of the surface is possible. Wang, Jones, et al. (2015) successfully synthesized the ultrathin BP nano-sheets by using water phase exfoliation. As shown in Fig. 4.3, is the process of BP exfoliation from the bulk BP samples in water. The further verification of nanosheets of BP samples is performed by the TEM and SEM characterization. The SEM image of the obtained bulk BP and the layered form of the BP are shown in Fig. 4.4a and Fig. 4.4b, respectively.

Fig. 4.5 shows the synthesis process of BPQDs and the size of 4.9 ± 1.6 nm and thickness of 1.9 ± 0.9 nm was achieved. BPQD images display the grid fringes of 0.34 and 0.52 nm, with high-resolution electron microscopy (HR-TEM) shown in Fig. 5.5(a), with the resulting levels of the (021) and (020) BP crystal. BP powders are usually applied to the NMP solvent and sonicated for 3 h in an ice bath. The supernatant containing the BPQDs has been gently decanting after 20 min centrifugation at 7,000 rpm. The lateral BPQD scale of 2.6 ± 1.8 nm and the thickness of 1.5 ± 0.6 nm was further improved by combining sample sonication and bath sunlight to promote exfoliation. The researchers produced large-scale solvothermal BPQDs with average side sizes of 2.1 nm (Fig. 4.6).

The degradation of oxidation leads to compositional changes and degraded optical and electronic properties. BP passivation has been performed by aluminum

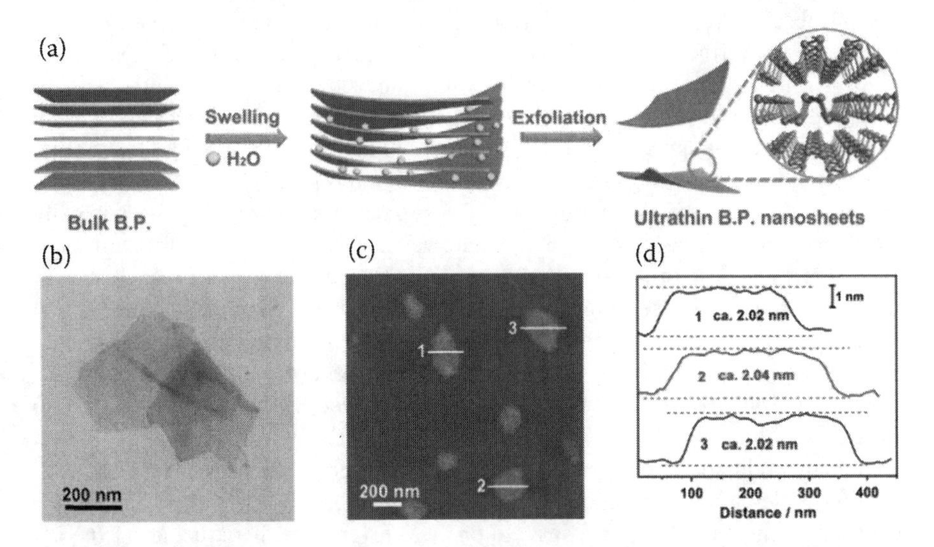

FIGURE 4.3 Ultrathin BP nano-sheets: preparation and morphology: a) Synthesis steps of ultrathin nanosheets using a water-exfoliation method from bulk BP; b) transmission electron microscopic (TEM) image; c) atomic force microscopic (AFM) image; d) height profile of BP nanosheet using AFM image. Reprinted with permission from Journal of the American Chemical Society. Copyright, 2015, American Chemical Society (Wang, Yang, et al. 2015).

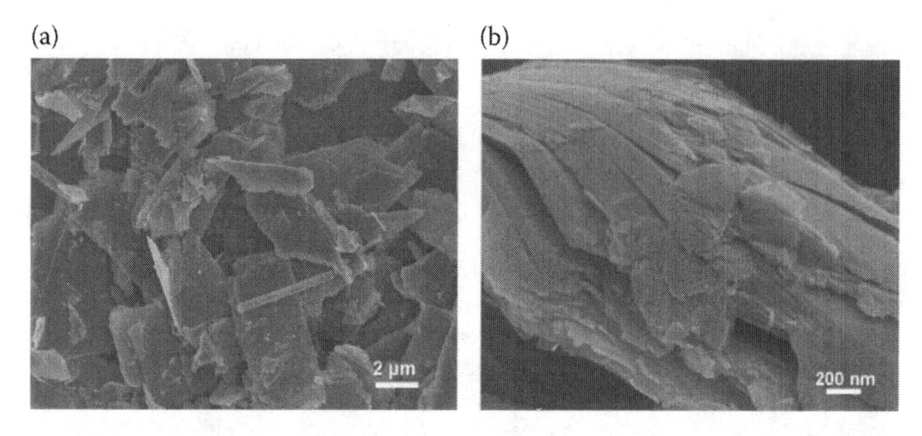

FIGURE 4.4 a) SEM images of the obtained bulk BP and b) showing layer-stacking morphology. Reprinted with permission from Journal of the American Chemical Society. Copyright, 2015, American Chemical Society (Wang, Yang, et al. 2015).

oxide (AlO_x) and it is also used to block molecules of water and oxygen to avoid degradation. Chemical modification can also change the synthesis of BP. Coordination and covalent modification are also useful methods to protect BP. The titanium benzene sulfonate ligand (TiL_4) is synthesized with an empty titanium orbital with heavy electronic benzene sulfonate absorption. TiL_4 coordinates with a BP lone-pair electron, observing nanosheets from bare BP, TiL_4@BP show great stability during long-term dispersion for air and water exposure, at the same time optical absorption and thermal degradation are minimal. The exfoliated BP nanosheets are further demonstrated to be stabilized through the formation of phosphorous–carbon covalent bonds with p-nitrobenzene diazonium and p-methoxybenzene diazonium salts. Also, after three weeks, the functionalized BP nanosheets are not apparent and the benzene ring improves the BP FET's mobility of electrons and switching ratio. Chen et al. (2017a) proposed BP nanosheets, using the dielectric layer of Al_2O_3 thin film for surface passivation, are mechanically exfoliated, and behave as a sensing/conductive channel in the field effect transistor (FET). The response of the sensor is determined by the electrical resistance shift in the BP following the use of antigens. The semi-conductor characterization method is used to characterize the sensor devices for electrical properties (transistor measurements and direct current) and to perform sensing tests such as dynamic measurements. The direct current is measured by recording the drain stream from –2.0 to +2.0 V with the drainage-source voltage V_{ds}. The FET-based BP biosensor shares common working principles, such as MoS_2 and reduced graphene oxide (rGO), with previously described semiconducting 2D materials. The efficiency of the BP sensor depends on the nanosheet thickness. In order to investigate the structure of the as-produced unit, the SEM on a BP sensor is performed and the SEM is shown in Fig. 4.7. The average size of a nanosheet is between many hundreds of nanometers and many microns. Since, the electrode space is 1.5 mm, the functions of the sensor can be bridged only by the nanosheets. High density of gold nanoparticles (NPs)

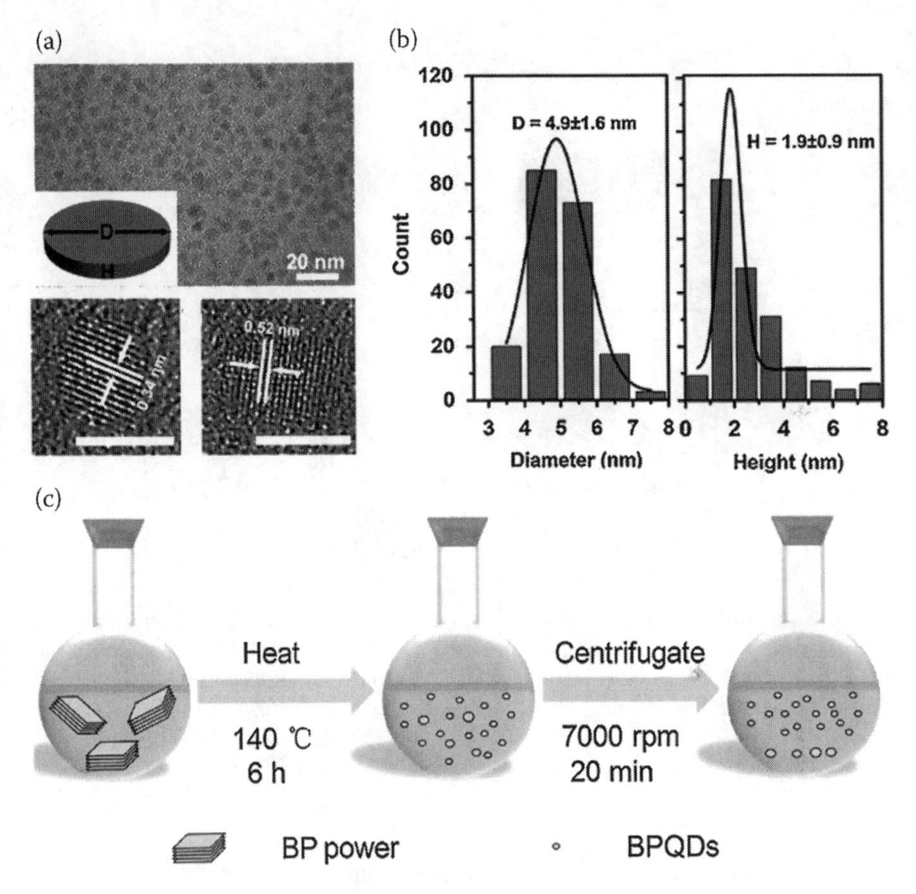

FIGURE 4.5 BPQD's liquid exfoliation: a) Top: BPQD TEM pictures. Bottom: HR-TEM pictures of (left) 0.34 nm and (right) 0.52 nm BPQDs with separate lattice frames. b) A (left) (D) diameter and (right) (H) thickness statistical analysis calculated in TEM images of 200 BPQDs. c) Schematic synthesis diagram of the BPQD production process using the NaOH NMP saturated solvothermal method. Reprinted with permission from Journal of Materials Science and Engineering: R: Reports. Copyright, 2017, Elsevier (Yi et al. 2017).

could lead to a high density of sensor antibodies that can improve the sensor response on the antigen-binding. The estimated BP nanosheet thickness is between 30 nm and 40 nm. Three Raman peaks at approximately 359 cm^{-1}, 434 cm^{-1}, and 461 cm^{-1} of a spectrum are reported, corresponding to Ag^2, Ag^1, and B_{2g} modes of pristine BP nanosheets, as shown in Fig. 4.7(b).

Late (2016) examined BP nanosheet fluid exfoliation, which functions as a sensor for humidity. The well-known fluid phase exfoliation method was synthesized with the simple ultrasonic solvent of BP nanosheets and nanoparticles, using N-methyl-2-pyrrolidone (NMP) as solvent, as this solvent results in a stable and condensed dispersions. Material can be characterized by UV-Vis-NIR absorption, Raman spectroscopy, and TEM. For AFM, the samples have been generated through the distribution of the BP nanosheets in ethanol and casting on clean Si

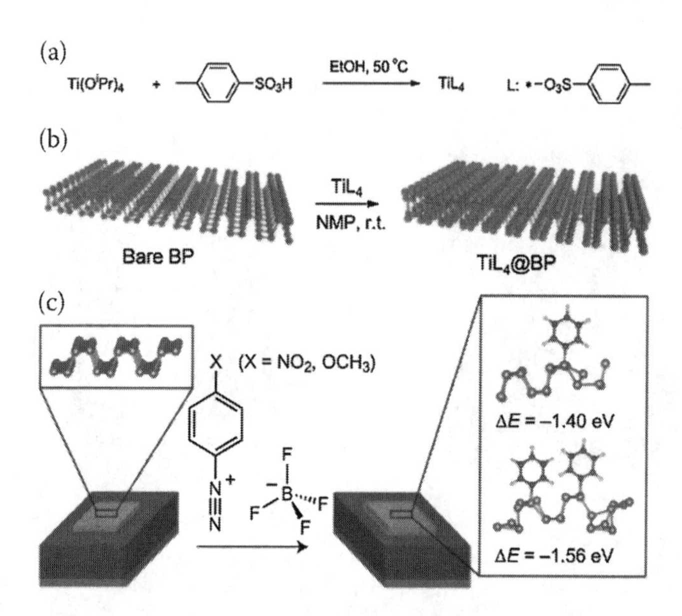

FIGURE 4.6 BP coordination and covalent changes: a) Synthesis and structural formula of TiL_4; b) surface coordination of TiL_4 to BP; b) TiL4@BP sheets on Si/SiO2 and exposed to the humid air at room temperature for 0 h (left), 12 h (middle), and 24 h (right). c) Benzene diazonium tetrafluoroborate derivatives reaction scheme and few-layer BP (light blue) mechanically exfoliated on an Si (grey)/SiO_2 (purple) substratum. Reprinted with permission from Journal of Materials Science and Engineering: R: Reports. Copyright, 2017, Elsevier (Yi et al. 2017).

substrates. Drop casting of the BP nanosheets/nanoparticles on a carbon-coated copper TEM grid was used to make samples for TEM characterization. The samples were cast in silicone substratum for Raman characterization and scattered into ethanol for UV-Vis-NIR absorption. Xu et al. (2019a), using three key methods to develop SiP, GeP, or As_xP_{1-x} single crystals, synthesized recent developments in BP and BP-analog materials, i.e., method of flux region, high pressure (HP) method of melting growth method, and method of chemical vapor transport (CVT). As a precursor in the flux area process, silicon (5 N)/Ge (5 N) and RP (6 N) have been used and Sn/Bi has become the flux for o-SiP and o-GeP synthesis. The oven was heated to 500°C slowly for SiP and maintained its temperature for 36 h. Then, the temperature was raised for heating at 1150°C for 12 h at that temperature. After the cooling process, o-SiP crystals were obtained. The CVT method is a standard universal way of synthesizing high-quality single crystals. The pure elements for SiP and GeP were first mixed. As_xP_{1-x} is a blend of grey arsenic with red phosphorus with a molar ratio of 9:1 to 2:8. The starting materials were supplemented by lead iodide (PbI_2) or tin iodide (SnI_4) and used as a mineralizer. The mixture was heated up to 550°C and maintained throughout 20–80 h at that temperature, and then slowly refreshed and received large crystals. Kuriakose et al. (2018) investigated plasma treatment effects of BP-layered electric and opto-electronic properties, and concluded that pristine BP flakes are thick, with the aid of atomic

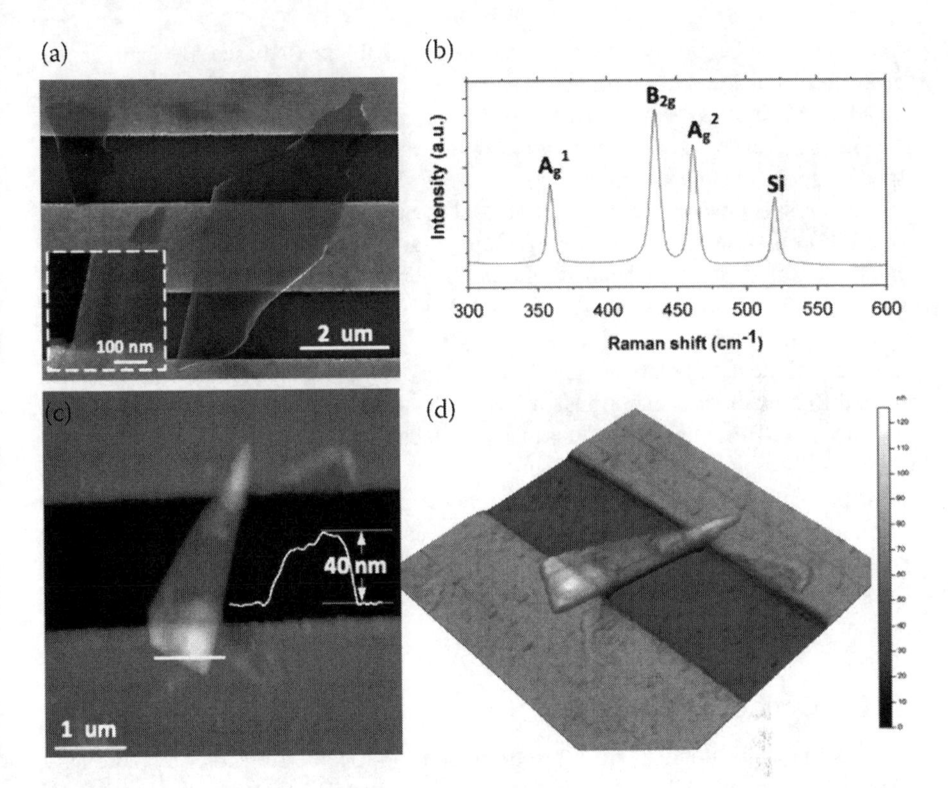

FIGURE 4.7 Characterization of the BP FET sensor. a) SEM picture on interdigitated gold electrodes of mechanically exfoliated BP nanosheets. The inset depicts the BP nanosheet's high-magnification image. On the surface of the BP nanosheet, dense and isolated dispersed gold nanoparticles are observed. b) BP nanosheet analyzed by Raman spectrum. The spectrum identifies three Raman peaks at approximately 359 cm^{-1}, 434 cm^{-1}, and 461 cm^{-1}. The pitch, 519 cm^{-1}, is of the Si substrate.(c–d) BP nanosheet AFM pictures. c) The nanosheet thickness is measured in the 30–40 nm range. d) A 3D view of the gold electrodes placed over the BP nanosheet. Reprinted with permission from Journal of Biosensors and Bioelectronics. Copyright, 2017, Elsevier (Chen et al. 2017a).

force microscopy (AFM). A variety of inductively coupling plasma (ICP) and radio frequency (RF) power conditions were initially examined at a fixed pressure of 10 mTorr in order to create an optimal plasma etching recipe that provided a controlled layer removal of BP without causing minimum damage to the crystal structure and morphology. In the light of their differing efficiencies in ionization, after a series of etching cycles followed by each AFM characterization, an optimal set of ICP and RF powers was defined for both O_2 and A_r etching.

A single atomic layer (monolayer) of BP is also called phosphorene. The thermodynamically most stable BP is one of the top three allotropes of phosphorus, as opposed to its red and white equivalent. In order to understand the properties of this novel material, physical characterization is necessary. A BP crystal is identified with Raman spectroscopy. The asymmetry and crystalline design of the BP is calculated by the reflectance microscope and a thickness-based BP bandgap is measured by using atomic force microscopy (Haratipour 2017).

After successfully synthesizing and preparing BP, it is necessary to characterize the sample. Hence, the most common techniques used to characterize BP are Raman spectroscopy, reflectance microscopy, and AFM.

Raman spectroscopy is an effective tool in characterizing 2D materials because details on the structure, electronic structure, thickness of the material, and vibrations can be revealed. There is detailed vibration information on each chemical bond in a substance. This knowledge can also be used as a fingerprint for the identification of the material structure.

A technique based on reflectance microscopy is that the optical absorption varies with different crystal directions in materials that have an anisotropic electrical conductivity. BP is an anisotropic material, meaning that the mass is efficient and thus mobility changes in various directions of transport. The crystal orientation must be regulated to regulate the electrical properties of BP MOSFETs. Angular Raman and reflective microscopy may also be used for the identification of BP flake crystal directions on the substrate. AFM is a high-resolution instrument to measure the sample's surface topography to extract the channel's thickness and obtain a total system topography.

4.2.2 Optical Properties

BP nanomaterial exhibits highly anisotropy optical properties. It shows extraordinary linear dichroism property. Due to virtues of the property, BP absorbs light signals with different polarization at dissimilar rates, especially frequencies near to the energy bandgap of the material (Carvalho et al. 2016; Dai and Zeng 2014; Du et al., n.d.; Sorkin et al. 2017). BP and phosphorene absorb an armchair direction polarization component of the light rays, but it is transparent for the light ray in a different direction, i.e., zigzag direction. Due to the changeover of the y-polarized light laterally, the crisscross route is prohibited by the regularities of the electron wave function in the valence and conduction bands. Thus, single or multiple layers of BP nanomaterial are normal optical photon polarizers suitable for optical device designing. This optoelectronic anisotropy property can be taken advantage of to show the orientation of the crystallographic axes. Additionally, bulk BP shows a large advancement in the optical conductivity that causes the inclination toward band-nesting (Yuan et al. 2015). Due to these properties, phosphorene is the best suitable material for the designing of the photodetector in the UV region (Carvalho et al. 2016; Sorkin et al. 2017; Tran, Soklaski, Liang, and Yang 2014), and also shows high responsivity in the UV region. A BP with few layers shows better responsivity than photoresponsivity of graphene-based devices, even 100 times more, when compared to the TMDCs, which show less responsivity. The BP shows responsivity in the range of 7.5–780 A W^{-1} (Lopez-Sanchez et al. 2013). But, a TMDC material-based photodetector has low mobility and reacts to light slower than phosphorene-based devices (Dai and Zeng 2014; Deng et al. 2014; Sorkin et al. 2017; Xia, Wang, and Jia 2014). The best thing about BP 2-D material is that it absorbed the light ray whose energy was near the energy bandgap of the material.

The most important parameter of BP that defines the optical absorption of 2D materials is the bandgap structure of that material. The band structure will be used to describe the optical properties of 2D materials, including those that can interfere with light. In the following section, the optical properties of BP are calculated theoretically and experimentally from the visible to mid-IR range.

4.2.2.1 Linear Properties

Anisotropic electronic mobility and optical reaction of BP are generated by the puckered honeycomb lattice (Çakir, Sahin, and Peeters 2014; Chen, Huang, Gua, and Xie 2016; Chen, Ponraj, Fan, and Zhang 2020; Han et al. 2017; Hemsworth et al. 2016; Lan, Rodrigues, Kang, and Cai 2016; Li et al. 2015). The layered structure facilitates mechanical exfoliation to enable extraction of thin BP from bulk crystals. The property of BP related to direct bandgap to its dependence on several layers makes it attractive for many applications. It has the bandgap for a single layer, double layer, and triple-layer as ~2.0 eV, ~1.3 eV, and ~0.8 eV, respectively, as mentioned in Fig. 4.2(b) due to its ability of bandgap tunability. The zigzag and armchair crystalline structure of BP is illustrated in Fig. 4.8(a) while, Fig. 4.8(b) shows the electron structures for bulk BP. The angle-resolve photoemission spectroscopy (ARPES) is used to verify the band structure that agrees with the results of the screened hybrid functional calculations (Li et al. 2014).

Its unique band structure, along with puckered honeycomb lattice, results in linear optical responses that are thickness and polarization-dependent. Fig. 4.9 exhibits the reflection spectra and linear absorption for a few layers of BP (Qiao et al. 2014; Tran, Soklaski, Liang, and Yang 2014). It is concluded that from absorption spectrum threshold values that it is layer dependent, as shown in Fig. 4.9b. From monolayer to bulk, the cut-off light energy becomes lower, indicating that the bandgap is inversely correlated with the thickness or the number of layers of BP. Most of the BP-based optical devices work on a broadband spectrum due to the bandgap tunable ability of BP, which varies from ultraviolet to mid-infrared. A detailed analysis has been done to calculate the absorption of BP and bulk BP for different layers (monolayer to pentalayer) along the zigzag axis and armchair axis, which shows the strong anisotropic properties in Fig. 4.9(c–d). The optical response provides an indication to depict dependence on polarization (Hong et al. 2014; Lan, Rodrigues, Kang, and Cai 2016; Low et al. 2014; Ribeiro et al. 2015; Wang, Jones, et al. 2015). The analysis shows that the absorption efficiency is less along the zigzag than the armchair direction. It is observed by Ribeiro et al. (2015) that the thickness as well as angular dependence, which is generated by photon interaction, electron, and band structure, is similar to Raman scattering. Different sizes of BP quantum dots have been used for high sensitivity photoacoustic imaging sensors due to their photoluminescence ability in the broadband spectrum (Chen et al. 2019; Gu et al. 2017; Shao et al. 2016).

4.2.2.2 Nonlinear Properties

Researchers and scientists globally have contributed to the work related to BP to unveil its outstanding properties. Besides excellent electrical (Li et al. 2014; Liu, Wang, et al. 2014), mechanical (Islam, Lee, and Feng 2018; Jiang and Park 2014;

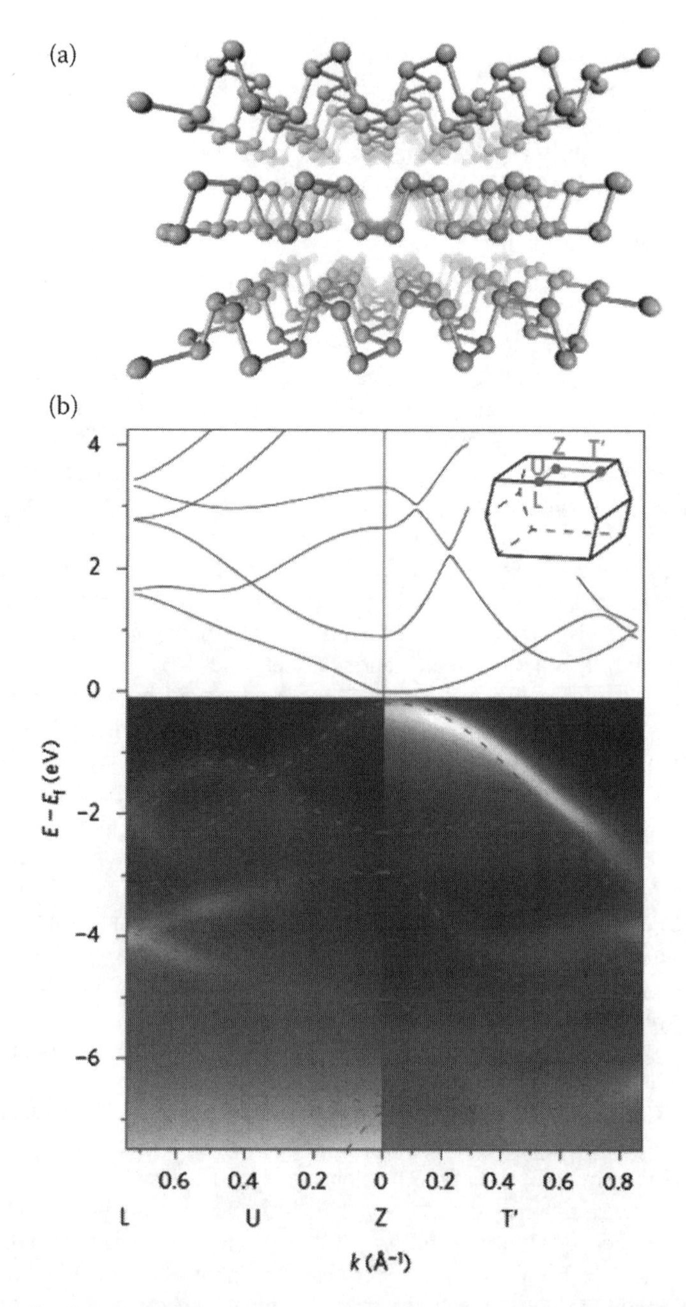

FIGURE 4.8 Crystal and electron structure of bulk BP is represented with the help of crystal structure and electron structure. a) BP atomic structure. b) Bulk BP band structure mapped out by ARPES measurements. Reprinted with permission from Journal of Nanoscale. Copyright, 2009, Royal Society of Chemistry (Chen, Ponraj, Fan, and Zhang 2020).

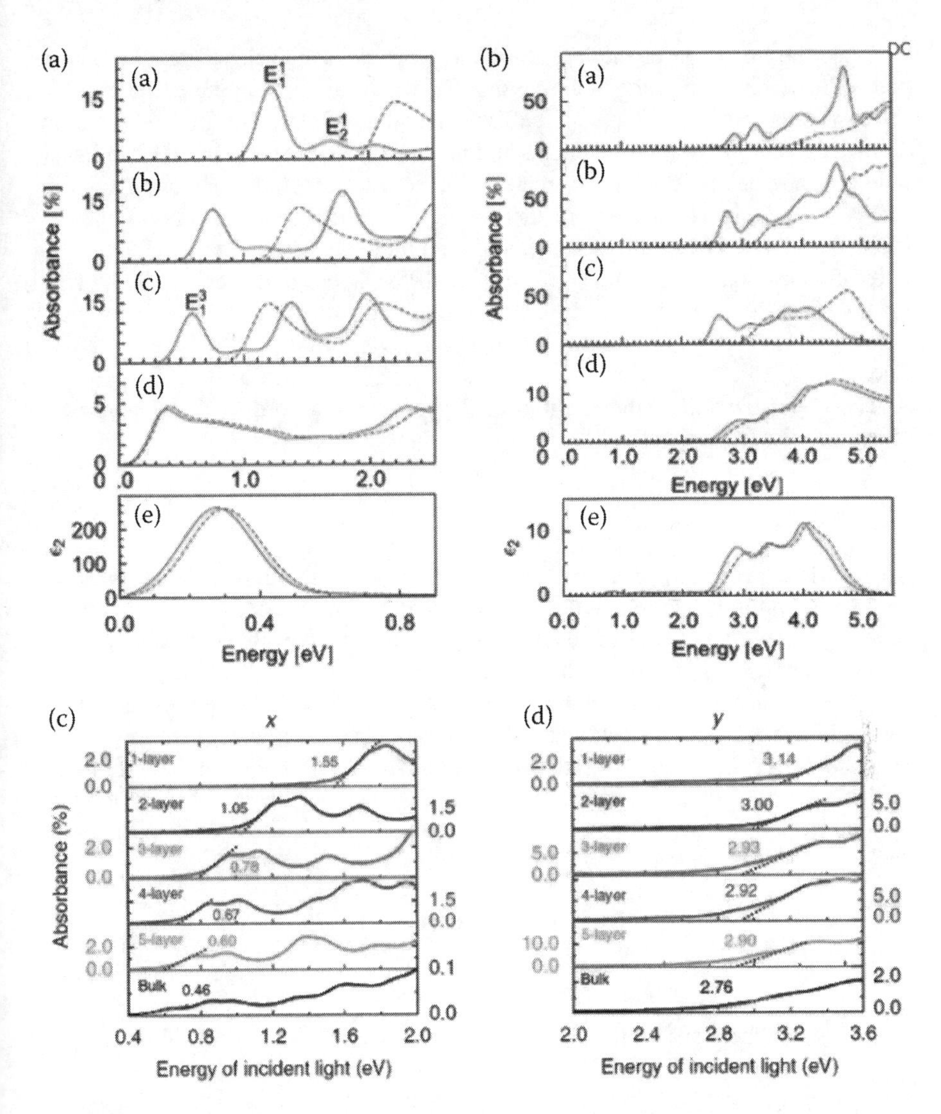

FIGURE 4.9 The absorption spectrum of BP for different thickness and orientation. a) and b) DFT analysis is used to calculate the values theoretically. c) and d) The experimental values of BP absorbance. Reprinted with permission from Journal of Nanoscale. Copyright, 2009, Royal Society of Chemistry (Chen, Ponraj, Fan, and Zhang 2020).

Moreno-Moreno et al. 2016), and thermal properties (Flores et al. 2015; Jiang 2015; Late 2015; Liu and Ruden 2017; Qin et al. 2015; Yoshizawa, Shirotani, and Fujimura 1986), BP has an amazing optical feature. In recent years, the nonlinear optical properties of 2D materials have been investigated. They found that graphene, TMDs, as well as BP, exhibit a unique nonlinear optical performance. Wang et al. (2015) observed the nonlinear response of graphene suspension under irradiation of a nanosecond laser with wavelengths of 532 nm and 1,064 nm.

Since the invention of the laser, nonlinear optics has developed rapidly. It has been already reported in the literature that BP nonlinear optical efficiencies include the absorbing saturable effect and the Kerr optical effect. In many ways, the Z-scan method is very easy, quite practical, and popularly used to examine the nonlinear optical properties of a particular material. The Z-scan method diagram shown in Fig. 4.9a is designed to measure the coefficient of nonlinear absorption and refractive index (RI) of a particular material with multiple wavelengths (Pálfalvi et al. 2009; Sheik-Bahae, Said, and Van Stryland 1989; Sheik-Bahae et al. 1990). This method was therefore adopted by many researchers to obtain and ascertain coefficients of nonlinear absorption of 2D materials (Jiang et al. 2017; Lu et al. 2015; Luo et al. 2015).

The P polarization in the medium could be written according to the nonlinear theory of susceptibility as follows:

$$P = \varepsilon_o \chi^{(1)} E + \varepsilon_o \chi^{(2)} E^2 + \varepsilon_o \chi^{(3)} E^3 + \ldots \tag{4.1}$$

where ε_o is the vacuum's dielectric constant, E is the electric field, and $\chi^{(n)}$ is the n-order susceptibility tensor. Optical parametric amplification, second-harmonic generation, sum frequency, and difference frequency are all connected by the second-order term. In the meantime, the third-order term is given in Eq. (4.1) and is associated with the third harmonic generation, two-photon absorption, saturation absorption (SA), Kerr effect, and other phenomena. The two effects of black phosphorus will be discussed in the subsections: the Kerr effect and the SA effect.

Saturated absorption is a material property in which the amount of absorption of light taken in decreases with the increase of light intensity. The following Eq. (4.2) describes the nonlinear BP absorption coefficient

$$\alpha(I) = \frac{\alpha_o}{1 + \frac{I}{I_s}} \tag{4.2}$$

where the nonlinear absorption coefficient is $\alpha(I)$, coefficient of linear absorption is α_o, the saturable intensity of light is I_s, and deplete light intensity is I.

When $I = I_s$, $\alpha(I) = \frac{\alpha_o}{2}$

The following is a formula for calculating and expressing the transmittance:

$$T = 1 - \frac{\alpha_s}{1 + \frac{I}{I_s}} - \alpha_{ns} \tag{4.3}$$

Saturable and non-saturable absorption are denoted by α_s and α_{ns}, respectively. I_s stands for saturable intensity, this is described as optical intensity, which is half that unbleached value for the optical absorbance. Fig. 4.9b depicts the saturable absorption coefficient diagram. The following Eq. (4.4) describes the relationship between propagation depth, intensity, and nonlinear coefficient of absorption with respect to the propagation depth (L):

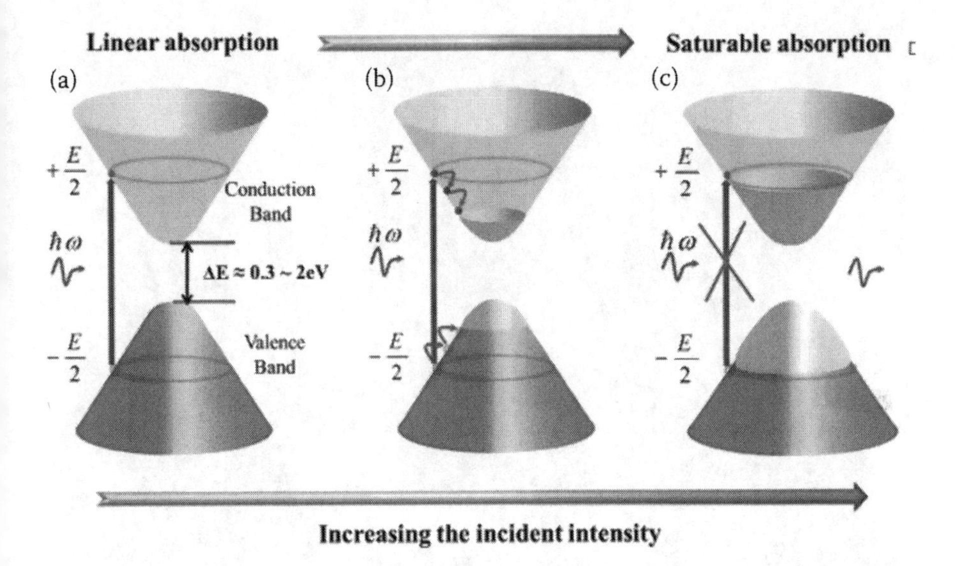

FIGURE 4.10 Representation of BP absorption in the saturation mode. Reprinted with permission from Journal of Nanoscale. Copyright, 2009, Royal Society of Chemistry (Chen, Ponraj, Fan, and Zhang 2020).

$$\frac{dI}{dL} = -\alpha(I)I \tag{4.4}$$

The absorption model is based on a single photon and energy-band design based on two stages and could explain the system of BP saturable absorption. The processes of excitation in linear and nonlinear light absorption are schematically depicted in Fig. 4.10 (Lu et al. 2015). As shown in Fig. 4.10(a), it is possible to excite electrons from the valence band (VB) that are sent to the conduction band (CB) by incident light having a photonic energy *E* greater than the bandgap of BP. Then, photo-excited, warm electrons heat up to build up a hot distribution Fermi-Dirac. As shown in Fig. 4.10(b), this process required the energy ($\kappa_B T_e$) to stop the new excitation of interband near the valance band. If this is a strong enough incident, the photo-excited electrons may cause the state to fill up almost half of the photon energy, blocking further absorption, as shown in Fig. 4.10(c).

4.3 APPLICATION OF BP IN SPR SENSORS

In addition to anisotropic behavior, very high carrier mobility, high upstream on/off ratios, and high operating frequencies, even an unfunctional BP has an intrinsically direct band difference. As per the unique optical, electronic, and mechanical properties of the BP, it is a highly suitable 2D material for different applications (Koenig et al. 2016; Li et al. 2014; Liu, Wang, et al. 2014; Qiao et al. 2014; Wang et al. 2014; Zhang et al. 2015), as shown in Fig. 4.11. These applications include thermal management (Ong, Cai, Zhang and Zhang 2014), nano-electromechanical

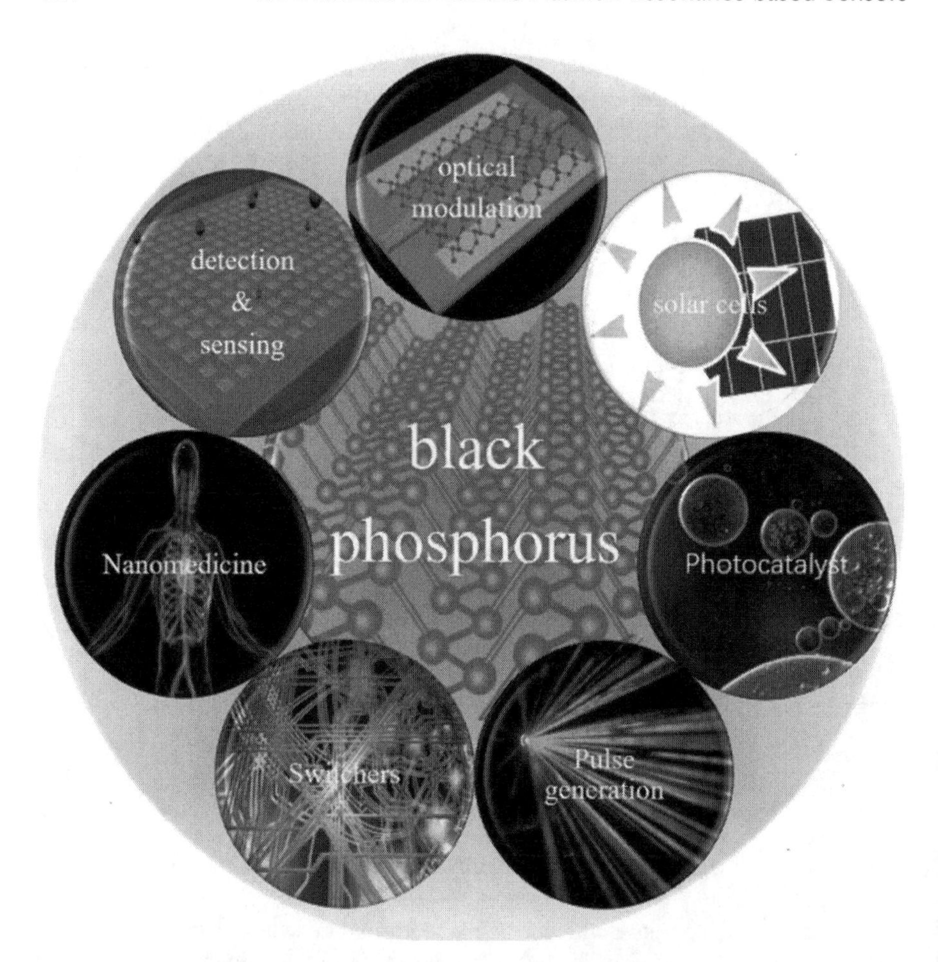

FIGURE 4.11 Schematic illustration of BP 2D nanomaterials applications including photocatalyst, pulse generation, switches, nanomedicines, detection and sensing, optical modulators, and solar cells. Reprinted with permission from Journal of Nanoscale. Copyright, 2009, Royal Society of Chemistry (Chen, Ponraj, Fan, and Zhang 2020).

oscillators (Park, Kim, Kim, and Sohn 2010), gas sensors (Kou, Frauenheim, and Chen 2014), and tunable infrared photo-detectors (Buscema et al. 2014; Engel, Steiner, and Avouris 2014), etc. Mainly, BP applications have been divided into two categories: (i) electronics and (ii) opto-electronics applications.

4.3.1 Electronics Applications

Due to the bandgap tunability and enhanced electronic performance of the BP, it is used in electronic devices to increase the performance of the devices, such as thin-film solar cells (Dai and Zeng 2014), electrode channel contacts material (Qiao et al. 2014), channel material in FETs (Li et al. 2014; Liu, Neal, et al. 2014; Xia, Wang, and Jia 2014), lithium-ion batteries (Park, Kim, Kim, and Sohn 2010; Sun et al.

2012), and thermoelectric devices. Lee et al. (2017) have designed and analyzed the BP-based H_2 sensor that is based on the field-effect transistor. The functionalized exfoliated BP sheet with platinum (Pt) nanoparticles is used to enhance the H_2-sensing efficiency of the fabricated sensor, as shown in Fig. 4.12. They obtained good reproducibility, fast response-decay, and better sensitivities at room temperature. There are various BP-based FET sensors for different applications demonstrated by other researchers, as shown in Table 4.1.

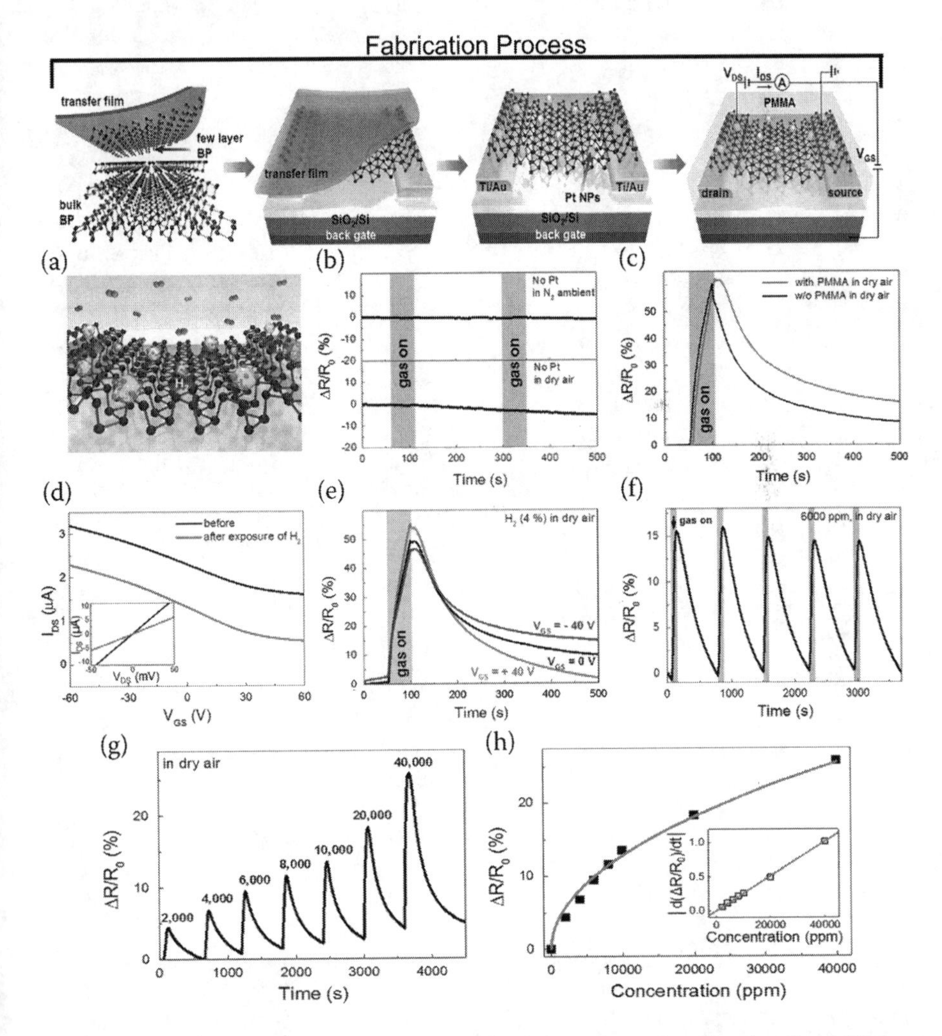

FIGURE 4.12 Schematic illustration of fabrication process of pt-functionalized BP hydrogen sensor: a) Sensing mechanism of the pt nanoparticle coated BP flakes hydrogen sensor. b) Sensor output in N_2 (upper) and dry air (lower) in the presence of only BP flakes for hydrogen sensing. c) Performance of the sensor with PMMA (red) and without PMMA (black) in the dry air w.r.t. time. d) Variation of drain current w.r.t. gate voltage before and after exposure of H_2 gas (e–h). Reprinted with permission from Applied Physics Letters. Copyright, 2017, AIP Publishing (Lee et al. 2017).

TABLE 4.1
The Performance Table of Field-Effect Transistors Based on BP With a Thickness (t), Channel Length (L). Reprinted With Permission from Materials Science and Engineering: R. Copyright, 2017, Elsevier (Yi et al. 2017).

t (nm)	*L* (μm)	Gate dielectric	μ_{FE} (cm^2 (V s)$^{-1}$)	On/Off ratio	Ref.
10	**1.0**	SiO_2 (back)	$\mu_h \sim 286$	10^4	(Liu, Neal, et al. 2014)
8	**1**	SiO_2 (back)	$\mu_h \sim 300$	10^4	(Xia, Wang, and Jia 2014)
2	**1**	SiO_2 (back)	$\mu_h \sim 50$	5×10^5	(Xia, Wang, and Jia 2014)
10	**–**	SiO_2 (back)	$\mu_h \sim 984$	10^5	(Li et al. 2014)
8	**2.6**	SiO_2 (back)	$\mu_h \sim 197$	10^5	(Li et al. 2014)
5	**4.5**	SiO_2 (back)	$\mu_h \sim 55$	10^5	(Li et al. 2014)
15	**3**	HfO_2 (top)	$\mu_h \sim 413$	2×10^3	(Liu et al. 2016)
1.9	**2**	SiO_2 (back)/ Al_2O_3 (top)	$\mu_h \sim 172$ μe ~ 38	2.7×10^4 (h) 4.4 × 103 (e)	(Das, Demarteau, and Roelofs 2014)
15	**2.7**	Polyimide (back)/ Al_2O_3 (top)	$\mu_h \sim 310$, $\mu_e \sim 89$	10^3–10^4	(Zhu et al. 2015)
13	**0.5**	Polyimide (back)/ Al_2O_3 (top)	$\mu_h \sim 233$	10^2	(Zhu et al. 2016)
<10	**0.02**	Al_2O_3 (top)	$\mu_h \sim 17$	$>10^2$	(Miao et al. 2015)
12		PMMA (top)/ SiO_2 (back)	$\mu_h \sim 1{,}150$	10^5	(Li et al. 2014)
10	**2.4**	BN (back)/Cu adatoms	$\mu_h \sim 690$, 1,780 (7 K) μ_e 80, 2,140 (7 K)	–	(Koenig et al. 2016)
5	**2**	SiO_2 (back)/Al adatoms/ Al_2O_3 (top)	$\mu_h > 2{,}150$ (120 K) $\mu_h > 1{,}495$ (260 K)	$>10^5$ (120 K) $>10^3$ (260 K)	(Prakash et al. 2017)
~43	**–**	PMMA/ MMA(top)/ SiO_2 (back)	$\mu_h \sim 900$ (0.3 K)	$> 10^5$ (180 K)	(Tayari et al. 2015)
~5	**2**	SiO_2 (back)/ionic liquid (top)	$\mu_h \sim 510$	1.4×10^4	(Saito and Iwasa 2015)
10.7 ± 0.8	**0.17–0.55**	HfO_2 (back)	$\mu_h \sim 44$	1.2×10^3	(Haratipour, Robbins, and Koester 2015)
13	**–**	SiO_2 (back)	$\mu_h \sim 950$, $\mu_e \sim 950$	$>10^5$	(Perello, Chae, Song, and Lee 2015)

TABLE 4.1 (Continued)
The Performance Table of Field-Effect Transistors Based on BP With a Thickness (t), Channel Length (L). Reprinted With Permission from Materials Science and Engineering: R. Copyright, 2017, Elsevier (Yi et al. 2017).

t (nm)	*L* (μm)	Gate dielectric	μ_{FE} (cm^2 (V s)$^{-1}$)	On/Off ratio	Ref.
5.6	–	graphite/ BN (back)	$\mu_h \sim 400, 3{,}900$ (1.5 K) $\mu_e \sim 83, 1{,}600$ (1.5 K)	–	(Li et al. 2015)
~10	–	graphite/BN (back)/BN (top)	$\mu_h \sim 920$ $\mu_h \sim 7{,}100$ (4 K)	–	(Li, Yang, et al. 2016)
~10	–	BN (back)/ BP (top)	$\mu_h \sim 400$ $\mu_h \sim 4{,}000$ (1.6 K)	10^5	(Gillgren et al. 2014)
~8	10	BN (back)/ BP (top)	$\mu_h \sim 1{,}350, 2{,}700$ (1.7 K)	10^5 (300 K) $>10^8$ (1.7 K)	(Chen et al. 2015)
–	–	BN (back)/ BP (top)	$\mu_h \sim 5{,}200, 45{,}000$ (2 K)	–	(Long et al. 2016)

BP nanosheets with a few layers that have been labeled with gold nanoparticle-antibody conjugates were used as a BP-based FET biosensor by Chen et al. (2017b). The mechanically exfoliated synthesis technique was used to develop BP nanosheets. It is used as the sensing/conducting path in the FET-based sensor and surface passivation antibody probes were paired with gold nanoparticles using an Al_2O_3 thin film as the dielectric sheet, as shown in Fig. 4.13. Surface functionalization sputtered on the BP, and the sensor response was determined by the difference in electrical resistance of the BP after antigens were added. Via complex antigen-antibody binding interactions, the adsorbed antigens are caused by an applied gate voltage, thus modifying the current of drain-source terminal. The BP biosensor as produced demonstrated specific selectivity for human immunoglobulin G and high sensitivity with a lower limit of detection of 10 ng ml^{-1}. The findings of this study show that BP performs exceptionally well as a sensing channel for FET biosensor applications.

4.3.2 Biosensing Applications

The rapid development of BP-based gas/small molecule sensors, as well as their superior sensing efficiency, has attracted attention in biomolecule detection. Chen et al. (2017b) developed a BP-based FET biosensor with sensitivity in the nanogram range for detecting antigen-antibody interactions in this biosensor, which is illustrated in Fig. 4.13. The extraordinary optical properties of the BP and sp^3

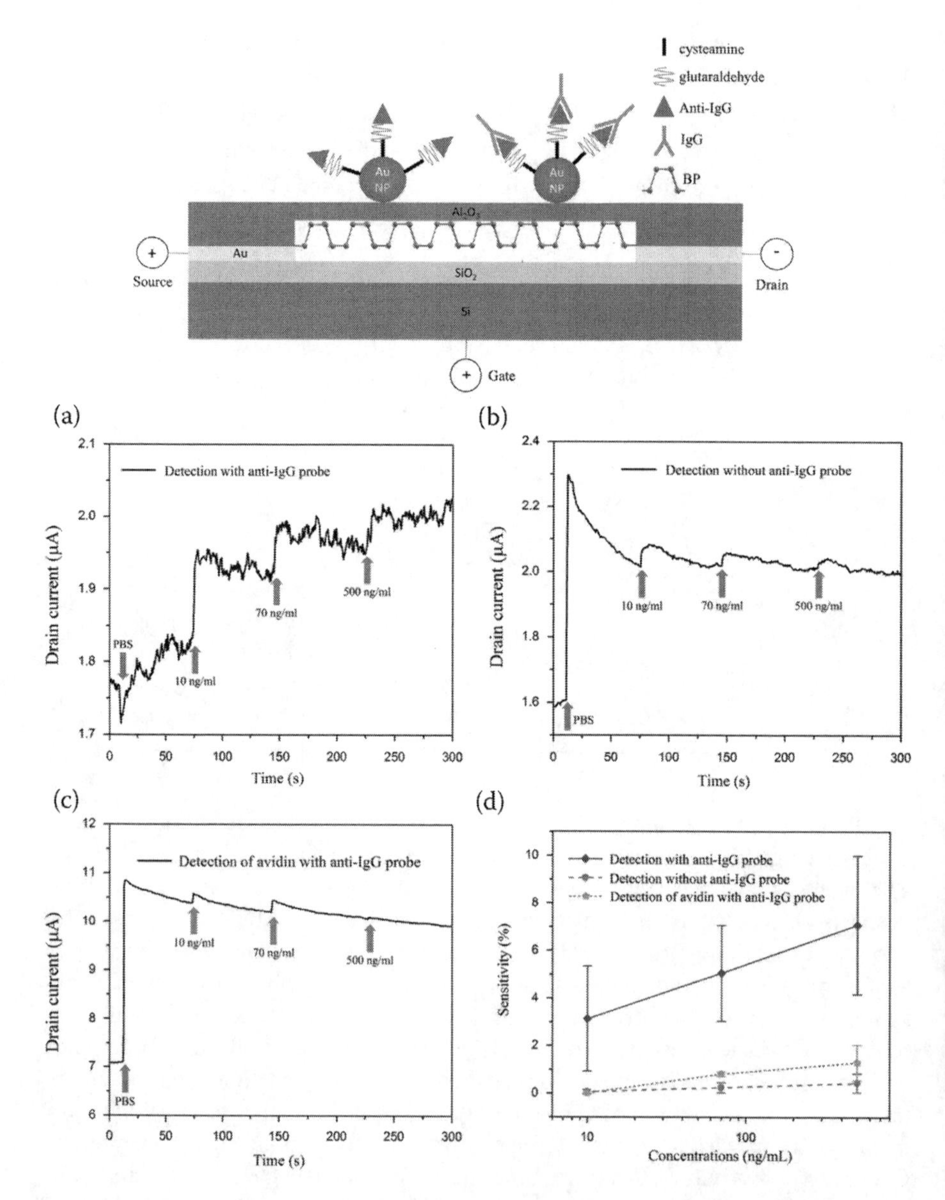

FIGURE 4.13 Illustrative diagram of the mechanically exfoliated BP-FET sensor: a) Response of drain current as a function of time with changing the concentrations of IgG at fixed source-drain voltage. b) Experimental analysis of non-specific binding to the BP sensor in the absence of antibody probes at fixed source-drain voltage. c) Experimental analysis of detection of a non-specific protein at fixed source-drain voltage. d) Sensitivity curve with respect to the concentration of target protein. Reprinted with permission from Journal of Biosensors and Bioelectronics. Copyright, 2017, Elsevier (Chen et al. 2017b).

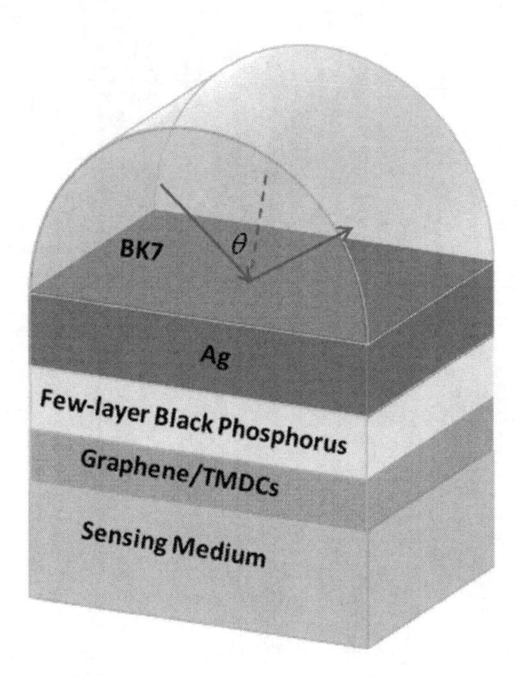

FIGURE 4.14 To improve sensitivity of the biochemical sensor, BP and heterostructures of graphene/TMDCs were used. Reprinted with permission from Journal of Sensors and Actuators B: Chemical. Copyright, 2017, Elsevier (Wu et al. 2017).

hybridization of the structure play a significant role in the SPR-based bio/gas sensors. Wu et al. (2017) proposed a prism-based biochemical sensor with SPR phenomenon, as shown in Fig. 4.14. In this biochemical sensor, heterostructures of a few BP layers with graphene/TMDCs are considered over the prism surface. The results of the proposed sensor show the maximum sensitivity of 279° RIU^{-1}, which is approximately 2.4 times higher as compared to a conventional SPR sensor, as mentioned in Fig. 4.15 and Fig. 4.16. The effect of BP on the performance of SPR-based biosensors is discussed theoretically by Pal, Verma, Prajapati, and Saini (2017). This paper used the Au as the SPR active material and BP as the biorecognition element (BRE) layer. The performance of the proposed sensor is estimated at the wavelength of 633 nm and finds the better performance, which is 1.42 times more than the conventional sensor and 1.4 times more than the graphene-based SPR biosensor.

Srivastava, Verma, Das, and Prajapati (2020) proposed a biosensor design using MXene nanomaterial. The design consists of BK7 prism, BP, gold, and transition metal dichalcogenides (TMDCs). MXene is used as a sensing material and the proposed design operating at 633 nm wavelength. The maximum sensitivity, 190.22 RIU^{-1}, is achieved and offers 1.52 times more penetration depth as compared to conventional biosensors. Kumar et al. (2020) theoretically demonstrated silicon layer-based SPR biosensors with high sensitivity. The proposed structure is based on a multilayer Kretschmann configuration, as shown in Fig. 4.17. It consists of silver, a silicon layer, hybrid nanostructure of BP and MXene nanomaterial. BK7

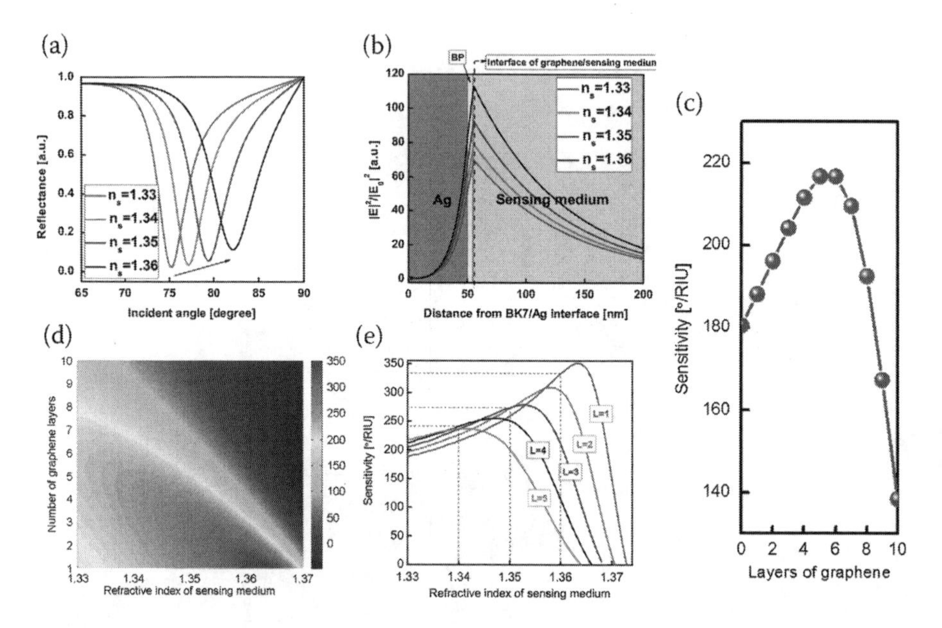

FIGURE 4.15 Performance analysis of the proposed biochemical sensor: a) SPR reflectance spectrum w.r.t. the incident angle and b) variation of electric field along with the normal distance of the metal-dielectric interface with few layers of BP and monolayer graphene heterostructure for different sensing analyte RI. c) Calculation of sensitivity as the number of graphene layers changing from 0 to 10 at fixed BP layer (5 nm). d) 2D color contour plot of the sensitivity w.r.t the refractive index changes from 1.33 RIU to 1.37 RIU and number of graphene layers varies from 1 to 10. e) Sensitivity curve w.r.t. the RI of sensing analyte with a different number of graphene layers for the proposed biochemical sensor. Reprinted with permission from Journal of Sensors and Actuators B: Chemical. Copyright, 2017, Elsevier (Wu et al. 2017).

prism is used as a coupling prism at the metal layer. The highest sensitivity 264° RIU^{-1} has been claimed. The proposed sensor is biocompatible and provides a hydrophilic surface that is a more suitable biosensing application.

Meshginqalam and Barvestani (2018) investigated ultrasensitive multi-layered biosensors. It consists of a BK7 prism for coupling, a bimetallic layer of gold and aluminum, and different 2-D materials like BP, WS_2, and graphene, as shown in Fig. 4.18a. An Si nanosheet is added to improve the performance of the proposed biosensor. BP is also used to enhance the sensitivity and biocompatibility with biomolecular interaction and medical diagnosis. The maximum sensitivity of the sensor, 342° RIU^{-1}, is obtained.

Das and Moirangthem (2021) have theoretically analyzed the dual-mode SPR sensor using a spectral interrogation technique, as shown in Fig. 4.19. This multilayer biosensor has hafnium oxide (HfO_2) as a dielectric spacer, which is a sandwich between two layers of Au and some layers of BP. The highest sensitivity, 1,692 nm RIU^{-1}, is obtained and compared to the proposed analysis with the conventional SPR sensor, as depicted in Fig. 4.20. This sensor can detect large molecules of the analyte.

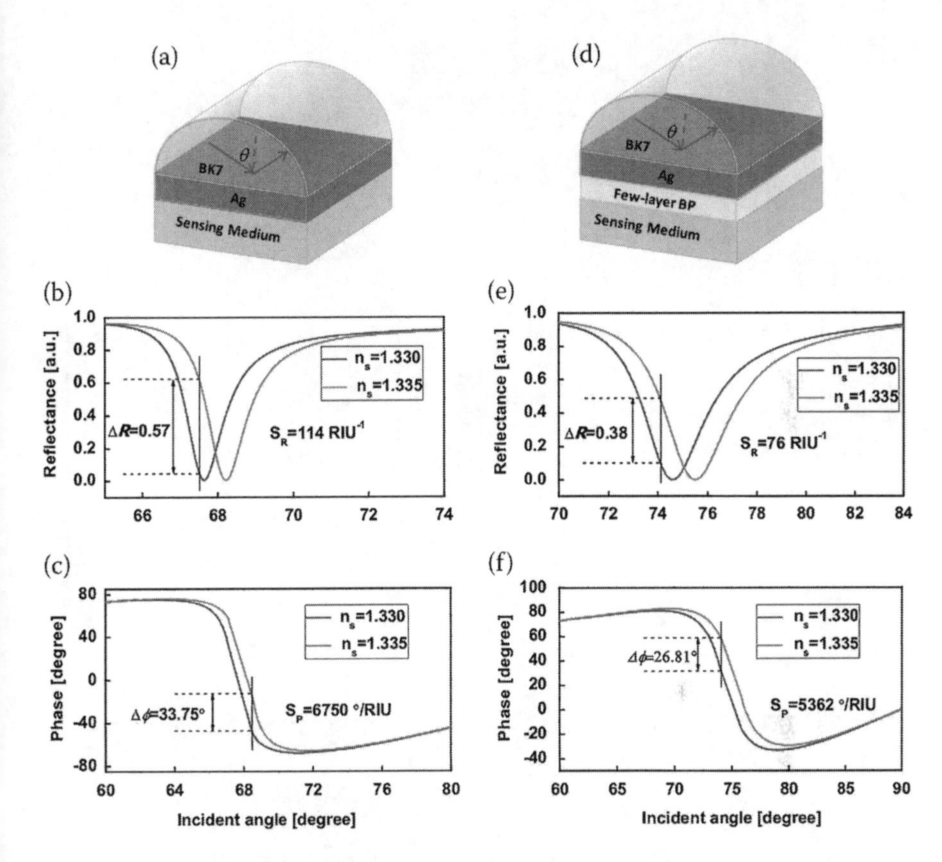

FIGURE 4.16 Comparative analysis of the intensity and phase sensitivity of the BP-based biochemical sensor with conventional SPR sensor. (a–c) For conventional SPR sensor with BK7/Ag/sensing layer, and (d–f) For 2D material based SPR sensor with BK7/Ag/few layer BP/sensing layer. Reprinted with permission from Journal of Sensors and Actuators B: Chemical. Copyright, 2017, Elsevier (Wu et al. 2017).

Srivastava and Jha numerically validate SPR biosensors for gas-sensing applications (Srivastava and Jha 2018). The proposed sensor has BP and few layers of 2D material such as graphene and molybdenum diselenide ($MoSe_2$). Here, BK7 is the coupling prism and silver (Ag) is used as the SPR active material. The sensitivity of 110° RIU^{-1} is achieved using this sensor. The proposed sensor is suitable for the detection of harmful volatile organic composites and dangerous gases. Maurya, Prajapati, Raikwar, and Saini (2018) demonstrated a multilayer SPR biosensor using silicon and BP, that is able to detect nitrogen dioxide (NO_2) gas. NO_2 gas has a high fitment property with the surface of the BP; hence, the proposed gas sensor is susceptible to the NO_2 gas detection. Pal, Verma, Saini, and Prajapati (2019) suggested a modified multiplayer configuration of an SPR biosensor using BP, WS_2, WSe_2 TMDC nanomaterial, silicon, and Au metal.

The performance parameters, like sensitivity, minimum reflectivity, and FWHM, are calculated. The maximum sensitivity, 163° RIU^{-1}, is obtained using BP and

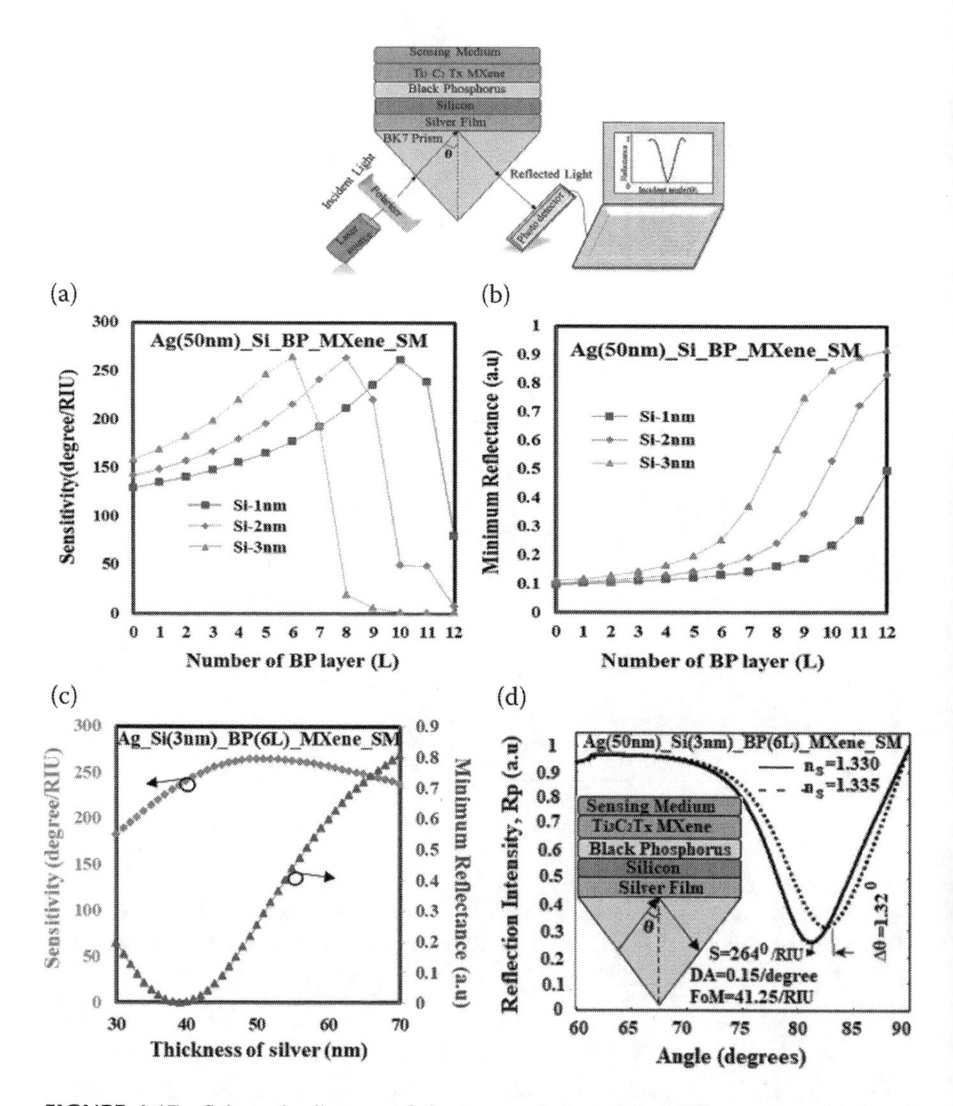

FIGURE 4.17 Schematic diagram of the proposed structure: a) Effects on sensitivity of numbers of BP layers, b) minimum reflectance w.r.t. the numbers of BP layers, c) calculation of sensitivity and minimum reflectance w.r.t. the silver layer, and d) reflectance curve for the optimized structure. Reprinted with permission from Journal of Superlattices and Microstructures. Copyright, 2020, Elsevier (Kumar et al. 2020).

WSe_2 and without WSe_2 184.6° RIU^{-1}. Due to the absorption property of TMDC nanomaterial, the decrement in the sensitivity occurred. Kumar et al. (2021) analyzed a biochemical sensor using BP, MXene, and a bimetallic layer of Cu and Ni using an angular interrogation technique. BP has the extraordinary property to enhance the light-matter interaction due to tunable and direct energy bandgap. This is an ultrasensitive biochemical sensor and shows a maximum sensitivity of 304.47°

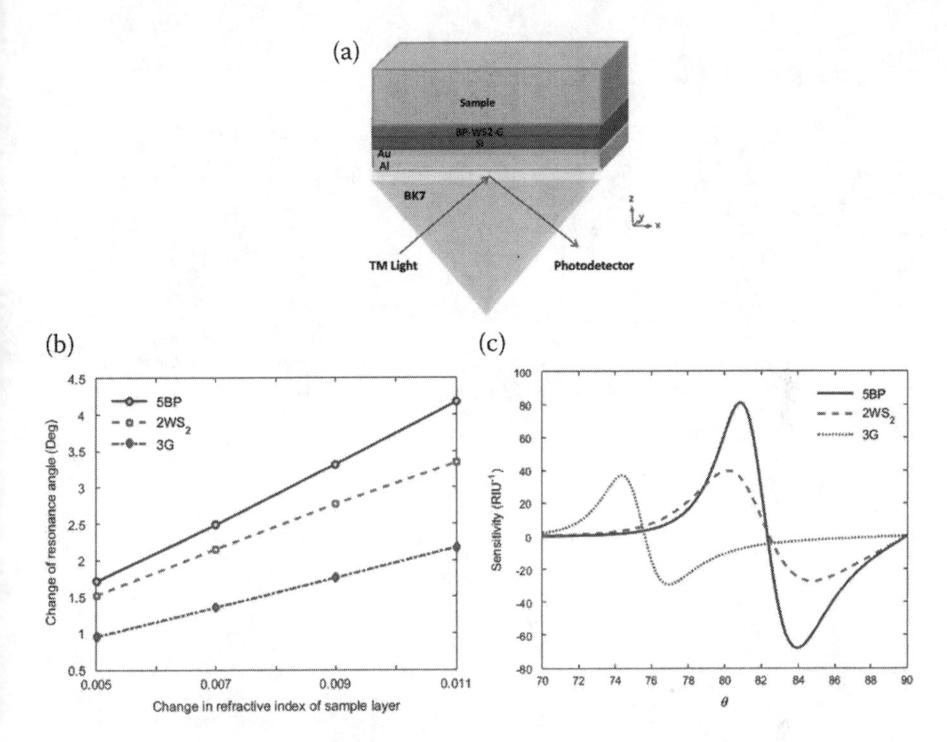

FIGURE 4.18 a) Proposed SPR biosensor structure, b) variation in resonance angle w.r.t. the change in RI of sensing medium for all three 2D material, and c) performance analysis of the structure. Reprinted with permission from Journal of Optical Materials. Copyright, 2018, Elsevier (Meshginqalam and Barvestani 2018).

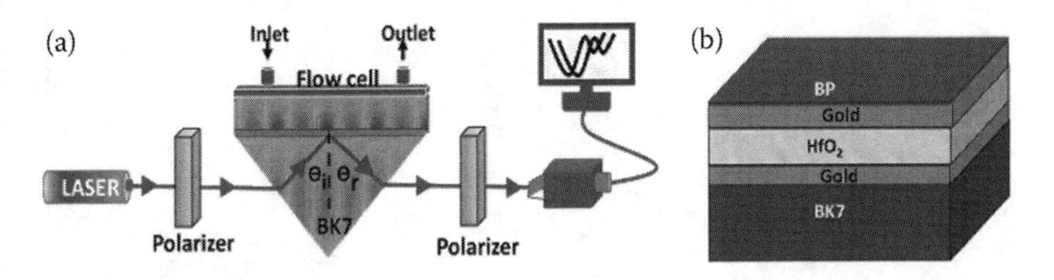

FIGURE 4.19 a) Schematic diagram of the experimental setup, and b) structure of the proposed prism-based SPR sensor. Reprinted with permission from Materials Today: Proceedings. Copyright, 2021, Elsevier (Das and Moirangthem 2021).

RIU^{-1}. Pal et al. demonstrated a sensor for the detection of DNA hybridization numerically (Pal et al. 2018). This is a modified multilayered configuration that has a few layers of BP, graphene, and a layer of gold. The phosphate-buffered saline (PBS) solution is used as a sensing medium with the highest sensitivity of 125° RIU^{-1}. Pal, Verma, Prajapati, and Saini (2020) proposed a modified Kretschmann multilayered configuration that contains BP, TMDC nanomaterials, and MXene material layers. The 2D surface structure of the BP, TMDC, and MXene makes a

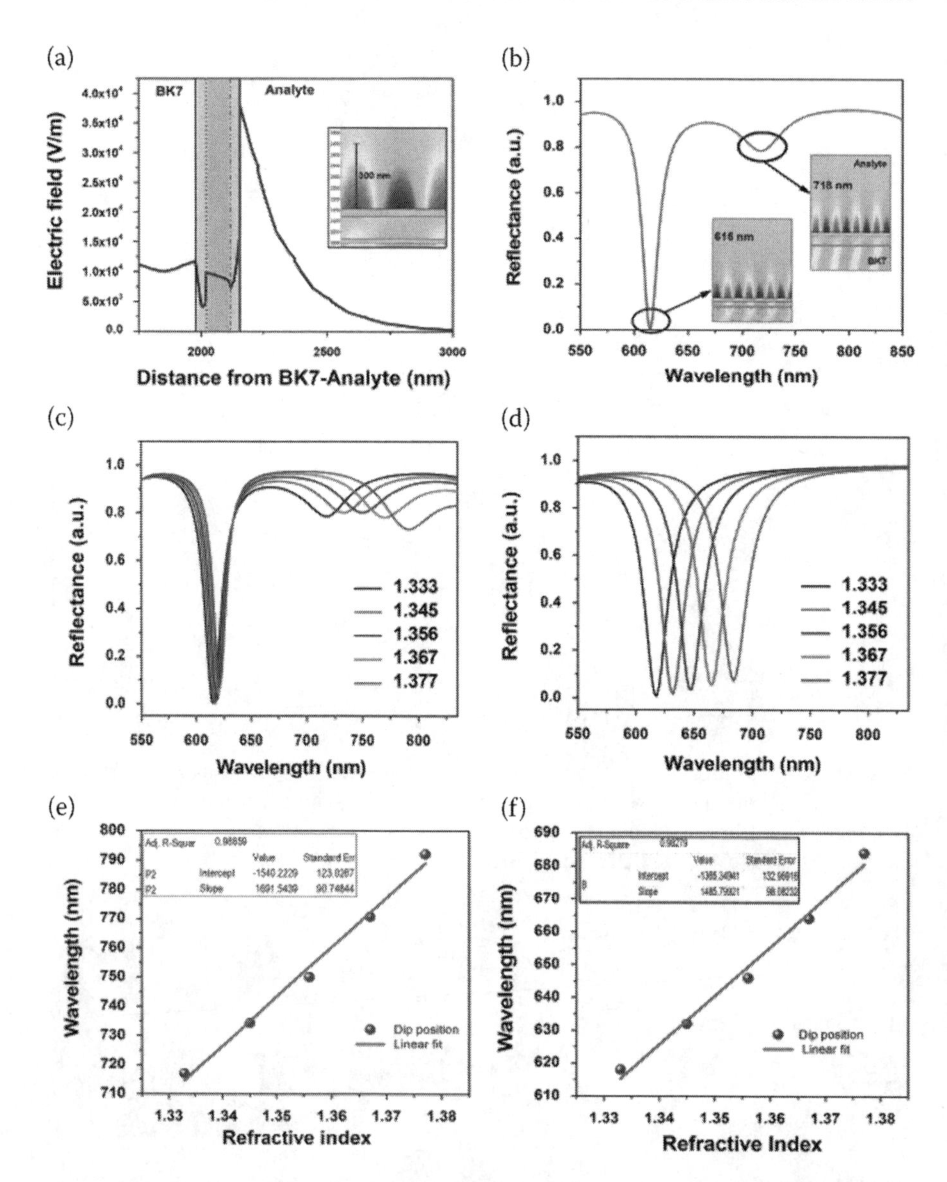

FIGURE 4.20 a) Field distribution (E⊥) profile across the suggested multilayer structure (BK7/Au/HfO_2/Au/BP/sensing layer). b) The reflectance curve with respect to the wavelength of incident light with sensing medium (water) for c) proposed structure (BK7/Au/HfO2/Au/BP/Analyte). d) Conventional structure (BK7/Au/Analyte) system (e) and (f) plots the resonance wavelength w.r.t. the RI of sensing analyte for proposed structure and conventional structure, respectively. Reprinted with permission from Materials Today: Proceedings. Copyright, 2021, Elsevier (Das and Moirangthem 2021).

van der Waals heterostructure by placing them in the vertical direction and showing extraordinary electronics and optical properties necessary for SPR sensing. The investigated sensor shows the highest sensitivity, 388° RIU^{-1}, and is useful for the

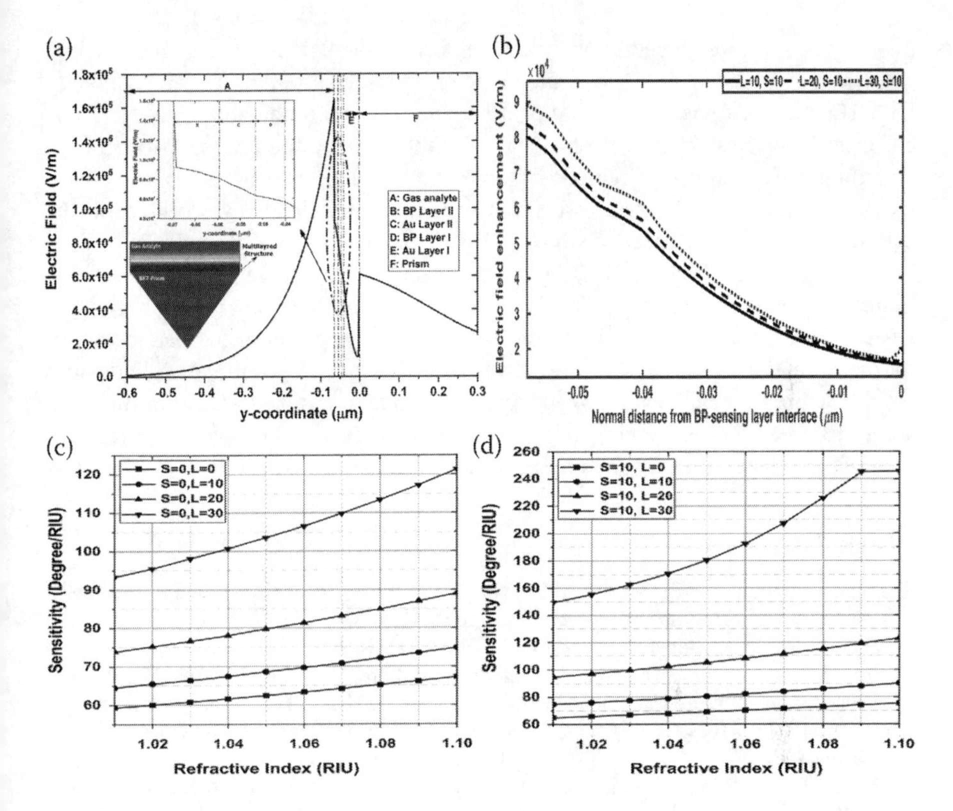

FIGURE 4.21 Performance analysis of the BP-based SPR sensor for gas sensing. a) Electric field distribution at the resonance condition. b) The effect of different BP layers on the electric field distribution. c) The sensitivity curve at different combinations of BP layers. Reprinted with permission from IEEE Sensors Letters. Copyright, 2019, IEEE (Singh and Raghuwanshi 2019).

detection of the biochemical. Singh and Raghuwanshi (2019) proposed a gas sensor using two layers of BP and two layers of gold, alternatively. The electric field distribution within the multilayer is also discussed, as shown in Fig. 4.21(a). Fig. 4.21(b) illustrates the effect of BP layers on the electric field. The highest sensitivity of 245.5° RIU^{-1} is achieved, as shown in Fig. 4.21 with the optimized structure. Singh, Paswan, and Raghuwanshi (2021) demonstrate an SPR sensor using a multilayered structure. The modified Kretschmann configuration consists of two Au layers, BP, and graphene. In this structure, BP is considered between two layers of gold, whereas the graphene is used as sensing layer. The highest sensitivity of 218° RIU^{-1} is achieved at an operating wavelength of 633 nm for the proposed sensor and the suggested sensor also. It is suitable for chemical/biochemical sensing.

4.4 SUMMARY

In fast-growing 2D materials, BP has extraordinary properties (physical, optical, mechanical, and electrical). BP is used in different applications, i.e., switches, photocatalysts, pulse generation, solar cells, nanomedicines, optical modulators, detection, and sensing at a large scale in the present scenario. All-optical technology with 2D materials has many significant benefits, including the ability to overcome electron delay, heat loss, and cross chat. It is the foundation of the next-generation communication network. In optical signal processing, the all-optical modulator plays an important role. Many research groups are continuously working on the synthesis and transfer mechanism of the BP nanosheets from the bulk BP. These nanosheets (single layer) of BP are also used in SPR sensing to detect the biomolecules/bio-chemicals/gases. Because of its tuneable bandgap with a single layer, bilayer, and trilayer, BP is used to optimize the performance parameters of the sensing devices. If the development of 2D materials continues, all devices would enable us to achieve the all-optical age, and it is feasible to fulfill the aim that we all foresee. With the emergence of nanotechnology and the advancement of 2D materials science, it is reasonable to expect the age of all-optical devices to arrive.

REFERENCES

Asahina, Hideo, and Akira Morita. 1984. "Band Structure and Optical Properties of Black Phosphorus." *Journal of Physics C: Solid State Physics* 17 (11): 1839.

Brent, Jack R., Nicky Savjani, Edward A. Lewis, Sarah J. Haigh, David J. Lewis, and Paul O'Brien. 2014. "Production of Few-Layer Phosphorene by Liquid Exfoliation of Black Phosphorus." *Chemical Communications* 50 (87): 13338–41.

Bridgman, P. W. 1914. "Two New Modifications of Phosphorus." *Journal of the American Chemical Society* 36 (7): 1344–63. 10.1021/ja02184a002.

Bridgman, P. W. 1935. "Polymorphism, Principally of the Elements, up to 50,000 kg/cm^2." *Physical Review* 48 (11): 893–906.

Bridgman, P. W. 1937. "Shearing Phenomena at High Pressures, Particularly in Inorganic Compounds." *Proceedings of the American Academy of Arts and Sciences* 71 (9): 387–460. http://www.jstor.org/stable/20023239.

Buscema, Michele, Dirk J. Groenendijk, Sofya I. Blanter, Gary A. Steele, Herre S. J. Van Der Zant, and Andres Castellanos-Gomez. 2014. "Fast and Broadband Photoresponse of Few-Layer Black Phosphorus Field-Effect Transistors." *Nano Letters* 14 (6): 3347–52.

Çakir, Deniz, Hasan Sahin, and François M. Peeters. 2014. "Tuning of the Electronic and Optical Properties of Single-Layer Black Phosphorus by Strain." *Physical Review B* 90 (20): 205421.

Carvalho, Alexandra, Min Wang, Xi Zhu, Aleksandr S Rodin, Haibin Su, and Antonio H Castro Neto. 2016. "Phosphorene: From Theory to Applications." *Nature Reviews Materials* 1 (11): 1–16.

Castellanos-Gomez, Andres. 2015. "Black Phosphorus: Narrow Gap, Wide Applications." *The Journal of Physical Chemistry Letters* 6 (21): 4280–91. 10.1021/acs.jpclett.5b01686.

Chen, Hao, Peng Huang, Dan Guo, and Guoxin Xie. 2016. "Anisotropic Mechanical Properties of Black Phosphorus Nanoribbons." *The Journal of Physical Chemistry C* 120 (51): 29491–97.

Chen, Xing, Joice Sophia Ponraj, Dianyuan Fan, and Han Zhang. 2020. "An Overview of the Optical Properties and Applications of Black Phosphorus." *Nanoscale* 12 (6): 3513–34.

Chen, Yantao, Ren Ren, Haihui Pu, Jingbo Chang, Shun Mao, and Junhong Chen. 2017a. "Field-Effect Transistor Biosensors with Two-Dimensional Black Phosphorus Nanosheets." *Biosensors and Bioelectronics* 89: 505–10. 10.1016/j.bios.2016.03.059.

Chen, Yantao, Ren Ren, Haihui Pu, Jingbo Chang, Shun Mao, Junhong Chen. 2017. "Field-Effect Transistor Biosensors with Two-dimensional Black Phosphorus Nanosheets." *Biosensors and Bioelectronics* 89 , Part 1, 505–510. https://doi.org/10.1016/j.bios.2016.03.059

Chen, Xiaolong, Yingying Wu, Zefei Wu, Yu Han, Shuigang Xu, Lin Wang, Weiguang Ye, et al. 2015. "High-Quality Sandwiched Black Phosphorus Heterostructure and Its Quantum Oscillations." *Nature Communications* 6 (1): 1–6.

Chen, Chen, Feng Chen, Xiaolong Chen, Bingchen Deng, Brendan Eng, Daehwan Jung, Qiushi Guo, et al. 2019. "Bright Mid-Infrared Photoluminescence from Thin-Film Black Phosphorus." *Nano Letters* 19 (3): 1488–93.

Coleman, Jonathan N., Mustafa Lotya, Arlene O'Neill, Shane D. Bergin, Paul J. King, Umar Khan, Karen Young, et al. 2011. "Two-Dimensional Nanosheets Produced by Liquid Exfoliation of Layered Materials." *Science* 331 (6017): 568–71.

Corbridge, D. E. C. 2016. *Phosphorus: Chemistry, Biochemistry and Technology*, 6th ed. Boca Raton: CRC Press. 10.1201/b12961.

Cui, Shumao, Haihui Pu, Eric C. Mattson, Zhenhai Wen, Jingbo Chang, Yang Hou, Carol J. Hirschmugl, and Junhong Chen. 2014. "Ultrasensitive Chemical Sensing through Facile Tuning Defects and Functional Groups in Reduced Graphene Oxide." *Analytical Chemistry* 86 (15): 7516–22. 10.1021/ac501274z.

Dai, Jun, and Xiao Cheng Zeng. 2014. "Bilayer Phosphorene: Effect of Stacking Order on Bandgap and Its Potential Applications in Thin-Film Solar Cells." *The Journal of Physical Chemistry Letters* 5 (7): 1289–93.

Das, Anindita, and Rakesh S. Moirangthem. 2021. "A Theoretical Approach to Develop a Black Phosphorous Coated Multilayer Plasmonic Sensor by Using Hafnium Oxide as Dielectric Spacer." *Materials Today: Proceedings* 46, Part 12, 5874–5877.

Das, Saptarshi, Marcel Demarteau, and Andreas Roelofs. 2014. "Ambipolar Phosphorene Field Effect Transistor." *ACS Nano* 8 (11): 11730–38.

Deng, Yexin, Zhe Luo, Nathan J. Conrad, Han Liu, Yongji Gong, Sina Najmaei, Pulickel M. Ajayan, Jun Lou, Xianfan Xu, and Peide D. Ye. 2014. "Black Phosphorus--Monolayer MoS2 van Der Waals Heterojunction p--n Diode." *ACS Nano* 8 (8): 8292–9.

Du, Haiwei, Xi Lin, Zhemi Xu, and Dewei Chu. 2015. "Recent Developments in Black Phosphorus Transistors." *Journal of Materials Chemistry C* 3 (34): 8760–75.

Du, Y., Z. Luo, H. Liu, X. Xu, and P. D. Ye. n.d. "Anisotropic Properties of Black Phosphorus." 2D Materials: Properties and Devices: 413–34. Cambridge: Cambridge University Press. 10.1017/9781316681619.023.

Engel, Michael, Mathias Steiner, and Phaedon Avouris. 2014. "Black Phosphorus Photodetector for Multispectral, High-Resolution Imaging." *Nano Letters* 14 (11): 6414–17.

Flores, Eduardo, Jose R. Ares, Andres Castellanos-Gomez, Mariam Barawi, Isabel J. Ferrer, and Carlos Sánchez. 2015. "Thermoelectric Power of Bulk Black-Phosphorus." *Applied Physics Letters* 106 (2): 22102.

Gillgren, Nathaniel, Darshana Wickramaratne, Yanmeng Shi, Tim Espiritu, Jiawei Yang, Jin Hu, Jiang Wei, et al. 2014. "Gate Tunable Quantum Oscillations in Air-Stable and High Mobility Few-Layer Phosphorene Heterostructures." *2D Materials* 2 (1): 11001.

Gu, Wei, Yinghan Yan, Xueyu Pei, Cuiling Zhang, Caiping Ding, and Yuezhong Xian. 2017. "Fluorescent Black Phosphorus Quantum Dots as Label-Free Sensing Probes for Evaluation of Acetylcholinesterase Activity." *Sensors and Actuators B: Chemical* 250: 601–7.

Han, F. W., W. Xu, L. L. Li, C. Zhang, H. M. Dong, and F. M. Peeters. 2017. "Electronic and Transport Properties of N-Type Monolayer Black Phosphorus at Low Temperatures." *Physical Review B* 95 (11): 115436.

Haratipour, Nazila. 2017. "Two-Dimensional Black Phosphorus for High Performance Field Effect Transistors." Retrieved from the University of Minnesota Digital Conservancy, https://hdl.handle.net/11299/200176

Haratipour, Nazila, Matthew C. Robbins, and Steven J. Koester. 2015. "Black Phosphorus P-MOSFETs with 7-Nm HfO2 Gate Dielectric and Low Contact Resistance." *IEEE Electron Device Letters* 36 (4): 411–3.

Hemsworth, Nicholas, Vahid Tayari, Francesca Telesio, Shaohua Xiang, Stefano Roddaro, Maria Caporali, Andrea Ienco, et al. 2016. "Dephasing in Strongly Anisotropic Black Phosphorus." *Physical Review B* 94 (24): 245404.

Hong, Tu, Bhim Chamlagain, Wenzhi Lin, Hsun-Jen Chuang, Minghu Pan, Zhixian Zhou, and Ya-Qiong Xu. 2014. "Polarized Photocurrent Response in Black Phosphorus Field-Effect Transistors." *Nanoscale* 6 (15): 8978–83.

Hultgren, Ralph, N. S. Gingrich, and B. E. Warren. 1935. "The Atomic Distribution in Red and Black Phosphorus and the Crystal Structure of Black Phosphorus." *The Journal of Chemical Physics* 3 (6): 351–5. 10.1063/1.1749671.

Islam, Arnob, Jaesung Lee, and Philip X.-L. Feng. 2018. "All-Electrical Transduction of Black Phosphorus Tunable 2D Nanoelectromechanical Resonators." In *2018 IEEE Micro Electro Mechanical Systems (MEMS)*: 1052–5

Jaramillo, Thomas F., Kristina P. Jørgensen, Jacob Bonde, Jane H. Nielsen, Sebastian Horch, and I. B. Chorkendorff. 2007. "Identification of Active Edge Sites for Electrochemical H2 Evolution from MoS2 Nanocatalysts." *Science* 317 (5834): 100–02. 10.1126/science.1141483.

Jiang, Jin-Wu. 2015. "Thermal Conduction in Single-Layer Black Phosphorus: Highly Anisotropic?" *Nanotechnology* 26 (5): 55701.

Jiang, Jin-Wu, and Harold S. Park. 2014. "Negative Poisson's Ratio in Single-Layer Black Phosphorus." *Nature Communications* 5 (1): 1–7.

Jiang, Qianqian, Lei Xu, Ning Chen, Han Zhang, Liming Dai, and Shuangyin Wang. 2016. "Facile Synthesis of Black Phosphorus: An Efficient Electrocatalyst for the Oxygen Evolving Reaction." *Angewandte Chemie* 128 (44): 14053–7.

Jiang, Xiao-Fang, Zhikai Zeng, Shuang Li, Zhinan Guo, Han Zhang, Fei Huang, and Qing-Hua Xu. 2017. "Tunable Broadband Nonlinear Optical Properties of Black Phosphorus Quantum Dots for Femtosecond Laser Pulses." *Materials* 10 (2): 210.

Kang, Joohoon, Joshua D. Wood, Spencer A. Wells, Jae-Hyeok Lee, Xiaolong Liu, Kan-Sheng Chen, and Mark C. Hersam. 2015. "Solvent Exfoliation of Electronic-Grade, Two-Dimensional Black Phosphorus." *ACS Nano* 9 (4): 3596–604.

Koenig, Steven P., Rostislav A. Doganov, Leandro Seixas, Alexandra Carvalho, Jun You Tan, Kenji Watanabe, Takashi Taniguchi, Nikolai Yakovlev, Antonio H. Castro Neto, and Barbaros Ozyilmaz. 2016. "Electron Doping of Ultrathin Black Phosphorus with Cu Adatoms." *Nano Letters* 16 (4): 2145–51.

Kou, Liangzhi, Thomas Frauenheim, and Changfeng Chen. 2014. "Phosphorene as a Superior Gas Sensor: Selective Adsorption and Distinct I--V Response." *The Journal of Physical Chemistry Letters* 5 (15): 2675–81.

Kumar, Rajeev, Sarika Pal, Alka Verma, Y. K. Prajapati, and J. P. Saini. 2020. "Effect of Silicon on Sensitivity of SPR Biosensor Using Hybrid Nanostructure of Black Phosphorus and MXene." *Superlattices and Microstructures* 145: 106591.

Kumar, Rajeev, Sarika Pal, Narendra Pal, Vimal Mishra, and Yogendra Kumar Prajapati. 2021. "High-Performance Bimetallic Surface Plasmon Resonance Biochemical Sensor Using a Black Phosphorus—MXene Hybrid Structure." *Applied Physics A* 127 (4): 1–12.

Kuriakose, Sruthi, Taimur Ahmed, Sivacarendran Balendhran, Gavin E. Collis, Vipul Bansal, Igor Aharonovich, Sharath Sriram, Madhu Bhaskaran, and Sumeet Walia. 2018. "Effects of Plasma-Treatment on the Electrical and Optoelectronic Properties of

Layered Black Phosphorus." *Applied Materials Today* 12: 244–9. 10.1016/j.apmt.2018.06.001.

Lan, Shoufeng, Sean Rodrigues, Lei Kang, and Wenshan Cai. 2016. "Visualizing Optical Phase Anisotropy in Black Phosphorus." *ACS Photonics* 3 (7): 1176–81.

Late, Dattatray J. 2015. "Temperature Dependent Phonon Shifts in Few-Layer Black Phosphorus." *ACS Applied Materials & Interfaces* 7 (10): 5857–62.

Late, Dattatray J. 2016. "Liquid Exfoliation of Black Phosphorus Nanosheets and Its Application as Humidity Sensor." *Microporous and Mesoporous Materials* 225: 494–503. 10.1016/j.micromeso.2016.01.031.

Lee, Geonyeop, Sunwoo Jung, Soohwan Jang, and Jihyun Kim. 2017. "Platinum-Functionalized Black Phosphorus Hydrogen Sensors." *Applied Physics Letters* 110 (24): 242103.

Li, Diao, Henri Jussila, Lasse Karvonen, Guojun Ye, Harri Lipsanen, Xianhui Chen, and Zhipei Sun. 2015. "Polarization and Thickness Dependent Absorption Properties of Black Phosphorus: New Saturable Absorber for Ultrafast Pulse Generation." *Scientific Reports* 5 (1): 1–9.

Li, Hong, Charlie Tsai, Ai Leen Koh, Lili Cai, Alex W. Contryman, Alex H. Fragapane, Jiheng Zhao, et al. 2016. "Activating and Optimizing MoS2 Basal Planes for Hydrogen Evolution through the Formation of Strained Sulphur Vacancies." *Nature Materials* 15 (1): 48–53.

Li, Likai, Fangyuan Yang, Guo Jun Ye, Zuocheng Zhang, Zengwei Zhu, Wenkai Lou, Xiaoying Zhou, et al. 2016. "Quantum Hall Effect in Black Phosphorus Two-Dimensional Electron System." *Nature Nanotechnology* 11 (7): 593–7.

Li, Likai, Guo Jun Ye, Vy Tran, Ruixiang Fei, Guorui Chen, Huichao Wang, Jian Wang, et al. 2015. "Quantum Oscillations in a Two-Dimensional Electron Gas in Black Phosphorus Thin Films." *Nature Nanotechnology* 10 (7): 608–13.

Li, Likai, Yijun Yu, Guo Jun Ye, Qingqin Ge, Xuedong Ou, Hua Wu, Donglai Feng, Xian Hui Chen, and Yuanbo Zhang. 2014. "Black Phosphorus Field-Effect Transistors." *Nature Nanotechnology* 9 (5): 372.

Liu, Han, Adam T. Neal, Zhen Zhu, Zhe Luo, Xianfan Xu, David Tománek, and Peide D. Ye. 2014. "Phosphorene: An Unexplored 2D Semiconductor with a High Hole Mobility." *ACS Nano* 8 (4): 4033–41.

Liu, Fei, Yijiao Wang, Xiaoyan Liu, Jian Wang, and Hong Guo. 2014. "Ballistic Transport in Monolayer Black Phosphorus Transistors." *IEEE Transactions on Electron Devices* 61 (11): 3871–6.

Liu, Xinke, Kah-Wee Ang, Wenjie Yu, Jiazhu He, Xuewei Feng, Qiang Liu, He Jiang, et al. 2016. "Black Phosphorus Based Field Effect Transistors with Simultaneously Achieved Near Ideal Subthreshold Swing and High Hole Mobility at Room Temperature." *Scientific Reports* 6 (1): 1–8.

Liu, Yue, and P. Paul Ruden. 2017. "Temperature-Dependent Anisotropic Charge-Carrier Mobility Limited by Ionized Impurity Scattering in Thin-Layer Black Phosphorus." *Physical Review B* 95 (16): 165446.

Long, Gen, Denis Maryenko, Junying Shen, Shuigang Xu, Jianqiang Hou, Zefei Wu, Wing Ki Wong, et al. 2016. "Achieving Ultrahigh Carrier Mobility in Two-Dimensional Hole Gas of Black Phosphorus." *Nano Letters* 16 (12): 7768–73.

Lopez-Sanchez, Oriol, Dominik Lembke, Metin Kayci, Aleksandra Radenovic, and Andras Kis. 2013. "Ultrasensitive Photodetectors Based on Monolayer MoS 2." *Nature Nanotechnology* 8 (7): 497–501.

Low, Tony, A. S. Rodin, A. Carvalho, Yongjin Jiang, Han Wang, Fengnian Xia, and A. H. Castro Neto. 2014. "Tunable Optical Properties of Multilayer Black Phosphorus Thin Films." *Physical Review B* 90 (7): 75434.

Lu, Ganhua, Leonidas E. Ocola, and Junhong Chen. 2009. "Reduced Graphene Oxide for Room-Temperature Gas Sensors." *Nanotechnology* 20 (44): 445502.

Lu, S. B., L. L. Miao, Z. N. Guo, X. Qi, C. J. Zhao, H. Zhang, S. C. Wen, D. Y. Tang, and D. Y. Fan. 2015. "Broadband Nonlinear Optical Response in Multi-Layer Black Phosphorus: An Emerging Infrared and Mid-Infrared Optical Material." *Optics Express* 23 (9): 11183–94.

Luo, Zhi-Chao, Meng Liu, Zhi-Nan Guo, Xiao-Fang Jiang, Ai-Ping Luo, Chu-Jun Zhao, Xue-Feng Yu, Wen-Cheng Xu, and Han Zhang. 2015. "Microfiber-Based Few-Layer Black Phosphorus Saturable Absorber for Ultra-Fast Fiber Laser." *Optics Express* 23 (15): 20030–39.

Maurya, J. B., Y. K. Prajapati, S. Raikwar, and J. P. Saini. 2018. "A Silicon-Black Phosphorous Based Surface Plasmon Resonance Sensor for the Detection of NO2 Gas." *Optik* 160: 428–33.

Meshginqalam, Bahar, and Jamal Barvestani. 2018. "Aluminum and Phosphorene Based Ultrasensitive SPR Biosensor." *Optical Materials* 86: 119–25.

Miao, Jinshui, Suoming Zhang, Le Cai, Martin Scherr, and Chuan Wang. 2015. "Ultrashort Channel Length Black Phosphorus Field-Effect Transistors." *ACS Nano* 9 (9): 9236–43.

Moreno-Moreno, Miriam, Guillermo Lopez-Polin, Andres Castellanos-Gomez, Cristina Gomez-Navarro, and Julio Gomez-Herrero. 2016. "Environmental Effects in Mechanical Properties of Few-Layer Black Phosphorus." *2D Materials* 3 (3): 31007.

Narayan Maiti, Uday, Won Jun Lee, Ju Min Lee, Youngtak Oh, Ju Young Kim, Ji Eun Kim, Jongwon Shim, Tae Hee Han, and Sang Ouk Kim. 2014. "Chemically Modified/Doped Carbon Nanotubes & Graphene for Optimized Nanostructures & Nanodevices." *Advanced Materials* 26: 40–67.

Ong, Zhun-Yong, Yongqing Cai, Gang Zhang, and Yong-Wei Zhang. 2014. "Strong Thermal Transport Anisotropy and Strain Modulation in Single-Layer Phosphorene." *The Journal of Physical Chemistry C* 118 (43): 25272–77.

Pal, Sarika, Alka Verma, Y. K. Prajapati, and J. P. Saini. 2017. "Influence of Black Phosphorous on Performance of Surface Plasmon Resonance Biosensor." *Optical and Quantum Electronics* 49 (12): 403.

Pal, Sarika, Alka Verma, Y. K. Prajapati, and J. P. Saini. 2020. "Sensitive Detection Using Heterostructure of Black Phosphorus, Transition Metal Di-Chalcogenides and MXene in SPR Sensor." *Applied Physics A* 126 (10): 1–10.

Pal, Sarika, Alka Verma, Jai Prakash Saini, and Yogendra Kumar Prajapati. 2019. "Sensitivity Enhancement Using Silicon-Black Phosphorus-TDMC Coated Surface Plasmon Resonance Biosensor." *Iet Optoelectronics* 13 (4): 196–201.

Pal, Sarika, Alka Verma, S. Raikwar, Y. K. Prajapati, and J. P. Saini. 2018. "Detection of DNA Hybridization Using Graphene-Coated Black Phosphorus Surface Plasmon Resonance Sensor." *Applied Physics A* 124 (5): 1–11.

Pálfalvi, L., B. C. Tóth, G. Almási, J. A. Fülöp, and J. Hebling. 2009. "A General Z-Scan Theory." *Applied Physics B* 97 (3): 679–85.

Park, Cheol-Min, Jae-Hun Kim, Hansu Kim, and Hun-Joon Sohn. 2010. "Li-Alloy Based Anode Materials for Li Secondary Batteries." *Chemical Society Reviews* 39 (8): 3115–41.

Perello, David J., Sang Hoon Chae, Seunghyun Song, and Young Hee Lee. 2015. "High-Performance n-Type Black Phosphorus Transistors with Type Control via Thickness and Contact-Metal Engineering." *Nature Communications* 6 (1): 1–10.

Prakash, Amit, Yongqing Cai, Gang Zhang, Yong-Wei Zhang, and Kah-Wee Ang. 2017. "Black Phosphorus N-Type Field-Effect Transistor with Ultrahigh Electron Mobility via Aluminum Adatoms Doping." *Small* 13 (5): 1602909.

Qiao, Jingsi, Xianghua Kong, Zhi-Xin Hu, Feng Yang, and Wei Ji. 2014. "High-Mobility Transport Anisotropy and Linear Dichroism in Few-Layer Black Phosphorus." *Nature Communications* 5 (1): 1–7.

Qin, Guangzhao, Qing-Bo Yan, Zhenzhen Qin, Sheng-Ying Yue, Ming Hu, and Gang Su. 2015. "Anisotropic Intrinsic Lattice Thermal Conductivity of Phosphorene from First Principles." *Physical Chemistry Chemical Physics* 17 (7): 4854–8.

Ribeiro, Henrique B., Marcos A. Pimenta, Christiano J. S. De Matos, Roberto Luiz Moreira, Aleksandr S. Rodin, Juan D. Zapata, Eunézio A. T. De Souza, and Antonio H. Castro Neto. 2015. "Unusual Angular Dependence of the Raman Response in Black Phosphorus." *ACS Nano* 9 (4): 4270–6.

Saito, Yu, and Yoshihiro Iwasa. 2015. "Ambipolar Insulator-to-Metal Transition in Black Phosphorus by Ionic-Liquid Gating." *ACS Nano* 9 (3): 3192–8.

Shao, Jundong, Hanhan Xie, Hao Huang, Zhibin Li, Zhengbo Sun, Yanhua Xu, Quanlan Xiao, et al. 2016. "Biodegradable Black Phosphorus-Based Nanospheres for in Vivo Photothermal Cancer Therapy." *Nature Communications* 7 (1): 1–13.

Sheik-Bahae, Mansoor, Ali A. Said, and Eric W. Van Stryland. 1989. "High-Sensitivity, Single-Beam n 2 Measurements." *Optics Letters* 14 (17): 955–7.

Sheik-Bahae, Mansoor, Ali A. Said, T-H. Wei, David J. Hagan, and Eric W. Van Stryland. 1990. "Sensitive Measurement of Optical Nonlinearities Using a Single Beam." *IEEE Journal of Quantum Electronics* 26 (4): 760–9.

Shirotani, I. 1982. "Growth of Large Single Crystals of Black Phosphorus at High Pressures and Temperatures, and Its Electrical Properties." *Molecular Crystals and Liquid Crystals* 86 (1): 203–11.

Singh, Yadvendra, Mohan Kumar Paswan, and Sanjeev Kumar Raghuwanshi. 2021. "Sensitivity Enhancement of SPR Sensor with the Black Phosphorus and Graphene with Bi-Layer of Gold for Chemical Sensing." *Plasmonics*, 1–10. 10.1007/s11468-020-01315-3.

Singh, Yadvendra, and Sanjeev Kumar Raghuwanshi. 2019. "Sensitivity Enhancement of the Surface Plasmon Resonance Gas Sensor with Black Phosphorus." *IEEE Sensors Letters* 3 (12): 1–4.

Smith, Joshua B., Daniel Hagaman, and Hai-Feng Ji. 2016. "Growth of 2D Black Phosphorus Film from Chemical Vapor Deposition." *Nanotechnology* 27 (21): 215602.

Sorkin, V., Yongqing Cai, Z. Ong, Gang Zhang, and Yong-Wei Zhang. 2017. "Recent Advances in the Study of Phosphorene and Its Nanostructures." *Critical Reviews in Solid State and Materials Sciences* 42 (1): 1–82.

Srivastava, Triranjita, and Rajan Jha. 2018. "Black Phosphorus: A New Platform for Gaseous Sensing Based on Surface Plasmon Resonance." *IEEE Photonics Technology Letters* 30 (4): 319–22.

Srivastava, Akash, Alka Verma, Ritwick Das, and Y. K. Prajapati. 2020. "A Theoretical Approach to Improve the Performance of SPR Biosensor Using MXene and Black Phosphorus." *Optik* 203: 163430.

Sun, Li-Qun, Ming-Juan Li, Kai Sun, Shi-Hua Yu, Rong-Shun Wang, and Hai-Ming Xie. 2012. "Electrochemical Activity of Black Phosphorus as an Anode Material for Lithium-Ion Batteries." *The Journal of Physical Chemistry C* 116 (28): 14772–9.

Takao, Yukihiro, Hideo Asahina, and Akira Morita. 1981. "Electronic Structure of Black Phosphorus in Tight Binding Approach." *Journal of the Physical Society of Japan* 50 (10): 3362–9.

Tan, Teng, Xiantao Jiang, Cong Wang, Baicheng Yao, and Han Zhang. 2020. "2D Material Optoelectronics for Information Functional Device Applications: Status and Challenges." *Advanced Science* 7 (11). 10.1002/advs.202000058.

Tayari, V., N. Hemsworth, I. Fakih, A. Favron, E. Gaufrès, G. Gervais, R. Martel, and T. Szkopek. 2015. "Two-Dimensional Magnetotransport in a Black Phosphorus Naked Quantum Well." *Nature Communications* 6 (1): 1–7.

Tran, Vy, Ryan Soklaski, Yufeng Liang, and Li Yang. 2014. "Layer-Controlled Band Gap and Anisotropic Excitons in Few-Layer Black Phosphorus." *Physical Review B* 89 (23): 235319.

Varghese, Seba S., Sunil Lonkar, K. K. Singh, Sundaram Swaminathan, and Ahmed Abdala. 2015. "Recent Advances in Graphene Based Gas Sensors." *Sensors and Actuators B: Chemical* 218: 160–83. 10.1016/j.snb.2015.04.062.

Varghese, Seba Sara, Saino Hanna Varghese, Sundaram Swaminathan, Krishna Kumar Singh, and Vikas Mittal. 2015. "Two-Dimensional Materials for Sensing: Graphene and Beyond." *Electronics* 4 (3): 651–87. 10.3390/electronics4030651.

Wang, Qing Hua, Kourosh Kalantar-Zadeh, Andras Kis, Jonathan N. Coleman, and Michael S. Strano. 2012. "Electronics and Optoelectronics of Two-Dimensional Transition Metal Dichalcogenides." *Nature Nanotechnology* 7 (11): 699–712.

Wang, Xiaomu, Aaron M. Jones, Kyle L. Seyler, Vy Tran, Yichen Jia, Huan Zhao, Han Wang, Li Yang, Xiaodong Xu, and Fengnian Xia. 2015. "Highly Anisotropic and Robust Excitons in Monolayer Black Phosphorus." *Nature Nanotechnology* 10 (6): 517–21.

Wang, Han, Xiaomu Wang, Fengnian Xia, Luhao Wang, Hao Jiang, Qiangfei Xia, Matthew L. Chin, Madan Dubey, and Shu-Jen Han. 2014. "Black Phosphorus Radio-Frequency Transistors." *Nano Letters* 14 (11): 6424–9.

Wang, Hui, Xianzhu Yang, Wei Shao, Shichuan Chen, Junfeng Xie, Xiaodong Zhang, Jun Wang, and Yi Xie. 2015. "Ultrathin Black Phosphorus Nanosheets for Efficient Singlet Oxygen Generation." *Journal of the American Chemical Society* 137 (35): 11376–82. 10.1021/jacs.5b06025.

Wu, Leiming, Jun Guo, Qingkai Wang, Shunbin Lu, Xiaoyu Dai, Yuanjiang Xiang, and Dianyuan Fan. 2017. "Sensitivity Enhancement by Using Few-Layer Black Phosphorus-Graphene/TMDCs Heterostructure in Surface Plasmon Resonance Biochemical Sensor." *Sensors and Actuators B: Chemical* 249: 542–8.

Xia, Fengnian, Han Wang, and Yichen Jia. 2014. "Rediscovering Black Phosphorus as an Anisotropic Layered Material for Optoelectronics and Electronics." *Nature Communications* 5 (1): 1–6.

Xu, Yijun, Zhe Shi, Xinyao Shi, Kai Zhang, and Han Zhang. 2019a. "Recent Progress in Black Phosphorus and Black-Phosphorus-Analogue Materials: Properties, Synthesis and Applications." *Nanoscale* 11 (31): 14491–527. 10.1039/c9nr04348a.

Xu, Yijun, Zhe Shi, Xinyao Shi, Kai Zhang, and Han Zhang. 2019b. "Recent Progress in Black Phosphorus and Black-Phosphorus-Analogue Materials: Properties, Synthesis and Applications." *Nanoscale* 11 (31): 14491–527. 10.1039/C9NR04348A.

Yang, Zhibin, Jianhua Hao, Shuoguo Yuan, Shenghuang Lin, Hei Man Yau, Jiyan Dai, and Shu Ping Lau. 2015. "Field-Effect Transistors Based on Amorphous Black Phosphorus Ultrathin Films by Pulsed Laser Deposition." *Advanced Materials* 27 (25): 3748–54.

Yang, Shengxue, Sefaattin Tongay, Yan Li, Qu Yue, Jian-Bai Xia, Shun-Shen Li, Jingbo Li, and Su-Huai Wei. 2014. "Layer-Dependent Electrical and Optoelectronic Responses of ReSe2 Nanosheet Transistors." *Nanoscale* 6 (13): 7226–31.

Yi, Ya, Xue Feng Yu, Wenhua Zhou, Jiahong Wang, and Paul K. Chu. 2017. "Two-Dimensional Black Phosphorus: Synthesis, Modification, Properties, and Applications." *Materials Science and Engineering R: Reports* 120: 1–33. 10.1016/j.mser.2017.08.001.

Yoshizawa, Masahito, Ichimin Shirotani, and Tadao Fujimura. 1986. "Thermal and Elastic Properties of Black Phosphorus." *Journal of the Physical Society of Japan* 55 (4): 1196–202.

Yuan, Hongtao, Xiaoge Liu, Farzaneh Afshinmanesh, Wei Li, Gang Xu, Jie Sun, Biao Lian, et al. 2015. "Polarization-Sensitive Broadband Photodetector Using a Black Phosphorus Vertical p–n Junction." *Nature Nanotechnology* 10 (8): 707–13.

Yuan, Wenjing, and Gaoquan Shi. 2013. "Graphene-Based Gas Sensors." *J. Mater. Chem. A* 1 (35): 10078–91. 10.1039/C3TA11774J.

Zhang, Xiao, Haiming Xie, Zhengdong Liu, Chaoliang Tan, Zhimin Luo, Hai Li, Jiadan Lin, et al. 2015. "Black Phosphorus Quantum Dots." *Angewandte Chemie International Edition* 54 (12): 3653–57.

Zhu, Weinan, Maruthi N. Yogeesh, Shixuan Yang, Sandra H. Aldave, Joon-Seok Kim, Sushant Sonde, Li Tao, Nanshu Lu, and Deji Akinwande. 2015. "Flexible Black Phosphorus Ambipolar Transistors, Circuits and AM Demodulator." *Nano Letters* 15 (3): 1883–90.

Zhu, Weinan, Saungeun Park, Maruthi N. Yogeesh, Kyle M. McNicholas, Seth R. Bank, and Deji Akinwande. 2016. "Black Phosphorus Flexible Thin Film Transistors at Gigahertz Frequencies." *Nano Letters* 16 (4): 2301–6.

5 MXene as a 2D Material for the Surface Plasmon Resonance Sensing

5.1 INTRODUCTION

In the next generation, compact, wearable, and high-performance optoelectronic applications and two-dimensional (2D) materials have huge promise due to their special architectures and outstanding optical and electrical properties. MXene is a recent class of carbides, carbonitride and nitrides of 2D transition metals. MXene was first discovered in 2011 by Barsomm and co-workers and was developed by exfoliating selective MAX levels. The MAX step refers to the $M_{n+1}AX_n$ formula and here is the range of n from 1 to 3, where M signifies primary d-block transition metals and a represents either C or N atoms of the key groups 13 and 14.

MXene is used for the sensing methods because it has a strong metallic conductivity, a medium region of diffusion, and a wide surface area. Through using $Ti_3C_2T_x$ MXene to excite the SPPs, the above topic has been demonstrated in SPR-based sensors. However, there is a downside that to provide a low detection accuracy (DA) and high figure of merit (FOM) in the SPR-based sensors and their reflectance curve is also large. We also suggested a new $Ti_3C_2T_x$ MXene-based long-range surface plasmon resonance (SPR) sensor to obtain high DA and FOM (Dai et al. 2019). 2D solids, which are described as crystals with large aspect ratios and atomic layer widths of a few hicks, have recently attracted a lot of attention. Novoselov et al. (2004) have given a brief description of graphene, although it has been known for last decades, and were able to isolate and describe a single graphene sheet and then interest in the material skyrocketed. In 2010, this function was awarded the Nobel Prize in Physics.

By selectively etching of MAX phases' A layers, it is possible to MAX phases exfoliation and create 2D layers of transition metal carbides or nitrides. At room temperature, aqueous hydrofluoric acid was used for the etching process. For this process, the 13 separate MXenes are produced, namely (1) Mo_2C (2) $(Ti_{0.50}, Nb_{0.50})_2$ C (3) Ta_4C_3 (4) Nb_4 C_3 (5) Ti_2C (6) Nb_2 C (7) V_2 C (8) $(V_{0.50}, Cr_{0.5})_3$ C_2 (9) Ti_3CN (10) $(Ti_{0.50}, V_{0.50})_2$ C (11) Ti_3C_2 (12) $(Ti_{0.50}, V_{0.50})_3$ C_2, and (13) $(Nb_{0.50}, V_{0.5})_4$ C_3. The synthesis of MXenes was followed by their termination through a mixture of OH, F, and O groups. Sonication of MXenes caused a slight separation of the stacked layers. When Ti_3C_2 was intercalated with dimethyl sulfoxide and then sonicated in water, substantial delamination occurred, resulting in aqueous colloidal solutions appropriate for the fabrication of MXene "paper."

MAX is extremely anisotropic and is caused by mechanical deformation of the phases of the material. Given the fact that the majority of the M–A bond represents

DOI: 10.1201/9781003190738-5

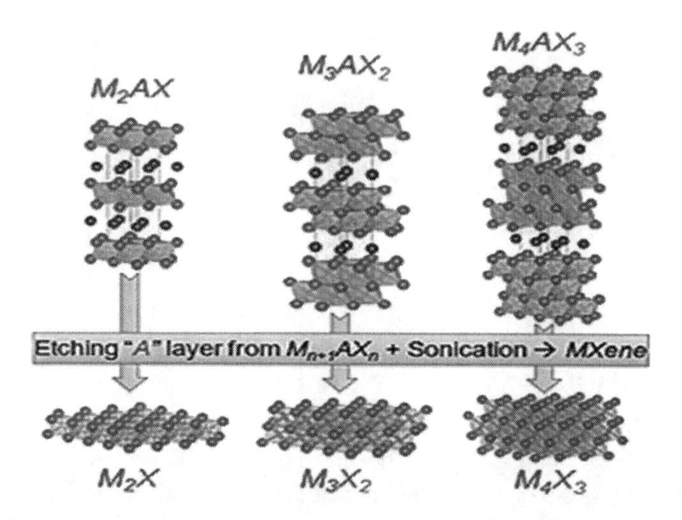

FIGURE 5.1 The phases of MAX structure and the resulting two-dimensional layers (after etching and sonicating the A layers) are depicted schematically. Reprinted with permission from Trends in Biotechnology. Copyright, 2020, Elsevier (Szuplewska et al. 2020).

graphene, as shown in Fig. 5.1, none of the phases of MAX has ever been exfoliated into a crystalline film a few nanometers thick.

There have been a few studies on the selective space or mildness of the A-group layers from the phases of MAX or their exfoliation. Hoffman et al. (2008) demonstrated that extracting both M and A to produce carbide-derived carbons, or CDCs, is a reasonably straightforward method. The first step is to exfoliate the MAX stages that are followed by selective etching of the A layers from MAX stages to form 2D layers of the transition metal. The resulting 2D $M_{n+1}X_n$ layers are called "ene" to emphasize the absence of the A group in the phases of MAX and "MXenes" to emphasize their 2D existence and relationship to graphene. The phrase "selective etching of A layers from MAX phases" refers to the method used to exfoliate the MAX phases. About 60 MAX phases have been observed to date (Sun 2011). By eliminating the A atom layers from MAX structure, more than 20 distinct "pure" $M_{n+1}X_n$ chemistries could be formed. Additionally, MAX solid solutions can be used to synthesize MXenes at the M and X sites, resulting in a variety of other 2D materials.

2D $Ti_3C_2T_x$ MXene is metallic, highly stable, solution-processable, and highly conductive, where T_x denotes (O, OH, or F) terminating groups (Szuplewska et al. 2020). MXene is suitable for low-contact resistance interconnects due to its high Fermi level density of states and electronic conductivity of up to 15,000 S cm^{-1}. MXene's electronic properties are anisotropic and tuneable because of their 2D-layered structure, which also allows for varying surface chemistry and layered spacing (Coleman et al. 2011). Despite that, hydrophilicity enables scaled and straightforward implementation.

Biocompatible (bio)sensors for analyte recognition have been identified for 2D materials; likewise "graphene" (Han et al. 2021) and "MoS_2" (Lawal 2018), and a novel group of 2D materials like MXene are shown in Table 5.1.

TABLE 5.1

The Parameters of MXene-based Sensors are Shown. Reprinted with Permission from Trends in Biotechnology. Copyright, 2020, Elsevier (Szuplewska et al. 2020)

Analyte	MXene	Detection System	Response Time	Linear Range	Limit of detection	Sensitivity	Refs
Nitrite	Ti_3C_2	Amperometric	<3 s	0.5–11,800 μM	0.12 μM	No data	(Sinha, Tan, et al. 2018)
H_2O_2	Ti_3C_2	Voltammetric	<3 s	0.1–260 μM	20 μM	No data	(Liu et al. 2015)
H_2O_2	Pristine and oxidized $Ti_3C_2\ T_x$ MXene	Chronoamperometric	10 s	No data	0.7 μM	596 mA cm^{-2} mM^{-1}	(Wang et al. 2014)
Glucose	Ti_3C_2 Tx	Amperometric	10 s	0.1–18 μM	5.9 μM	4.2 mA M^{-1} cm^{-2}	(Lorencova et al. 2017)
Volatile organic compounds	Ti_3C_2 Tx	Resistive	No data	No data	9.27 ppm	No data	(Rakhi, Nayak, Xia, and Alshareef 2016)
Single nucleotide	Ti_3C_2 Tx	Electrochemiluminescent	No data	No data	1 μM	No data	(Lee et al. 2017)
Tripropylamine	Ti_3C_2 Tx	Electrochemiluminescent	No data	1.0×10^{-8} to 1.0×10^{-3}	5 μM	No data	(Fang et al. 2018)
Dopamine	Ti_3C_2	Conductometric	No data	100 nM to 50 μM	100 μM	No data	(Xu et al. 2016)
Human bending release activities	Ti_3C_2	Resistive	<10 ms	No data	351 Pa	Gauge factor 180.1	(Ma et al. 2017)
Humidity	Ti_3C_2	Resistive, gravimetric	100 s	0–85% relative humidity	0.8% relative humidity	3%	(Zhou et al. 2017)
Organophosphorus pesticides	Ti_3C_2 Tx	Amperometric	No data	1×10^{-14} to 1×10^{-8} M	0.3×10^{-14} M	No data	(Zhou et al. 2017)

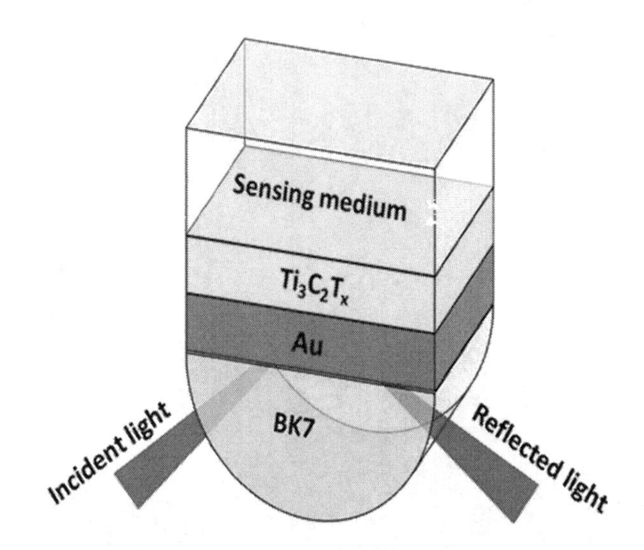

FIGURE 5.2 Using MXene, a schematic diagram of an SPR biosensor is shown. Reprinted with permission from Sensors and Actuators B: Chemical. Copyright, 2018, Elsevier (Wu et al. 2018).

The novel SPR biosensor with a few layers of MXene proposed by Wu et al. (2018) is depicted in Fig. 5.2. The proposed SPR-based biosensor with a few layers of MXene was investigated by coating the surface of thin-metal films with a few layers of $Ti_3C_2T_x$ MXene. At a 633 nm wavelength, the proposed structure, when considering the 4 layers of Au, 7 layers of Ag, 12 layers of Al, and 9 layers of Cu, exhibit sensitivity is increased by 16.8%, 28.4%, 46.3%, and 33.6%, respectively. Additionally, with monolayer MXene at a 532-nm wavelength, it was investigated that the sensitivity can obtain 224.5°/RIU.

$Ti_3C_2T_x$ MXene with a few layers was discovered to have possible applications, and MXene and other 2D MXene nanomaterials (NMs) are expected to find auspicious applications in SPR biosensors.

Fig. 5.3 shows that the theoretical investigation of an SPR-based structure is investigated with a single layer of Au-based biosensor, the sensitivity obtained is 137°/RIU and, using the Au and four layers of the MXene, the sensitivity found was 160°/RIU (Wu et al. 2018).

Fig. 5.4 represents the other BK7 prism-based SPR sensor for better sensitivity; in this, a few layers of the MXene-based SPR sensor are theoretically investigated and found the FOM as 270°/RIU, which is very high compared to conventional Au film (50 nm) (Dai et al. 2019).

Similarly, in another sensor structure (BK7/Ag/Si/MXene/sensing medium) proposed by Kumar, Pal, Prajapati, and Saini (2021), the SPR biosensor has been theoretically investigated using MXene, using a MXene layer with the highest sensitivity achieved by the proposed SPR sensor as 231°/RIU.

As shown in Fig. 5.5, the SPR biosensor (BK7/Silver/Silicon/Black Phosphorus/MXene-$Ti_3C_2T_x$/sensing medium) proposed by Kumar et al. (2020) found that the maximum sensitivity accomplished is 264°/RIU at wavelength 633 nm, the RI of

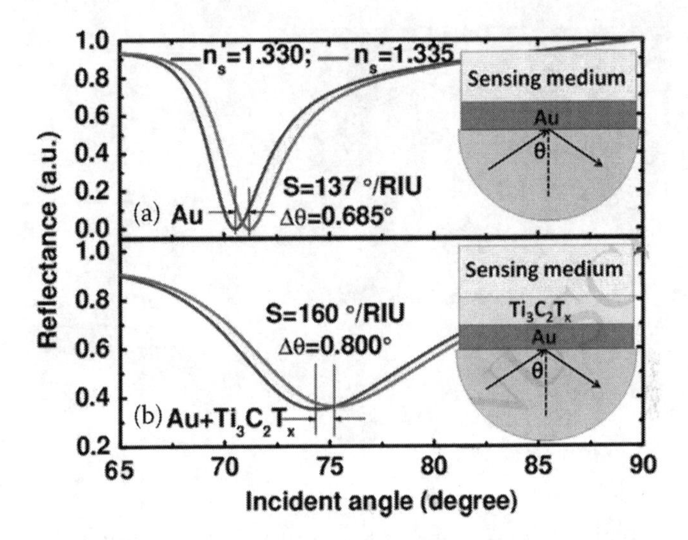

FIGURE 5.3 The difference in reflectance with an angle of incidence for a) standard biosensing based on a particular Au film and b) novel biosensing based on a few-layer $Ti_3C_2T_x$ MXene. Reprinted with permission from Sensors and Actuators B: Chemical. Copyright, 2018, Elsevier (Wu et al. 2018).

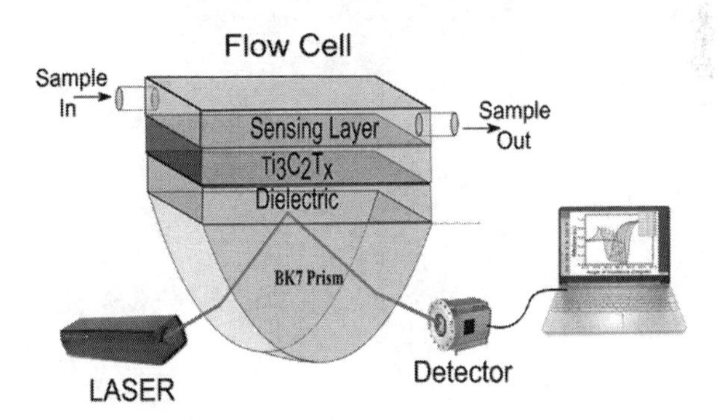

FIGURE 5.4 Shows the schematics diagram of the MXene/dielectric/BK7-based SPR sensor.

the BK7 is 1.5151, Ag as per the Drude model, Si = 3.916, Black Phosphorus = 3.5 + 0.01 × i, and for MXene it is 2.38 + 1.33 × i.

Similarly, Srivastava, Verma, Das, and Prajapati (2020) proposed the SPR biosensor as shown in Fig. 5.6, by the proposed structure the highest sensitivity achieved is 190.22°/RIU at 633 nm wavelength. The parameters considered for this structure are given in Table 5.2.

Surface plasmon excitation has been demonstrated to be beneficial in several areas, like biological, environmental, and optical sensors, and also in surface-enhanced Raman spectroscopy (SERS) (Homola, Yee, and Gauglitz 1999; Satheeshkumar et al. 2016; Willets and Van Duyne 2007; Zeng, Baillargeat, Ho,

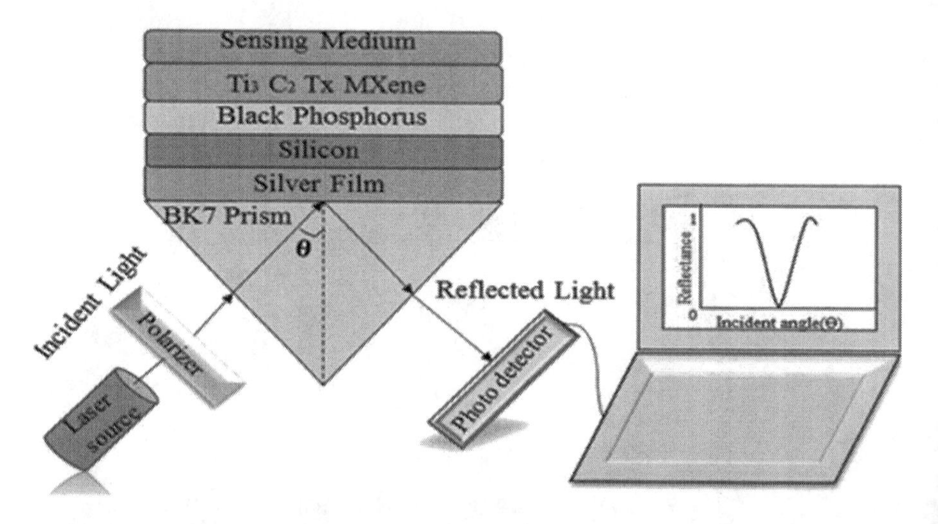

FIGURE 5.5 Schematic diagram of the MXene/BP/Silicon/Ag-based SPR-based biosensor. Reprinted with permission from Superlattices and Microstructures. Copyright, 2020, Elsevier (Kumar et al. 2021).

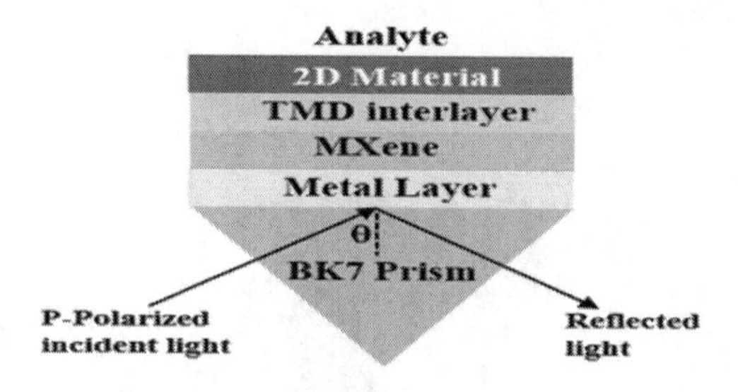

FIGURE 5.6 Using MXene, a schematic diagram of an SPR biosensor is shown. Reprinted with permission from Optik. Copyright, 2020, Elsevier (Srivastava, Verma, Das, and Prajapati 2020).

and Yong 2014). Using electron energy loss spectroscopy (EELS) of high-resolution transmission, it is possible to tune the surface plasmon frequency of the Ti_3C_2 MXene layer in the mid-infrared region (Mauchamp et al. 2014).

According to various studies, MXenes exhibit extraordinary plasmonic properties. MXene $Ti_3C_2T_x$ nanosheets combined with new metals like silver and gold can be used to attain SERS with such an improvement for both the methylene blue dye (Satheeshkumar et al. 2016). Raman signal enhancement substrates of spray-coated MXene are also fabricated and useful to detect Rhodamine 6G without the need of the novel metals.

As illustrated in Fig. 5.7, the transverse SP mode and the interband transition are dispersed similarly in monolayer and multilayer flakes. Also, as shown in Fig. 5.7(a), the longitudinal multipolar modes were strongly dependent on the flake's morphological features in terms of energy and spatial distribution, as illustrated in Fig. 5.7(b).

TABLE 5.2
Shows the Sensor Structure Design Parameters. Reprinted with Permission from Optik. Copyright, 2020, Elsevier (Srivastava, Verma, Das, and Prajapati 2020)

Layers	Materials Used	Refractive Index (RI) for λ = 633 nm		Thickness (nm)
		Real part (n)	Imaginary part (k)	
I	Prism BK7 (Palik 1998)	1.5151	-	
II	Metal (Au) layer (Pal, Verma, Prajapati, and Saini 2017)	0.19683	3.0905	45
III	MXene (Wu et al. 2018; Xu, Ang, Wu, and Ang 2019)	2.38	1.33	0.993
IV	WS2 monolayer (Xu, Hsieh, Wu, and Ang 2018)	4.9	0.3124	0.80
V	Black Phosphorus (Maurya, Prajapati, Raikwar, and Saini 2018; Pal, Verma, Prajapati, and Saini 2017)	3.5	0.01	0.53
VI	Sensing layer (ns)	1.330 to 1.335	–	–

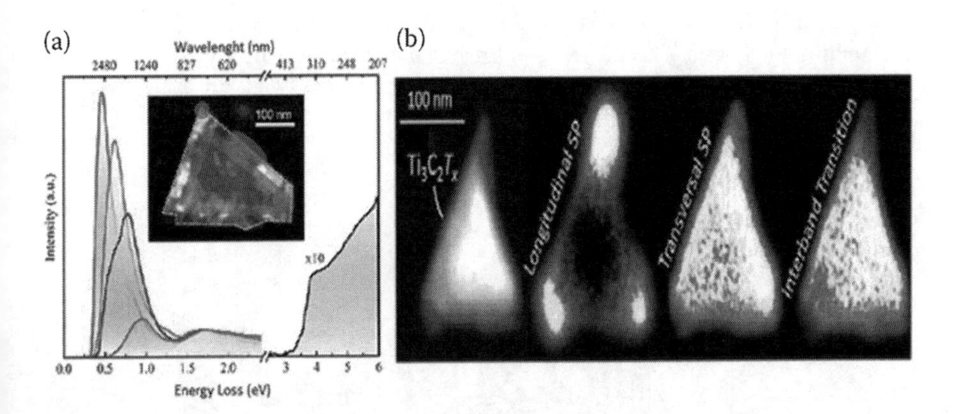

FIGURE 5.7 a) Electron energy loss spectral analysis of a triangular $Ti_3C_2T_x$ flake with no energy loss. The inset displays a HAADF-STEM. b) At the same, $Ti_3C_2T_x$ flake excited longitudinal surface plasmon is shown, transversal SP is shown, and interband transition distributions. Reprinted with permission from Physics Reports. Copyright, 2020, Elsevier (Jiang et al. 2020).

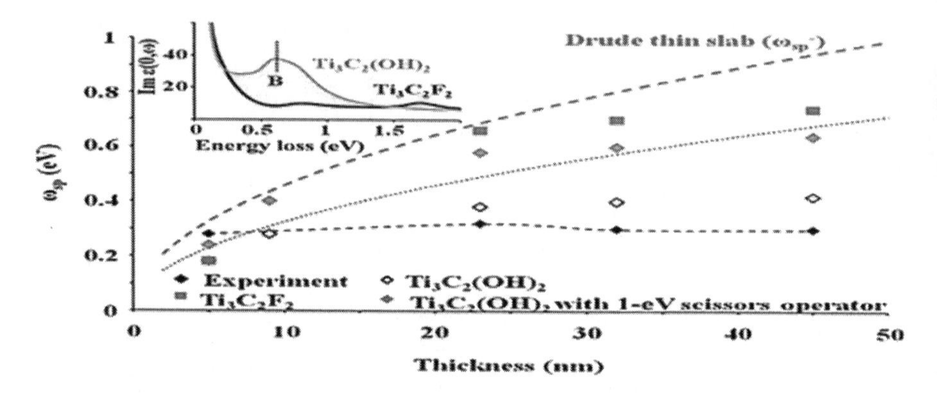

FIGURE 5.8 ω_{SP} (eV) vs MXene thickness (nm) calculated from experimental data and measurements as a function of. Inset: Im$\varepsilon(0, \omega)$ for $Ti_3C_2F_2$ & $Ti_3C_2(OH)_2$. Reprinted with permission from Physics Reports. Copyright, 2020, Elsevier (Jiang et al. 2020).

The metallic free electron density of $Ti_3C_2T_x$ flakes is compounded by the desorption of fluorine (F) termination at annealing temperatures greater than 550°C.

As shown in Fig. 5.8, MXene layers do not much affect the bulk plasmon excitation due to weak coupling between the $Ti_3C_2T_x$ sheets in the stack; it shows that MXene layers in a stack do not affect the probability of bulk plasmon. Edge and central longitudinal multipolar modes were discovered in $Ti_3C_2T_x$ flakes using ultra-high-resolution EELS (El-Demellawi et al. 2018).

MXene's functional groups exist in a variety of energy states, each of which results in a charge transfer to maintain a certain level, as shown in Fig. 5.9 (Jiang et al. 2020). Based on the Kretschmann design (Sambles, Bradbery, and Yang 1991; Sarycheva et al. 2017), a prism-based SPR sensor has also been proposed as an effective optical sensor for a biological or chemical analyte. It has been investigated that the sensitivity parameter is mainly dependent on the thickness of MXene and metal film thickness. The Al metal film enhanced the sensitivity by 46% when coated with 12 atomic layers of $Ti_3C_2T_x$ (Wu et al. 2018). Xu, Ang, Wu, and Ang

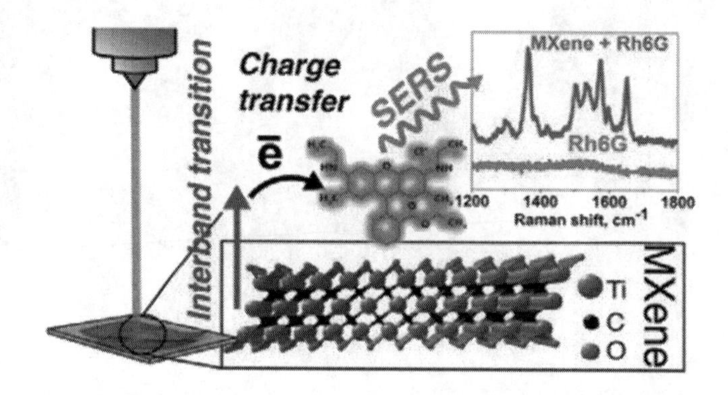

FIGURE 5.9 Schematic diagram of interband transition. Reprinted with permission from Physics Reports. Copyright, 2020, Elsevier (Jiang et al. 2020).

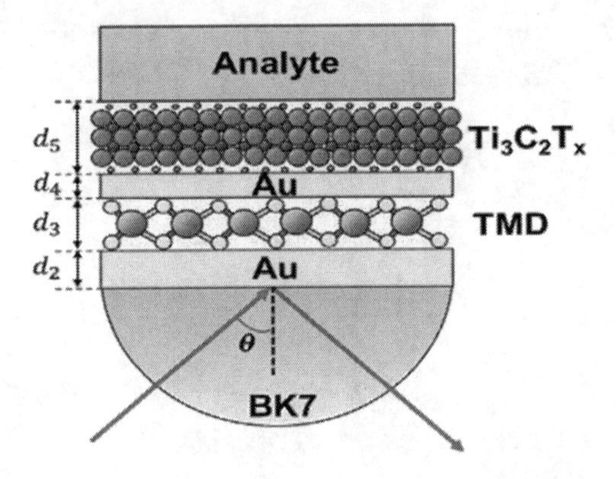

FIGURE 5.10 Proposed hybrid structure of Au–$Ti_3C_2T_x$-Au TMDs based on SPR sensor. Reprinted with permission from Physics Reports. Copyright, 2020, Elsevier (Jiang et al. 2020).

(2019) proposed the SPR-based hybrid structure with the Au-Ti_3C_2Tx-Au-TMDs with a performance improvement of 41% over the conventional Au metal SPR sensors (see Fig. 5.10) (Jiang et al. 2020). MXenes that are not terminated have recently gained attention as a potential alternative plasmonic material for fabricating metasurfaces. Chaudhuri et al. (2018) demonstrated the photon spin Hall effect in optical metasurfaces composed of TiN and ZrN. A metasurface composed of a Ti_3C_2Tx nanodisk array on an Au/Al_2O_3 substrate exhibited strong heterogeneous SPR at near-infrared frequencies, enabling visible interband transitions. This metasurface exhibits high absorption efficiency (90%) over a 1.55 μm broadband wavelength window, which may be advantageous for energy extraction from light, biomedical imaging, and sensing (Wang et al. 2018) (Fig. 5.11).

Pandey (2020) proposed a sensor structure as shown in Fig. 5.12. The highest angular sensitivity accomplished by the proposed structure is 322.46°/RIU and FWHM (full width half maximum) is obtained as 4.512° with FOM of 55.59/RIU for the three-layer monolayer-based SPR biosensor.

The R_{min} is obtained as 5.214×10^{-4} at $t_{Ag} = 44$ nm for the monolayer $Ti_3C_2T_x$ MXene monolayer BP over Ag at 633 nm wavelength and refractive index of sensing layer taken 1.320 as shown in Fig. 5.13.

TEM image shows a mono-layered $Ti_3C_2T_x$ slice that is thin and transparent; in the proposed experiment, the lateral size is considered 1 μm as shown in Fig. 5.14(a). The findings show that MXene nanosheets were effectively extracted from the MAX process. The AFM image is shown in Fig. 5.14(b) as-synthesized nanosheets. It demonstrates that a flat surface of $Ti_3C_2T_x$ nanoflakes with a lateral size of more than 1 μm was obtained, with a thickness of around 3 nm for the selected nanosheets. It is estimated that there are about 2 ± 1 layers of MXene based on the adsorption of surface functional groups (Wang et al. 2015). XRD patterns show that after the mixed etchant treatment, the peak near 38 vanishes (Fig. 5.14c), indicating that Al atoms are separated from MAX-Ti_3AlC_2 (Luo et al. 2016). Additionally, after etching, MXene flakes have been effectively delaminated (Wang et al. 2016).

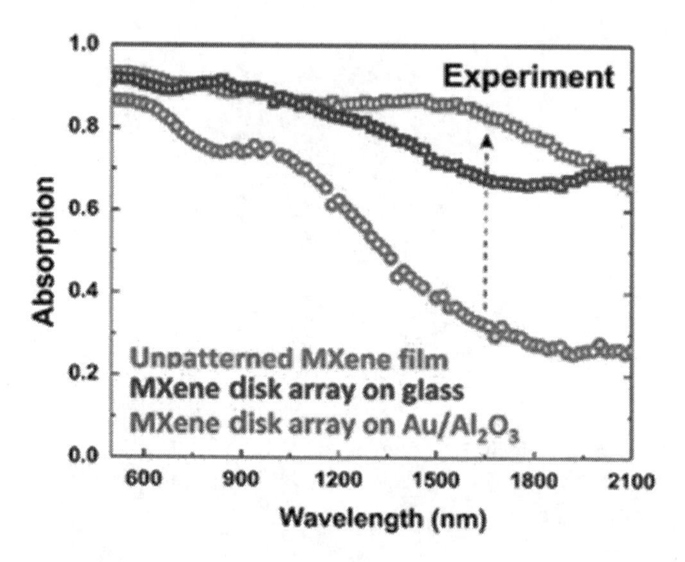

FIGURE 5.11 Broadband plasmonic absorbers made by MXene. Amid infrared regimes, MXene disc arrays are patterned that exhibit a strong absorption enhancement. Reprinted with permission from Physics Reports. Copyright, 2020, Elsevier (Jiang et al. 2020).

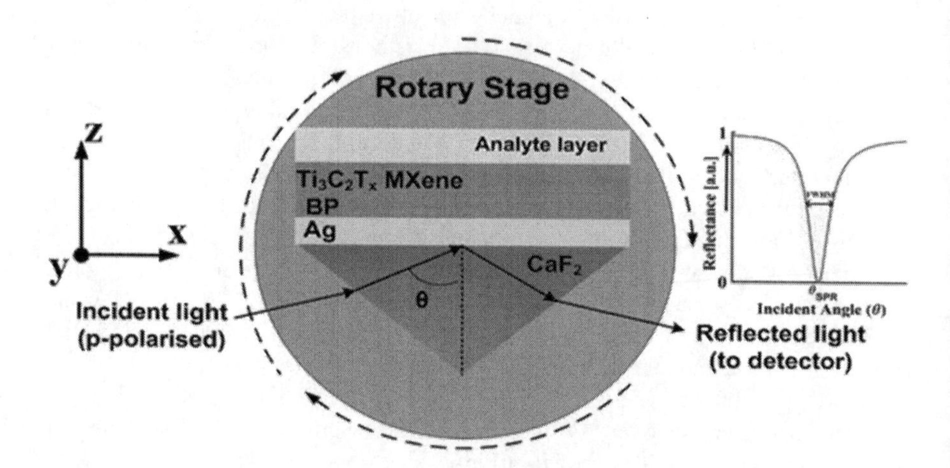

FIGURE 5.12 Schematic of BP and $Ti_3C_2T_x$ MXene monolayers are proposed. The width of the SPR curve at full width half maximum (FWHM). Reprinted with permission from Photonics and Nanostructures—Fundamentals and Applications. Copyright, 2020, Elsevier (Pandey 2020).

In Fig. 5.14(d), the Raman spectrum of as-prepared $Ti_3C_2T_x$ nanosheets are shown, with the peaks at 125 cm^{-1} and 202 cm^{-1} corresponding to in-plane and out-of-plane Ti atom vibrations, respectively. The peaks occur in the spectrum at 290 cm^{-1}, 372 cm^{-1}, and 630 cm^{-1}, respectively, as reflections of and vibrations of the surface groups. The 719 cm^{-1} peak results from the movement of carbon atoms out of a substrate away from the layer (Lioi et al. 2019; Sarycheva and Gogotsi

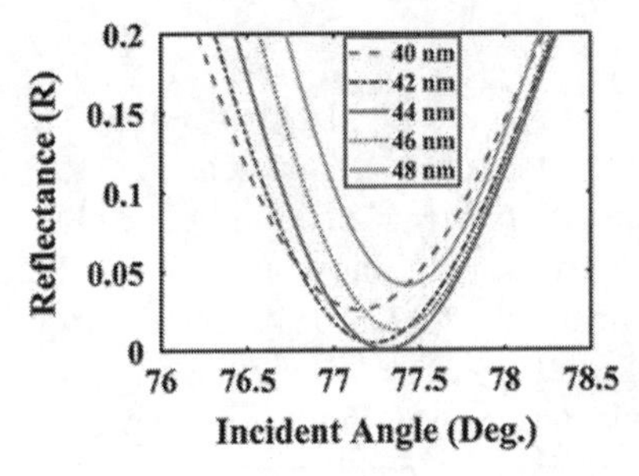

FIGURE 5.13 Different Ag layer thickness the variation of incident angle vs reflectance. Reprinted with permission from Photonics and Nanostructures—Fundamentals and Applications. Copyright, 2020, Elsevier (Pandey 2020).

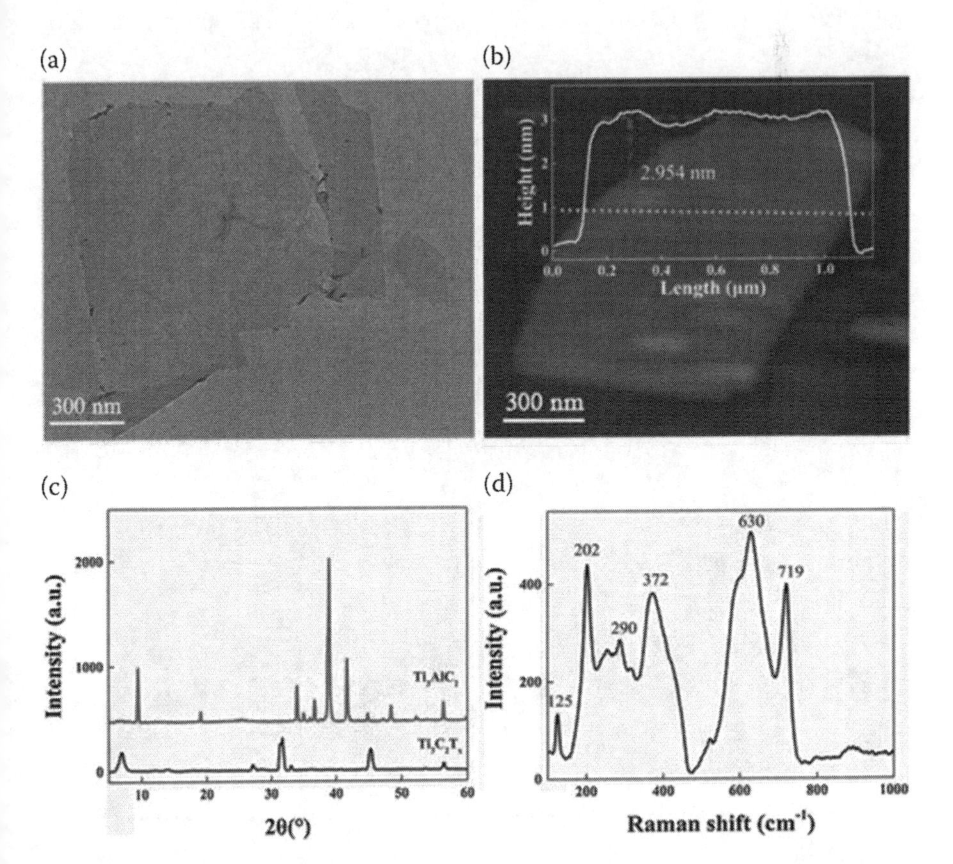

FIGURE 5.14 Characterization of MXene nanosheets that have been synthesized. a) Monolayer $Ti_3C_2T_x$ sheet TEM; b) AFM; c) MAX-Ti_3AlC_2 and MXene-$Ti_3C_2T_x$ XRD patterns; d) $Ti_3C_2T_x$ nanosheet Raman shift spectra. Reprinted with permission from Sensors and Actuators B: Chemical. Copyright, 2021, Elsevier (Liu et al. 2015).

2020). The Raman spectra, in particular, indicate that functional groups on the MXene surface are absorbed. The materials were successfully prepared and exhibit a robust morphology consistent with all previously described properties of $Ti_3C_2T_x$ MXene nanosheets.

5.2 BACKGROUND

MXenes can be synthesized through specific acid etching of MAX or non-MAX parents (typically from the periodic table groups 13 and 14), chemical transformations, either bottom-up construction techniques. In terms of viability, yields, controllability, and cost-effectiveness, the first approach currently outperforms the others. MAX ternary carbides and nitrides occur in over 70 various configurations (Anasori, Lukatskaya, and Gogotsi 2017), indicating that the MXene family has a great deal of diversity. Experiments have yielded over 30 different types of MXenes, with many more predicted based on theoretical predictions (Frey et al. 2019; Pan, Lany, and Qi 2017). Fig. 5.15 depicts a brief timeline of the growth of MXenes.

2D materials have gained attention in the scientific community. MXenes are used for many practical applications due to high conductivity, hydrophilicity, ion permeability, and wide surface area. It is found in the form of composites like intermetallic carbides, nitrides, and carbonitride.

In Fig. 5.16, salt mixtures of KF, NaF, and LiF were employed as etchants to etch Ti_4Alp-primers to make Al etchants at 550°C. Ti_2AlN is immersed in a KF-HCl

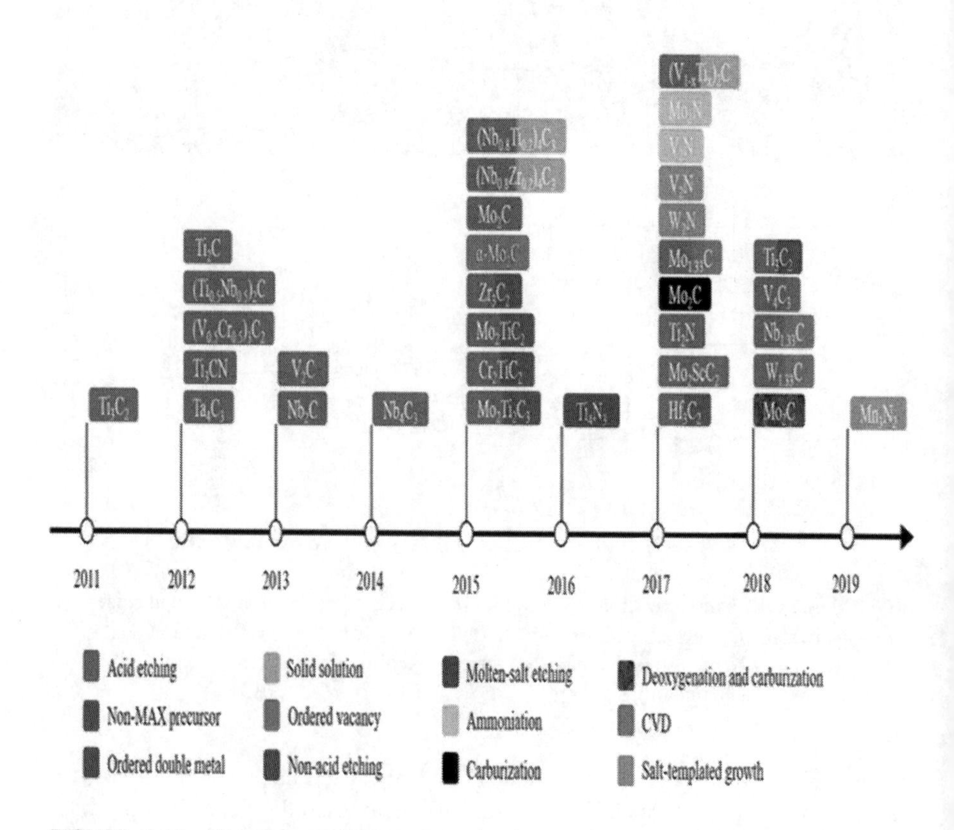

FIGURE 5.15 Timeline of MXene investigation progress. Reprinted with permission Physics Reports, Copyright, 2020, Elsevier (Jiang et al. 2020).

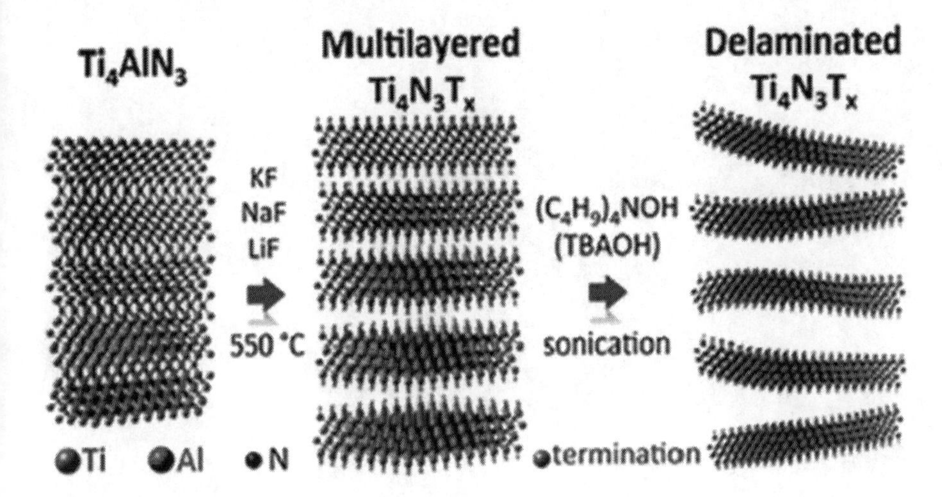

FIGURE 5.16 The synthesis of $Ti_4N_3T_x$ involves molten salt treatment of Ti_4AlN_3. Reprinted with permission, Chemical Engineering Journal. Copyright, 2021, Elsevier (Wu et al. 2021).

mixture to achieve selective etching of Al and layer intercalation. When exposed to the HF-HCl mixture and KF, MXenes, on the other hand, lose their layered morphology (Wu et al. 2021).

2D titanium carbide has many peculiar properties, such as large conductivity, thermal stability, and other properties; due to these properties, MXene is used in a variety of applications. Murugan et al. (2021) elaborated on the development of mixed-phase 2D MXene used as a catalytic material for the simultaneous detection of biomolecules, as shown in Fig. 5.17. The T_iCT_x mixed-phase MXene was synthesized and characterized using XRD, Raman spectroscopy, HR-SEM, and HR-TEM.

The selective etching synthesis approaches in Fig. 5.18(a) show MAX/non-MAX wet acid etching with HCl and HCl+LiF/KF. Fig. 5.18(b) shows an acid-free anodic etching process for MXene preparation in an electrolyte in water. Fig. 5.18(c) shows the molten-salt etching method for Ti_4N_3Tx MXene preparation at 550°C under Ar. Fig 5.18(d) shows high-temperature ammoniation of Mo_2C and V_2C to produce their transition metal nitrides Mo_2C and V_2N. Fig. 5.18(e) shows the thermal annealing of MoS_2 to Mo_2C MXene in the presence of CH_4 and H_2. Fig. 5.18(f) shows the M_oC to Mo_2C MXene deoxygenation and carburization process. Constructions built from the ground up are shown in Fig 5.18(g). Growth of a large-area Mo_2C thin layer on a Cu substrate by chemical vapor deposition is shown in Fig 5.18(h). Mn_3N_2 MXene synthesis uses a salt-templated growth process with an ammonia flow. Due to the simplicity, low cost, and high yield of this process, it's become the primary method for fabricating MXenes (Fig. 5.18).

Fig. 5.19 shows the process of obtaining the 2D MXene by the interleaving of the A atoms and also it has been shown how the 2D MXene sheets are obtained by the etching of A layers and HF treatment (Sinha, Dhanjai, et al. 2018).

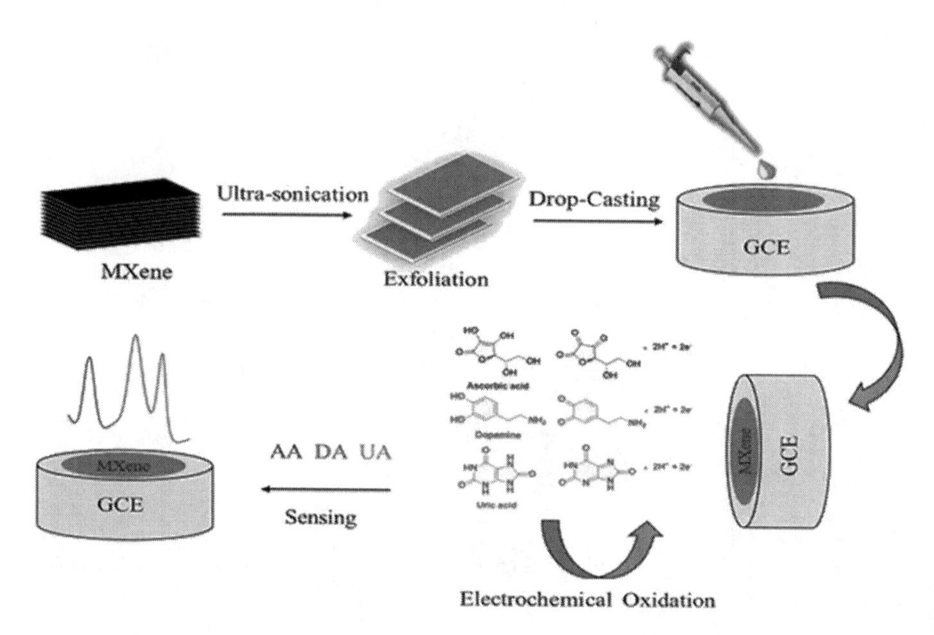

FIGURE 5.17 Different synthesis methods of MXene for the detection of molecules. Reprinted with permission, Journal of Material Science & Technology. Copyright, 2021, Elsevier (Murugan et al. 2021).

5.2.1 Materials and Methods

In this section is a summary of the MAX phrases that are used, as well as experimental information on techniques for synthesis and characterization of MXenes, preparation of electrodes, assembly of information on LIBs, and electrochemical testing.

5.2.1.1 Synthesis of MAX Phases

Powders with distinct properties were mixed, utilizing zirconia balls in plastic vessels for 12–18 h. Powders were combined and then put in alumina crucibles and heated to soaking temperature in a tube furnace using constant argon flow. The furnace was maintained at the soak T (°C) in Ar movement for the soaking time. Materials were then cooled to room temperature, or RT, in the furnace. The result was a powder compact that was crushed with titanium nitride-coated milling bits to remove powders for further testing. The compacts Ti_3AlCN, Ti_2SnC, Hf_2SnC, and Ti_3SnC_2 were ground to a powder with a pestle and mortar. The resulting powder was dissolved using a 400-mesh filter to a particle size of 35 m in both cases.

Titanium, Ti, and graphite, C, powders were combined for 18 h in a ball mill for the former, and after that, the metal indium, In, was applied. Shots were added to the mixture until it reached the desired temperature and was evenly distributed with a spatula. Ta_4AlC_3 was cold-pressed at a pressure of 500 MPa into cylindrical discs with a diameter of 25 mm and a height of up to 10 mm, rather than heated directly in the furnace. The discs were put together in a tube furnace. The remainder of the treatment followed a similar pattern to that previously mentioned.

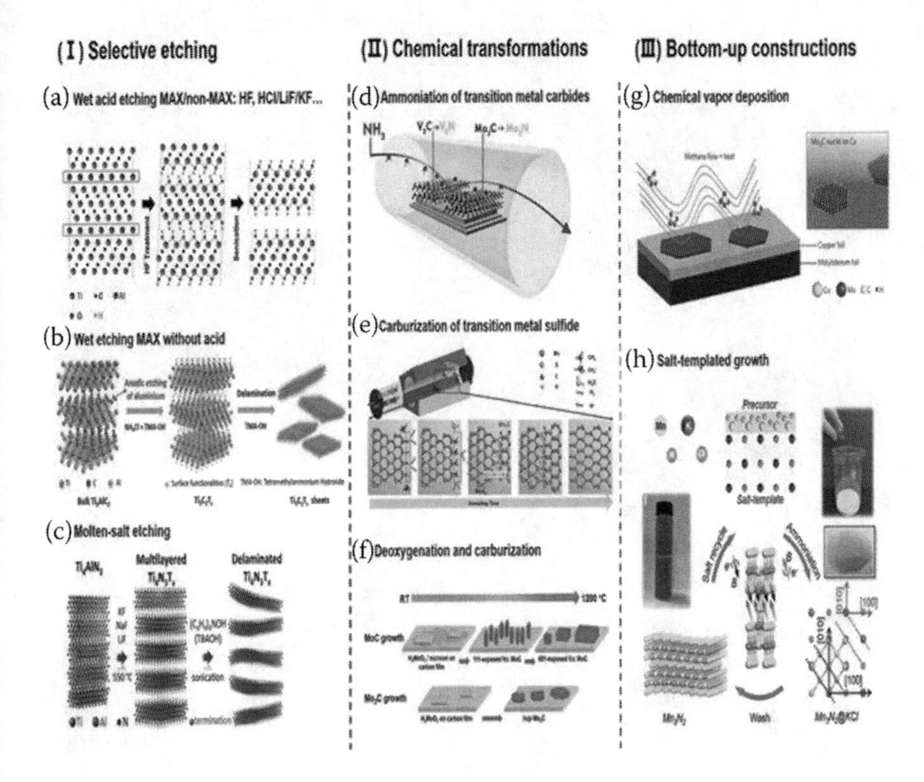

FIGURE 5.18 Schematic of the MXene synthesis techniques. Reprinted with permission, Physics Report. Copyright, 2020, Elsevier (Jiang et al. 2020).

5.2.1.2 Chemical Treatments of Phases of MAX

In order to etch the A-group layers, various MAX phases were subjected to a variety of chemical treatments. They can be divided into three categories: molten salts, non-aqueous fluorination, and wet chemical treatments.

5.2.1.2.a Molten Salt Treatments

In the experiment, powder and bulk Ti_2AlC samples were used. A mixture of Ti, Al_4C_3, and graphite powders were blended and warm-pressed under vacuum at 1600°C for 4 h and the bulk sample was subjected to a load equal to a stress of 40 MPa. As a result, the sample was dense and almost entirely a single phase. In platinum, Pt, crucible, 10 g Ti_2AlC powder was mixed with 30 g LiF and heated to 900°C for 2 h in the air. XRD Siemens D500 diffractometer was used to test the resulting powder after it had cooled. To achieve the desired orientation, 1-GPa powders were cold-pressed into 300 m thin and 2.5 cm diameter discs. Bulk samples were also immersed in molten LiF for 2 h at 900°C. Until cooling, the high-volume samples were washed by carefully grinding off the solidified salt, and their surfaces were examined using x-ray diffraction. The high-volume samples were also cross-sectioned, assembled, crushed, polished, and examined under an optical microscope to a thickness of 1 m. Using potassium fluoride, KF, and powder, the

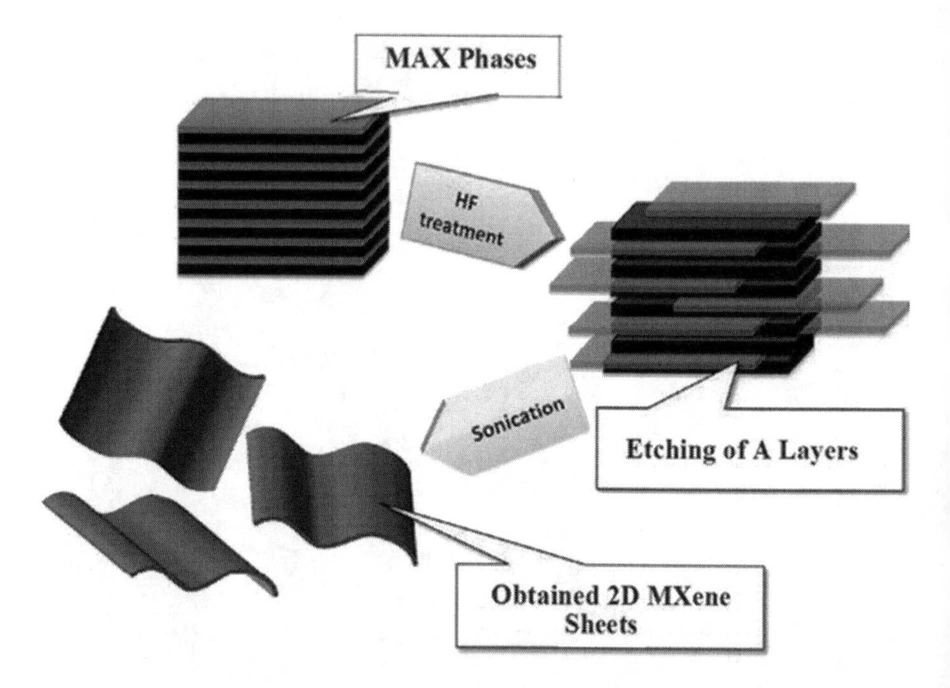

FIGURE 5.19 The systematic preparation of graphite, such as 2D MXenes. Reprint with permission. TrAC Trends in Analytical Chemistry. Copyright 2018. Elsevier (Murugan et al. 2021).

same procedures were carried out on bulk and powdered Ti_2AlC at the same temperature. Molten sodium hexa-fluoroaluminate, Na_3AlF_6, was used to treat bulk samples of Ti_3SiC_2 by immersing them in the latter for 2 or 4 h in air at 1000°C. In a Pt crucible, the heating took place. After cooling to room temperature, the samples were cleaned and prepared for characterization using the same technique as in the LiF event.

5.2.1.2.b Fluorination Treatments

At different temperatures, liquid anhydrous hydrogen fluoride (A-HF) was used to treat Ti_2AlC, Ti_2SC, Ti_3AlC_2, and Ti_3SiC_2 as the phases of MAX. Another collection of samples was treated at 200°C with 5% of overall fluorine, F2, gas (the rest is nitrogen), under this condition the Ti_3AlC_2 powders were synthesized. Each event utilized 325 mesh powders. Ten grams of each MAX phase powder were located in a reactor prior to being immersed in a liquid A-HF at the molar ratio specified for the A-HF treatment. Reactors were preheated to the times and temperatures before initiating the reaction. The emergence temperatures were determined by steadily increasing the temperature above room temperature while monitoring the evolution of the gas.

5.2.1.2.c Wet Chemical Treatments

In some situations, reagents could be used at their original concentrations and in others, reagents are diluted with deionized (DI) water. Whenever a list of as-received concentrations was provided, the aggregate of those samples was rounded down an integer and taken as the concentration of the compound. For example, when it comes to HF, a concentration of 50% by weight was used to define the as-obtained concentration. Additionally, some MAX phases are handled with a 1:3 molar concentration mixture of HNO_3 and HCl (aqua regia).

For the wet chemical treatments, phase of MAX was immersed in a reaction vessel. The latter was a plastic container with a hole in the lid for venting gas-containing objects. To prevent reaction product overflow, the reaction vessel also wasn't filled to more than ⅓ of the capacity. During the experiments, the reaction vessels were placed in another container in compliance with safety regulations. Some of the reactions were particularly violent due to their high exothermic nature. As a result, a small quantity with a phase of MAX powder (0.1 g) was applied to the solution when determining the aggressiveness of a new reaction. To avoid unnecessary heat accumulation, tiny amounts of powder are added slowly at intervals greater than 1 min. By weight, the whole phase of the MAX reagents ratio was 1:10. After adding the phase of MAX powder to the reaction vessel, besides, a magnetic stir bar was attained and the vessel was closed and located on the stirrer for a predetermined time, at which point the contents of the reaction vessel were moved to plastic centrifuging bottles and centrifuged for 1 h at 3,500 rpm using a centrifuging machine. The supernatant was then decanted from the vials.

The vials were then vigorously shaken to re-suspend the settled sediment, and the centrifuging and decanting procedures were repeated until the pH of the decanted liquid was less than five. After this, the wet sediment was immersed in ethanol and dried in room temperature air. Weight-loss percentage is determined by:

$$\text{Wt. loss } \% = \frac{\text{wt. before any treatment} - \text{wt. after the treatment}}{\textit{wt. before any treatment}} \times 100$$

5.3 PRACTICAL APPLICATIONS IN SPR SENSORS

MXenes are conductive and stable in aqueous environments, a rare mixture with enormous potential in a diverse range of applications, including energy storage, sensors, and catalysts, biosensor MXene has huge applications in many areas. As shown in Fig. 5.20, it has been used mainly for electrochemical biosensors, gas sensors, wearable sensors, luminescent sensors, and many more areas.

MXene can also be used as a patch antenna array with integrated feeding circuits on a conformal surface that performs similarly to a copper antenna array at 28 GHz, a practical 5G target frequency. MXene antennas are a promising candidate for integrated RF components in a variety of modular electronic devices due to their flexibility, scalability, and ease of solution processing. It has huge applications like lithium-ion batteries, brain electrodes, electrochromic devices, biomedical, etc.

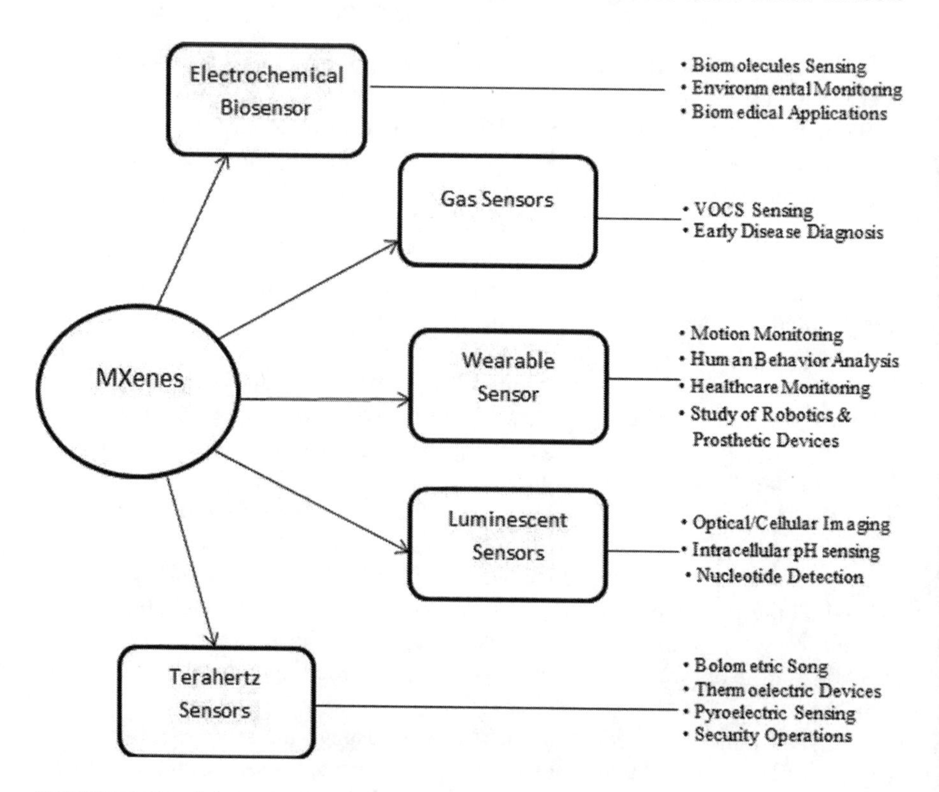

FIGURE 5.20 Schematic diagram for application of the MXenes.

After extensively testing its thermal stability in various atmospheres, vacuum calcination significantly improved the electrochemical properties of MXene Ti_3C_2 multilayer for Li-ion batteries (Kong et al. 2018) and biomedicine (Lin, Chen, and Shi 2018). The 2D MXene substance inks and their printing applications are a gateway to revolutionizing the health industry by preventing diseases and treating pain in time by continued non-invasive and real-time health surveillance in the health sector, as cost-effective printed smart devices (Sreenilayam, Ahad, Nicolosi, and Brabazon 2021). Lorencova, Sadasivuni, Kasak, and Tkac (2020) investigated that the MXene-based biosensor is able to detect cancer biomarkers mixed with blood; also, it is useful in gas sensing, pressure sensors, and in the biomedical field for biological imaging and antibacterial agents.

Similarly, it has also proposed that a MXene-based SPR biosensor is very effective in detecting the carcinoembryonic antigen (CEA). The process to detect the CEA is shown in Fig. 5.21; the materials used for the proposed biosensor are bovine serum albumin (BSA), hydrogen tetrachloroauratehydrate ($HAuCl_4$), sodium borohydride ($NaBH_4$). It has been investigated that the detection of the cancer biomarker with a Ti_3C_2-MXene-based SPR sensor has a significant advantage (Wu et al. 2019).

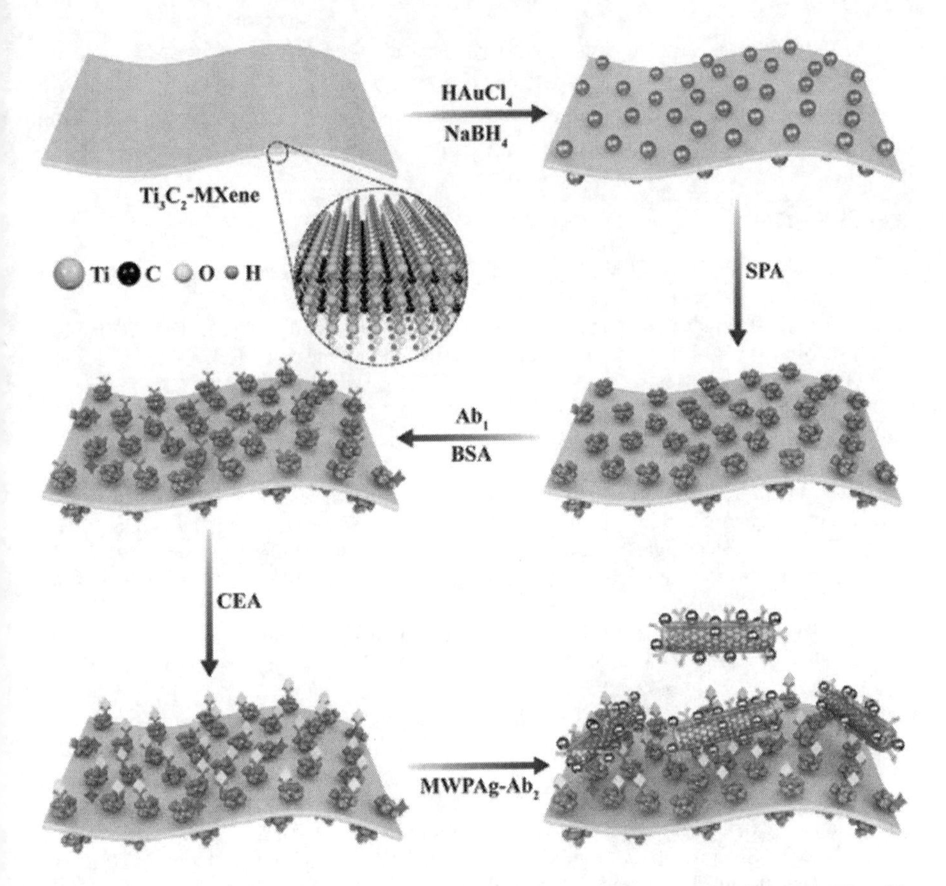

FIGURE 5.21 Schematics diagram of the process of the detection of the CEA. Reprint with permission. Biosensors and Bioelectronics. Copyright 2019. Elsevier (Wu et al. 2019).

5.4 SUMMARY

The term "MXenes" was coined to refer to this novel class of 2D materials to highlight their 2D structure and graphene-like properties. M_2AX, M_3AX_2, M_4AX_3, and solid M-site solutions have all been synthesized as MXenes from a variety of sub-MAX phase groups. Electrically conductive, hydrophilic, and highly stable in water, MXenes have demonstrated enormous potential in a wide range of applications. MXene is known as one of the most diligent materials in a wide variety of applications due to its superior optical, electrical, thermal, and mechanical properties. Numerous MXene-based NMs with extraordinary properties have been proposed, prepared, and used as a catalyst due to their 2D structure, extremely large surface area, ease of decoration, and high adsorption efficiency. $Ti_3C_2T_x$ dispersions of water with several layers and single layers can be regarded as two distinct colloidal structures. At various concentration regimes, they show markedly different viscous and viscoelastic properties. Additionally, 2D MXene flakes have been applied in a variety of applications, most notably sensors and microscale energy storage systems such as micro-supercapacitors. MXenes are 2D structures that exhibit properties similar to those of graphene. They are conductive, hydrophilic, and have high water stability. They have application potential in a variety of fields, ranging from energy storage to sensing.

REFERENCES

Anasori, Babak, Maria R. Lukatskaya, and Yury Gogotsi. 2017. "2D Metal Carbides and Nitrides (MXenes) for Energy Storage." *Nature Reviews Materials* 2 (2): 1–17.

Chaudhuri, Krishnakali, Amr Shaltout, Deesha Shah, Urcan Guler, Aveek Dutta, Vladimir M. Shalaev, and Alexandra Boltasseva. 2018. "Photonic Spin Hall Effect in Robust Phase Gradient Metasurfaces Utilizing Transition Metal Nitrides." *ACS Photonics* 6 (1): 99–106.

Coleman, Jonathan N, Mustafa Lotya, Arlene O'Neill, Shane D Bergin, Paul J King, Umar Khan, Karen Young, et al., 2011. "Two-Dimensional Nanosheets Produced by Liquid Exfoliation of Layered Materials." *Science* 331 (6017): 568–71.

Dai, Xiaoyu, Chunmei Song, Chunyan Qiu, Leiming Wu, and Yuanjiang Xiang. 2019. "Theoretical Investigation of Multilayer $Ti_3C_2T_x$ MXene as the Plasmonic Material for Surface Plasmon Resonance Sensors in Near Infrared Region." *IEEE Sensors Journal* 19 (24): 11834–8.

El-Demellawi, Jehad K., Sergei Lopatin, Jun Yin, Omar F. Mohammed, and Husam N. Alshareef. 2018. "Tunable Multipolar Surface Plasmons in 2D $Ti_3C_2T_x$ MXene Flakes." *ACS Nano* 12 (8): 8485–93.

Fang, Yanfeng, Xuecheng Yang, Tao Chen, Gengfang Xu, Mengli Liu, Jingquan Liu, and Yuanhong Xu. 2018. "Two-Dimensional Titanium Carbide (MXene)-Based Solid-State Electrochemiluminescent Sensor for Label-Free Single-Nucleotide Mismatch Discrimination in Human Urine." *Sensors and Actuators B: Chemical* 263: 400–7.

Frey, Nathan C., Jin Wang, Gabriel Iván Vega Bellido, Babak Anasori, Yury Gogotsi, and Vivek B. Shenoy. 2019. "Prediction of Synthesis of 2D Metal Carbides and Nitrides (MXenes) and Their Precursors with Positive and Unlabeled Machine Learning." *ACS Nano* 13 (3): 3031–41.

Han, Meikang, Yuqiao Liu, Roman Rakhmanov, Christopher Israel, Md Abu Saleh Tajin, Gary Friedman, Vladimir Volman, Ahmad Hoorfar, Kapil R Dandekar, and Yury Gogotsi. 2021. "Solution-Processed $Ti_3C_2T_x$ MXene Antennas for Radio-Frequency Communication." *Advanced Materials* 33 (1): 2003225.

Hoffman, Elizabeth N., Gleb Yushin, Tamer El-Raghy, Yury Gogotsi, and Michel W. Barsoum. 2008. "Micro and Mesoporosity of Carbon Derived from Ternary and Binary Metal Carbides." *Microporous and Mesoporous Materials* 112 (1–3): 526–32.

Homola, Jiví, Sinclair S. Yee, and Günter Gauglitz. 1999. "Surface Plasmon Resonance Sensors." *Sensors and Actuators B: Chemical* 54 (1–2): 3–15.

Jiang, Xiantao, Artem V. Kuklin, Alexander Baev, Yanqi Ge, Hans Ågren, Han Zhang, and Paras N Prasad. 2020. "Two-Dimensional MXenes: From Morphological to Optical, Electric, and Magnetic Properties and Applications." *Physics Reports* 848: 1–58.

Kong, Fanyu, Xiaodong He, Qianqian Liu, Xinxin Qi, Yongting Zheng, Rongguo Wang, Yuelei Bai. 2018. "Improving the Electrochemical Properties of MXene Ti_3C_2 Multilayer for Li-Ion Batteries by Vacuum Calcination." *Electrochimica Acta* 265: 140–50.

Kumar, Rajeev, Sarika Pal, Y. K. Prajapati, and J. P. Saini. 2021. "Sensitivity Enhancement of MXene Based SPR Sensor Using Silicon: Theoretical Analysis." *Silicon* 13:1887–94.

Kumar, Rajeev, Sarika Pal, Alka Verma, Y. K. Prajapati, and J. P. Saini. 2020. "Effect of Silicon on Sensitivity of SPR Biosensor Using Hybrid Nanostructure of Black Phosphorus and MXene." *Superlattices and Microstructures* 145: 106591.

Lawal, Abdulazeez T. 2018. "Progress in Utilisation of Graphene for Electrochemical Biosensors." *Biosensors and Bioelectronics* 106: 149–78.

Lee, Eunji, Armin Vahid Mohammadi, Barton C. Prorok, Young Soo Yoon, Majid Beidaghi, and Dong-Joo Kim. 2017. "Room Temperature Gas Sensing of Two-Dimensional Titanium Carbide (MXene)." *ACS Applied Materials & Interfaces* 9 (42): 37184–90.

Lin, Han, Yu Chen, and Jianlin Shi. 2018. "Insights into 2D MXenes for Versatile Biomedical Applications: Current Advances and Challenges Ahead." *Advanced Science* 5 (10): 1800518.

Lioi, David B., Gregory Neher, James E. Heckler, Tyson Back, Faisal Mehmood, Dhriti Nepal, Ruth Pachter, Richard Vaia, and W. Joshua Kennedy. 2019. "Electron-Withdrawing Effect of Native Terminal Groups on the Lattice Structure of $Ti_3C_2T_x$ MXenes Studied by Resonance Raman Scattering: Implications for Embedding MXenes in Electronic Composites." *ACS Applied Nano Materials* 2 (10): 6087–91.

Liu, Hui, Congyue Duan, Chenhui Yang, Wanqiu Shen, Fen Wang, and Zhenfeng Zhu. 2015. "A Novel Nitrite Biosensor Based on the Direct Electrochemistry of Hemoglobin Immobilized on MXene-Ti_3C_2." *Sensors and Actuators B: Chemical* 218: 60–6.

Lorencova, Lenka, Tomas Bertok, Erika Dosekova, Alena Holazova, Darina Paprckova, Alica Vikartovska, Vlasta Sasinkova, et al. 2017. "Electrochemical Performance of Ti3C2Tx MXene in Aqueous Media: Towards Ultrasensitive H_2O_2 Sensing." *Electrochimica Acta* 235: 471–9.

Lorencova, Lenka, Kishor Kumar Sadasivuni, Peter Kasak, and Jan Tkac. 2020. "Ti 3 C 2 MXene-Based Nanobiosensors for Detection of Cancer Biomarkers." In *Novel Nanomaterials*. London, United Kingdom: IntechOpen.

Luo, Jianmin, Xinyong Tao, Jun Zhang, Yang Xia, Hui Huang, Liyuan Zhang, Yongping Gan, Chu Liang, and Wenkui Zhang. 2016. "Sn4+ Ion Decorated Highly Conductive Ti_3C_2 MXene: Promising Lithium-Ion Anodes with Enhanced Volumetric Capacity and Cyclic Performance." *ACS Nano* 10 (2): 2491–9.

Ma, Yanan, Nishuang Liu, Luying Li, Xiaokang Hu, Zhengguang Zou, Jianbo Wang, Shijun Luo, and Yihua Gao. 2017. "A Highly Flexible and Sensitive Piezoresistive Sensor Based on MXene with Greatly Changed Interlayer Distances." *Nature Communications* 8 (1): 1–8.

Mauchamp, Vincent, Matthieu Bugnet, Edson P. Bellido, Gianluigi A. Botton, Philippe Moreau, Damien Magne, Michael Naguib, Thierry Cabioc'h, and Michel W. Barsoum. 2014. "Enhanced and Tunable Surface Plasmons in Two-Dimensional Ti_3C_2 Stacks: Electronic Structure versus Boundary Effects." *Physical Review B* 89 (23): 235428.

Maurya, J. B., Y. K. Prajapati, S. Raikwar, and J. P. Saini. 2018. "A Silicon-Black Phosphorous Based Surface Plasmon Resonance Sensor for the Detection of NO_2 Gas." *Optik* 160: 428–33.

Murugan, Nagaraj, Rajendran Jerome, Murugan Preethika, Anandhakumar Sundaramurthy, and Ashok K. Sundramoorthy. 2021. "2D-Titanium Carbide (MXene) Based Selective Electrochemical Sensor for Simultaneous Detection of Ascorbic Acid, Dopamine and Uric Acid." *Journal of Materials Science & Technology* 72: 122–31.

Novoselov, Kostya S., Andre K. Geim, Sergei V. Morozov, Dingde Jiang, Yanshui Zhang, Sergey V. Dubonos, Irina V. Grigorieva, and Alexandr A. Firsov. 2004. "Electric Field Effect in Atomically Thin Carbon Films." *Science* 306 (5696): 666–9.

Pal, Sarika, Alka Verma, Y. K. Prajapati, and J. P. Saini. 2017. "Influence of Black Phosphorous on Performance of Surface Plasmon Resonance Biosensor." *Optical and Quantum Electronics* 49 (12): 403.

Palik, Edward D. 1998. *Handbook of Optical Constants of Solids*. Vol. 3. United States: Academic Press.

Pan, Jie, Stephan Lany, and Yue Qi. 2017. "Computationally Driven Two-Dimensional Materials Design: What Is Next?" *ACS Nano* 11 (8): 7560–4.

Pandey, Ankit Kumar. 2020. "Plasmonic Sensor Utilizing Ti3C2Tx MXene Layer and Fluoride Glass Substrate for Bio-and Gas-Sensing Applications: Performance Evaluation." *Photonics and Nanostructures-Fundamentals and Applications* 42: 100863.

Rakhi, R. B., Pranati Nayak, Chuan Xia, and Husam N. Alshareef. 2016. "Novel Amperometric Glucose Biosensor Based on MXene Nanocomposite." *Scientific Reports* 6 (1): 1–10.

Sambles, J. R., G. W. Bradbery, and Fuzi Yang. 1991. "Optical Excitation of Surface Plasmons: An Introduction." *Contemporary Physics* 32 (3): 173–83.

Sarycheva, Asia, and Yury Gogotsi. 2020. "Raman Spectroscopy Analysis of the Structure and Surface Chemistry of Ti3C2T x MXene." *Chemistry of Materials* 32 (8): 3480–8.

Sarycheva, Asia, Taron Makaryan, Kathleen Maleski, Elumalai Satheeshkumar, Armen Melikyan, Hayk Minassian, Masahiro Yoshimura, and Yury Gogotsi. 2017. "Two-Dimensional Titanium Carbide (MXene) as Surface-Enhanced Raman Scattering Substrate." *The Journal of Physical Chemistry C* 121 (36): 19983–8.

Satheeshkumar, Elumalai, Taron Makaryan, Armen Melikyan, Hayk Minassian, Yury Gogotsi, and Masahiro Yoshimura. 2016. "One-Step Solution Processing of Ag, Au and Pd@MXene Hybrids for SERS." *Scientific Reports* 6 (1): 1–9.

Sinha, Ankita, Dhanjai, Huimin Zhao, Yujin Huang, Xianbo Lu, Jiping Chen, and Rajeev Jain. 2018. "MXene: An Emerging Material for Sensing and Biosensing." *TrAC—Trends in Analytical Chemistry* 105: 424–35. 10.1016/j.trac.2018.05.021.

Sinha, Ankita, Bing Tan, Yujin Huang, Huimin Zhao, Xueming Dang, Jiping Chen, Rajeev Jain, et al. 2018. "MoS_2 Nanostructures for Electrochemical Sensing of Multidisciplinary Targets: A Review." *TrAC Trends in Analytical Chemistry* 102: 75–90.

Sreenilayam, Sithara P., Inam Ul Ahad, Valeria Nicolosi, and Dermot Brabazon. 2021. "Mxene Materials Based Printed Flexible Devices for Healthcare, Biomedical and Energy Storage Applications." *Materials Today* 43: 99–131.

Srivastava, Akash, Alka Verma, Ritwick Das, and Y. K. Prajapati. 2020. "A Theoretical Approach to Improve the Performance of SPR Biosensor Using MXene and Black Phosphorus." *Optik* 203: 163430.

Sun, Z. M. 2011. "Progress in Research and Development on MAX Phases: A Family of Layered Ternary Compounds." *International Materials Reviews* 56 (3): 143–66.

Szuplewska, Aleksandra, Dominika Kulpińska, Artur Dybko, Agnieszka Maria Chudy Michałand Jastrz kebska, Andrzej Olszyna, and Zbigniew Brzózka. 2020. "Future Applications of MXenes in Biotechnology, Nanomedicine, and Sensors." *Trends in Biotechnology* 38 (3): 264–79.

Wang, Fen, ChenHui Yang, CongYue Duan, Dan Xiao, Yi Tang, and JianFeng Zhu. 2014. "An Organ-like Titanium Carbide Material (MXene) with Multilayer Structure Encapsulating Hemoglobin for a Mediator-Free Biosensor." *Journal of The Electrochemical Society* 162 (1): B16.

Wang, Xuefeng, Xi Shen, Yurui Gao, Zhaoxiang Wang, Richeng Yu, and Liquan Chen. 2015. "Atomic-Scale Recognition of Surface Structure and Intercalation Mechanism of Ti_3C_2X." *Journal of the American Chemical Society* 137 (7): 2715–21.

Wang, Hongbing, Jianfeng Zhang, Yuping Wu, Huajie Huang, Gaiye Li, Xin Zhang, and Zhuyin Wang. 2016. "Surface Modified MXene Ti3C2 Multilayers by Aryl Diazonium Salts Leading to Large-Scale Delamination." *Applied Surface Science* 384: 287–93.

Wang, Zhuoxian, Krishnakali Chaudhuri, Mohamed Alhabeb, Xiangeng Meng, Shaimaa I. Azzam, Alexander Kildishev, Young L. Kim, Vladimir M. Shalaev, Yury Gogotsi, and Alexandra Boltasseva. 2018. "MXenes for Plasmonic and Metamaterial Devices." In *CLEO: QELS_Fundamental Science*, FM2G--7.

Willets, Katherine A., and Richard P. Van Duyne. 2007. "Localized Surface Plasmon Resonance Spectroscopy and Sensing." *Annual Review of Physical Chemistry* 58: 267–97.

Wu, Leiming, Qi You, Youxian Shan, Shuaiwen Gan, Yuting Zhao, Xiaoyu Dai, and Yuanjiang Xiang. 2018. "Few-Layer Ti3C2Tx MXene: A Promising Surface Plasmon Resonance Biosensing Material to Enhance the Sensitivity." *Sensors and Actuators B: Chemical* 277: 210–5.

Wu, Qiong, Ningbo Li, Ying Wang, Yanchao Xu, Shuting Wei, Jiandong Wu, Guangri Jia, et al. 2019. "A 2D Transition Metal Carbide MXene-Based SPR Biosensor for Ultrasensitive Carcinoembryonic Antigen Detection." *Biosensors and Bioelectronics* 144: 111697.

Wu, You, Xiaoming Li, Hui Zhao, Fubing Yao, Jiao Cao, Zhuo Chen, Xiaoding Huang, Dongbo Wang, and Qi Yang. 2021. "Recent Advances in Transition Metal Carbides and Nitrides (MXenes): Characteristics, Environmental Remediation and Challenges." *Chemical Engineering Journal* 418: 129296.

Xu, Bingzhe, Minshen Zhu, Wencong Zhang, Xu Zhen, Zengxia Pei, Qi Xue, Chunyi Zhi, and Peng Shi. 2016. "Ultrathin MXene-Micropattern-Based Field-Effect Transistor for Probing Neural Activity." *Advanced Materials* 28 (17): 3333–9.

Xu, Yi, Chang-Yu Hsieh, Lin Wu, and L. K. Ang. 2018. "Two-Dimensional Transition Metal Dichalcogenides Mediated Long Range Surface Plasmon Resonance Biosensors." *Journal of Physics D: Applied Physics* 52 (6): 65101.

Xu, Yi, Yee Sin Ang, Lin Wu, and Lay Kee Ang. 2019. "High Sensitivity Surface Plasmon Resonance Sensor Based on Two-Dimensional MXene and Transition Metal Dichalcogenide: A Theoretical Study." *Nanomaterials* 9 (2): 165.

Zeng, Shuwen, Dominique Baillargeat, Ho-Pui Ho, and Ken-Tye Yong. 2014. "Nanomaterials Enhanced Surface Plasmon Resonance for Biological and Chemical Sensing Applications." *Chemical Society Reviews* 43 (10): 3426–52.

Zhou, Liya, Xiaoning Zhang, Li Ma, Jing Gao, and Yanjun Jiang. 2017. "Acetylcholinesterase/Chitosan-Transition Metal Carbides Nanocomposites-Based Biosensor for the Organophosphate Pesticides Detection." *Biochemical Engineering Journal* 128: 243–9.

6 Introduction to Carbon Nanotube 2D Layer Assisted by Surface Plasmon Resonance Based Sensor

6.1 INTRODUCTION

Chemical and gas sensors are gaining prominence due to the varied applications of environmental surveillance, industry, space travel, bio-medicine, and pharmacology. Gas sensors are used to measure explosive gases as well as to detect toxic or pathogenic gases in real time, with quite a large selectiveness and precision. People also realized the need for monitoring and regulating the atmosphere, particularly when it relates to global warming. The National Aeronautics and Space Administration (NASA) is researching how to identify atmospheric components on different planets using high-performance gas sensors. Furthermore, the discovery of nerve agents for homeland security is a hot subject (Meyyappan 2004). Commonly, there are many specific requirements for a successful gas-sensing device including sensitivity and selectivity, quick reaction time, low analyst consumption, temperature independence, stability in efficiency, and low operative temperature. Generally used gas-sensor materials include porous-structured materials like porous silicon, semiconductor metal oxides, and vapor-sensitive polymers (Fenner and Zdankiewicz 2001; Rittersma 2002; Traversa 1995). While adsorption and desorption of gas molecules explains much about gas-sensing theory on the surface, by the increment in the number of contact interfaces, the sensitivity can be improved.

The advancement of nanotechnology has contributed to the portable sensor production with high sensitivity, low power consumption, and compact scale. The hollow structure and essentially high ratio of surface-to-volume nanomaterials (NMs) help them easily absorb and store gases. Therefore, nanomaterials like carbon-nanotubes, i.e., CNTs, nanowires, or nanoparticles (NPs), have been studied to build gas sensors. These dimensional (D) NMs can be classified into 1D materials made of nanowires and nanotubes, 2D materials that include graphene and nanoclays, and 3D NPs, for example, cubical NPs and spherical NPs (Bhattacharya 2016; Rauti et al. 2019). The range of potential carbon hybridization (sp^1, sp^2, and sp^3) yields amorphous carbon, graphite, and diamond. Carbon nanotubes (CNTs), first proposed by Ichiro Iijima in 1991 (Iijima 1991), received the most scientific attention due to

DOI: 10.1201/9781003190738-6

their great geometry, properties, and structure. Their physical and chemical properties are researched thoroughly (like thermal, electronic, mechanical, and optical properties), as well as their applications in different fields. Simulation, as well as theoretical work, have also been performed to acknowledge the materials belonging to nano-scaled and associated phenomenon (Meyyappan 2004).

CNTs belong to the fullerene nanostructure family (Aqel, Abou El-Nour, Ammar, and Al-Warthan 2012). Fullerenes are a category of allotropes of carbon, shown in Fig. 6.1, mostly made of carbon molecules, are a hollow-shaped sphere, tube-shaped, or an ellipsoidal. CNTs, also known as buckytubes, are cylindrical fullerenes. Fullerenes have a graphite-like composition, consisting of a layer of hexagonal connected rings, except that they conceive of pentagonal (or heptagonal) rings that prevent planarization. A single-layer graphene sheet, as shown in Fig. 6.2, can be rolled in several directions to produce various forms of CNTs. Because of its symmetric composition, a normal CNT does have a hexagonal

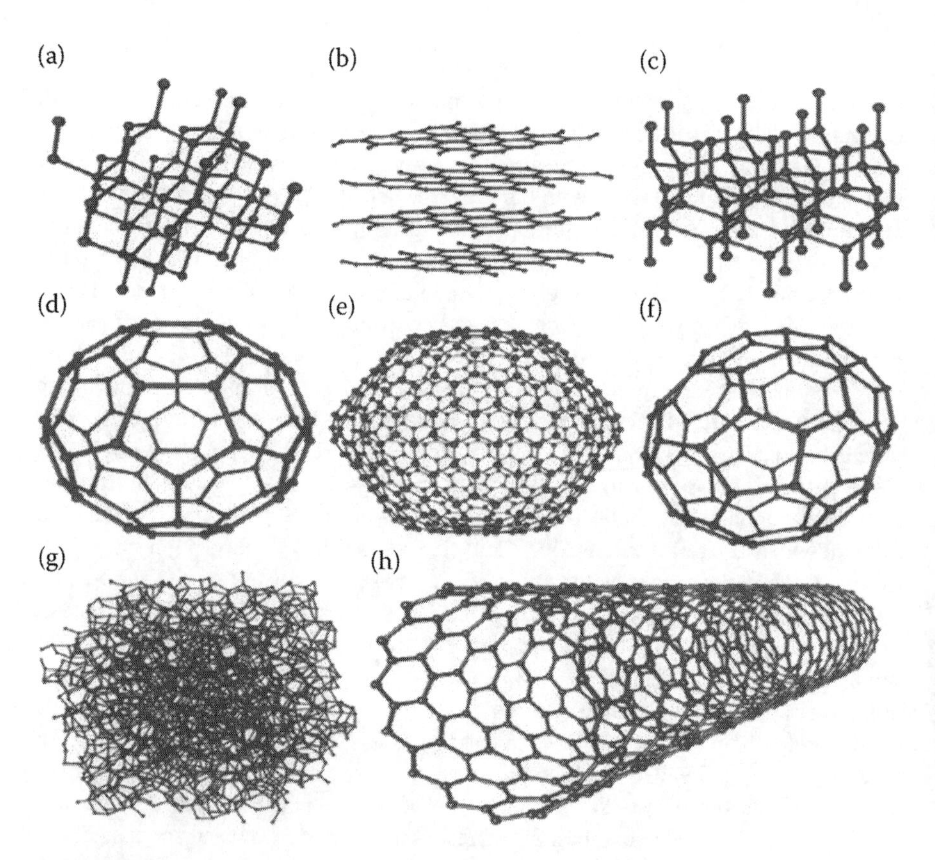

FIGURE 6.1 The eight carbon allotropic structures: a) diamond, b) graphite (single of graphite), c) Lonsdaleite, d) C60 (Bukyball or Buckminsterfullerene), e) C540 Fullerene, f) C70 Fullerene, g) Amorphous carbon, and h) SWCNT. Reprinted with permission from Arabian Journal of Chemistry. Copyright, 2012, Elsevier (Aqel, Abou El-Nour, Ammar, and Al-Warthan 2012).

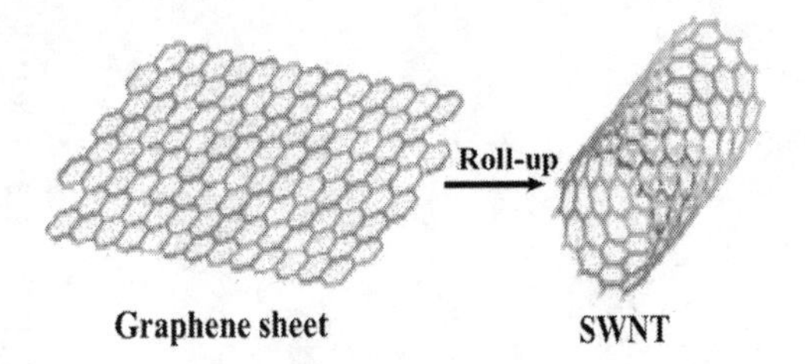

FIGURE 6.2 A single layer of graphite sheet is rolled onto single-walled CNTs. Reprinted with permission from Arabian Journal of Chemistry. Copyright, 2012, Elsevier (Aqel, Abou El-Nour, Ammar, and Al-Warthan 2012).

array of carbon atoms in a tube shape and has extraordinary properties. Their activity depends solely on the helix's nature, and therefore they behave as a metal or as a semiconductor.

The detailed classification of CNTs are grouped into three categories according to their synthesis techniques: (a) *Single-walled CNTs* (SWCNTs): This can be thought of as a single graphene layer (SWCNTs are a one-atom-thick film of rolled-up graphite with a diameter between 1 and 100 nm and a thickness from 1 to 100 μm). It is an interesting kind of CNT because they have important electric properties that the multi-walled CNT forms don't have. Electric wire is the most essential component of these devices, and this can be excellent conductor. Production of SWCNTs is still extremely costly, and the advancement of more cost-effective synthesis techniques is essential for the future of carbon nanotechnology. (b) *Double-walled CNTs* (DWCNTs): Coaxial nano-arms made of two SWCNTs with one encapsulated on another. It is a special case of MWCNTs that must be highlighted because they have very similar properties and morphology to SWCNTs. (c) *Multi-walled CNTs* (MWCNTs): MWCNTs wrap several layers of graphite around a structure to create a cylinder with the same central axis. This can be regarded as a layer of nanotubes within another layer of nanotubes (Iijima 1991; Kim et al. 2015; Mubarak, Abdullah, Jayakumar, and Sahu 2014). The interlayer gap between MWCNT layers is about 3.3 Å, as in the case of graphite. All of these types of CNT structures are highlighted in Fig. 6.3.

Literature (Mubarak, Abdullah, Jayakumar, and Sahu 2014) indicates that the SWCNT path seems to have taken place two years after the MWCNTs and are generally divided into three types, such as a zigzag, armchair, and chiral, depending on the size. CNTs may have metallic or semiconducting electrical properties, which depend on the tube's diameter and chirality (the orientation where the graphite sheet is folded to build the tube). Normally, the chirality is expressed as an integer pair (n, m). The metallic nanotubes have $n-m = 3j$ (j being a non-zero integer), while others are semiconducting. Based on the values of n or m, its formation varies, as shown in Fig. 6.4. These configurations rely primarily on the precise roll-up of the graphene board. A vector diagram of how to roll a hexagonal graphite sheet into a CNT is depicted in Fig. 6.5.

Zhang and Li (Koziol, Boskovic, and Yahya 2010) have documented the various forms and tubular morphologies of CNTs, as shown in Fig. 6.6. In addition, they

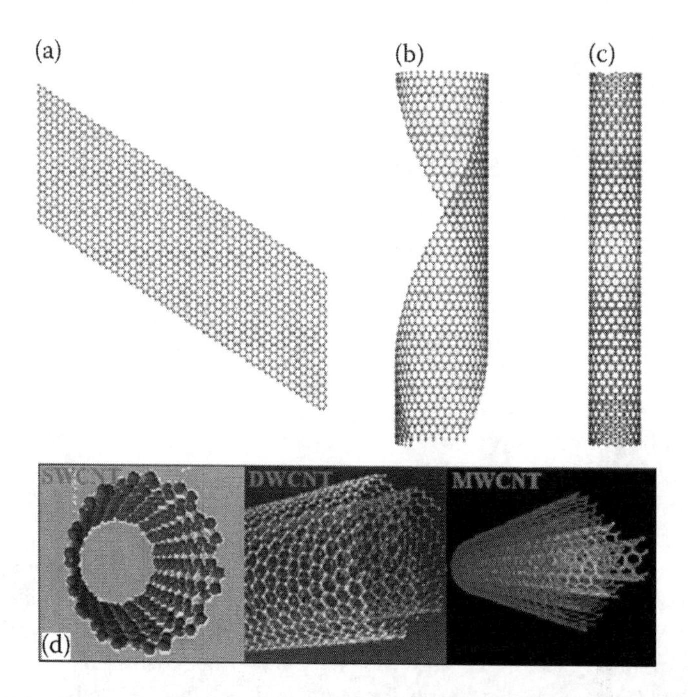

FIGURE 6.3 Schematic pictures illustrating the various forms of CNT as well as other carbon products: a) A graphite flat sheet, (b) a partially rolled graphite sheet, c) single-walled CNT, and d) structures of three different kinds of CNTs: SWCNTs, DWCNTs, and MWCNTs, respectively. Reprinted with permission from Arabian Journal of Chemistry. Copyright, 2012, Elsevier (Aqel, Abou El-Nour, Ammar, and Al-Warthan 2012).

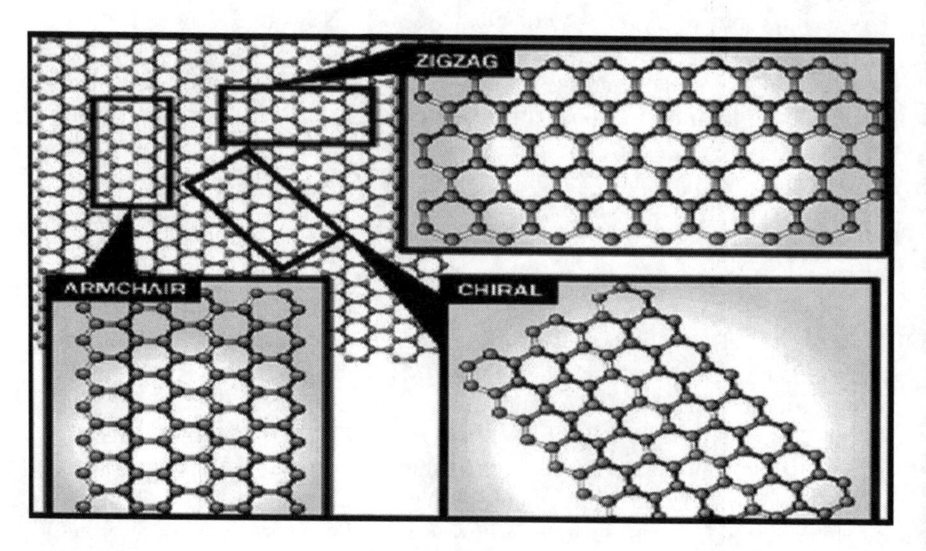

FIGURE 6.4 Schematic graphic represents how the various CNT chirality structures are formed. Reprinted with permission from Arabian Journal of Chemistry. Copyright, 2012, Elsevier (Aqel, Abou El-Nour, Ammar, and Al-Warthan 2012).

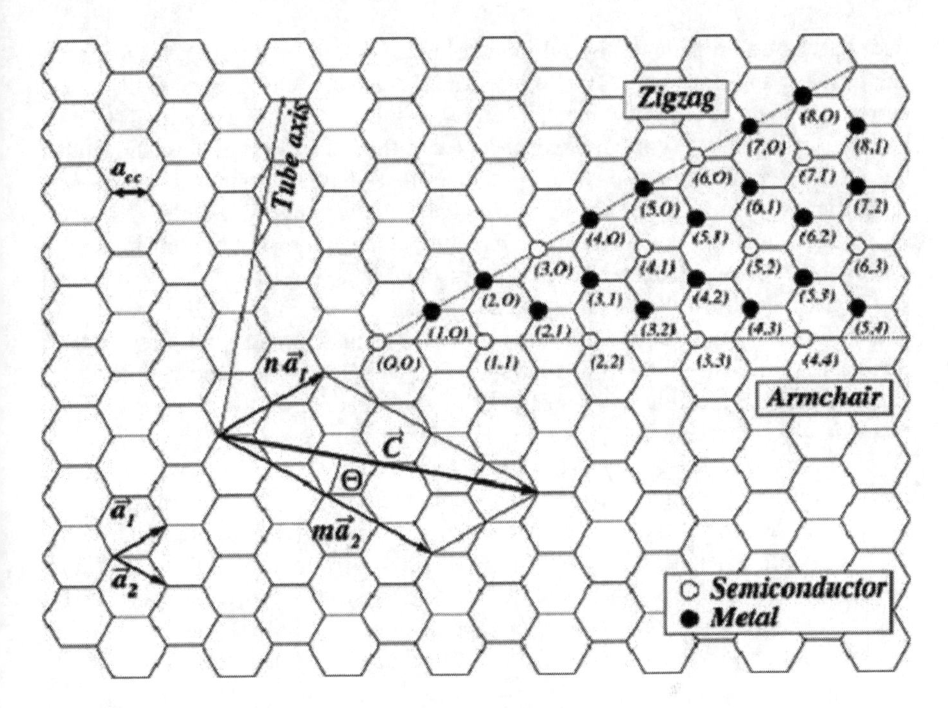

FIGURE 6.5 The diagram of a two-dimensional graphene sheet displaying a vector form description used to classify the structure of CNTs. Reprinted with permission from Arabian Journal of Chemistry. Copyright, 2012, Elsevier (Aqel, Abou El-Nour, Ammar, and Al-Warthan 2012).

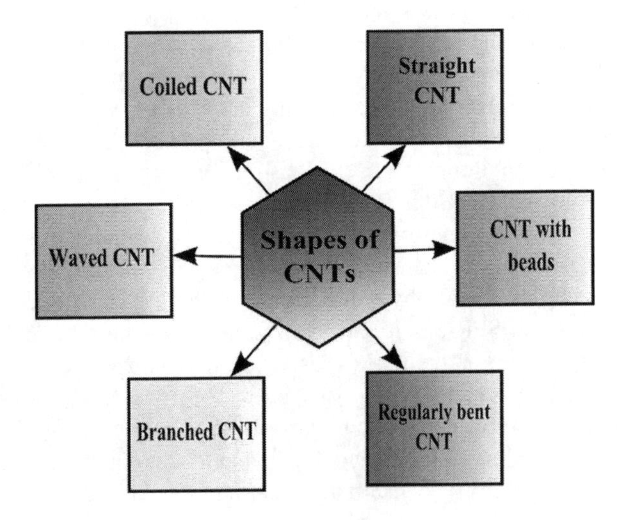

FIGURE 6.6 Different shapes of carbon nanotubes (CNTs).

find that, based on their forms and morphology, CNTs have unique characteristics and possible applications. CNTs are mechanically the best and most rigid fibers, are currently known for the C–C bond. Thermally, in both vacuum and air, CNTs give high thermal stability. Various experimental and theoretical efforts have shown that the CNTs (both single-walled and multi-walled) have outstanding mechanical properties, such as a higher aspect ratio ($\sim 10^4$), lightweight (~ 2 g cm^{-3}) (Xie et al. 2000), high electrical conductivity, high melting temperature, outstanding durability (excellent) of many biochemical processes (de Menezes et al. 2019; Radhamani 2018), greater corrosion resistance (Samuel Ratna Kumar, Smart, and Alexis 2017), and different optical properties (Samuel Ratna Kumar, Smart, and Alexis 2017; Shah and Tali 2016; Wang et al. 2018). CNTs have remarkable electrical properties and can be both metallic and semiconducting, depending on their design (i.e., on their diameter and helicity) (Saito, Fujita, Dresselhaus, and Dresselhaus 1992). A lot of excitement is caused by CNTs due to their excellent properties, which include near-infrared triggering, nanoelectronics, bio-material sensing, pharmaceuticals, medical applications, water filtration, and smart materials (Chakoli, He, and Huang 2018; Chowdhury et al. 2018; Eatemadi et al. 2014; Ferreira et al. 2019; Zahid et al. 2018). Significant use of CNT is in the reinforcement of lightweight materials for aircraft, vehicle, sporting, and medical applications. Several measurable, real-life implementations have also seen the positive impact of their use of CNTs and composites.

However, optical gas and chemical sensors depend on the mechanism of substantial changes in the optoelectronic characteristics of the surface required for sensing coated on an optical layer, such as grating, prism, or fiber-optic cable, as a result of disclosure to target molecules of gas in a specific spectrum, which exhibits itself as an observable and readable optical signal. Because the significant monitoring signals are of an optical type, then the risk of a spark and subsequent explosions in the sensing field is significantly reduced, endorsing the optical sensor secure to use inflammable and polluted ambiance. Furthermore, the use of optical sensors provide convenient and smoother adsorption over the substrate and a high level of interaction of sensing output, resulting in a productive measurement of gas species. Optical sensors also allow for fast and reversible detection in both high- and low-temperature environments, as well as maintaining stable performance over a large temperature scale. In this view, optical sensors have huge commercial potential for detecting and monitoring various gases and chemicals. In the last few decades, surface plasmon resonance (SPR) has proven to be extremely helpful to the methods of sensing different physical parameters, biological materials, chemical entities, and gases. An SPR-based fiber-optic sensor was initially deployed with a prism that has a high refractive index (RI) acting as an optical structure using the Kretschmann and Raether design, the substitution of a prism sensor configuration by a fiber optic, has proven enormous compactness, reliability, and efficient arrangements (Abdulhalim, Zourob, and Lakhtakia 2008; Gupta and Kant, 2018; Homola, Yee, and Gauglitz 1999; Sharma, Jha, and Gupta 2007).

The advancement of metals and carbon nanostructure sensors in comparison with other individually integrated structures or, in combination with nanocomponents moves a step forward in sensor-based gas-sensing fiber-optic sensor configurations.

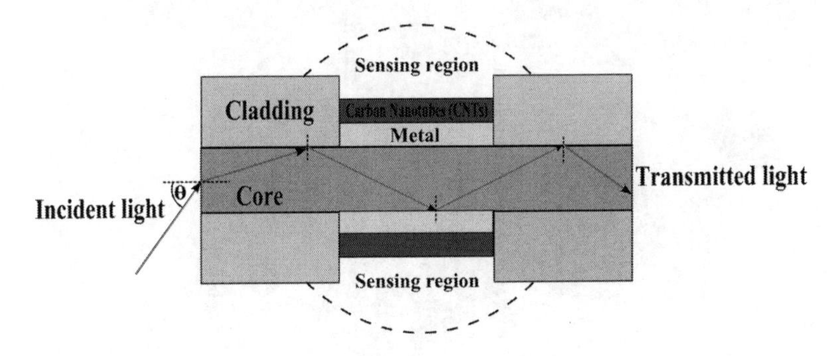

FIGURE 6.7 Schematic diagram of metal coated with CNT-based sensor.

Fig. 6.7 depicts the schematic diagram of a metal coated with a CNT-based sensor. The contact of a coated sensing surface across an unclad fiber core with targeted molecules of gas, which results in a shift in the sensing surface's RI and the SPR spectrum creation, which is the guiding SPR's principle to a fiber-optic gas sensor.

A fiber-optic SPR gas sensor, which is widely used in wavelength interrogation modules, the changes in target concentration of gas in the periphery of the fiber-optic probe's sensing region cause a difference in the resonance wavelength (Hinman, McKeating, and Cheng 2018; Mishra and Mishra 2012; Tabassum, Mishra, and Gupta 2013; Yin et al. 2018). The nanocomposite adapts the properties of all the groups and exploits their individual properties on top of that. This creates enough levels of defect in the sensing surface of a nanocomposite, producing the number of active sites on the sensor surface for efficient gas molecule adsorption and maximizing the interaction of those molecules. This greatly increases the sensitivity of the gas sensor and extends its available area. In addition, it guarantees sensor selectivity by using materials that are highly precise for the gaseous substituents undergoing investigation in the nanocomposite framework. Commonly, nanocomposite structures are generally prepared for fiber-optic gas sensors using a conducting polymer in graphene or metal oxides (Mishra, Kumari, and Gupta 2012). These polymers are chemically effective, thereby providing favorable surfaces for the production of gas-sensing surfaces. Thus, the coupling between CNTs and metal-based NPs is intended to combine the reactivity of CNTs to dangerous gases like H_2, NO_2, and CO with the localized SPR of metal NPs, which is considered to be highly sensitive to the change in the surrounding medium's dielectric properties, an operating method used by sensor applications.

6.2 GROWTH METHOD OF THIN LAYER

This section outlines the development made in the past few decades in the growth of CNTs by different strategies such as plasma torch (Chen et al. 2003; Hong and Uhm 2005), liquid electrolysis (Chen et al. 1998; Zhang, Gamo, Xiao, and Ando 2002), arc-discharge, laser ablation, and chemical vapor deposition (CVD), as shown in Fig. 6.8. Out of these, the utilization of the plasma torch process is more recent compared to the last three procedures. This technique has been in progress for the

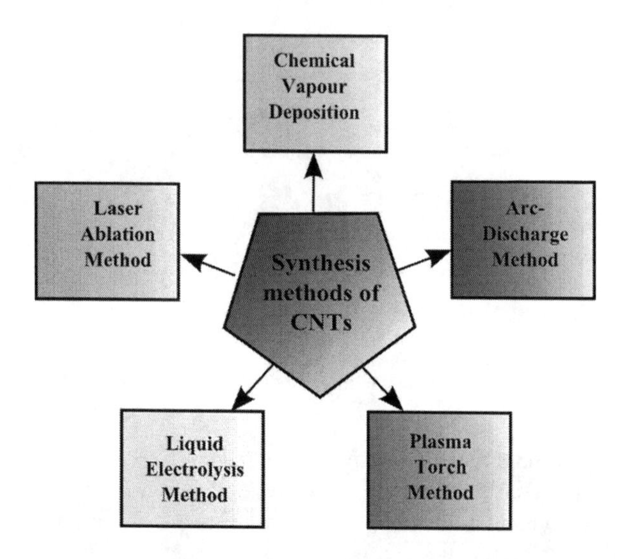

FIGURE 6.8 Schematic graph showing the processes currently employed for the CNT synthesis.

last decade and now can be used with silicon-based substrates to synthesize high-yield nanotubes (Chen et al. 2003; Hong and Uhm 2005; Jašek et al. 2006), whereas the liquid electrolysis technique has long been used for the development of CNTs. Research efforts have focused on this approach to identify the various molten salts utilized for high CNT production (Hsu et al. 1995).

However, the most common and basic methods for the growth techniques utilize mostly the last three processes. In the past 10 years, arc-discharge and laser ablation techniques were both used in the manufacture of nanotubes. Both techniques include carbon condensation from solid carbon source evaporation. The temperatures of these techniques are near the graphite melting point, 3000–4000°C. In *arc-discharge*, the carbon atoms evaporate by helium gas plasma, sparked by high currents passing through opposite carbon anode and cathode. The schematic diagram of arc-discharge setup is shown in Fig. 6.9, which has become an outstanding tool for manufacturing both superior kinds of SWNTs and MWNTs. MWNTs can be generated besides monitoring the growing conditions in the arcing current and discharge chambers, such as the inert gas pressure. In 1992, Ebbes and Ajayan first reported gram-level high-scale growth and purification of MWNTs by using an arc-discharge (Ebbesen and Ajayan 1992). The synthesized MWNTs have 10 microns of length and 5–30 nm of diameter. Also, the van der Waals interactions normally bind the nanotubes together into strong bundles. MWNTs formed by an arc-discharge have high visibility due to their unique structure. As evolved tubes, there are a few imperfections, including pentagons or heptagons on the sidewalls. Multi-layered graphite NPs (in polyhedron shapes) form as a by-product of the arc-discharge growing process. Purification of MWNTs can be accomplished in an oxygen atmosphere by heating the developed product to eliminate the graphene particles (Ebbesen and Ajayan 1992). Even if the oxidation rate of polyhedronitric

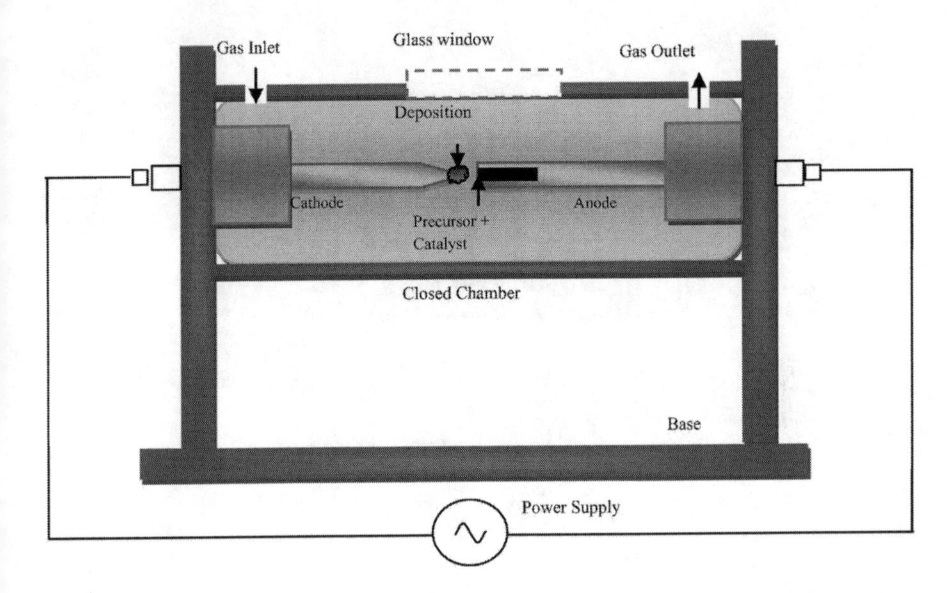

FIGURE 6.9 Schematic of an arc-discharge setup. Reprinted with permission from Diamond and Related Materials Journal. Copyright, 2014, Elsevier (Arora and Sharma 2014).

particles is higher than MWNTs, a small portion of nanotubes is still removed in the process. However, for an increase in SWNTs' growth, an arc-discharge metal catalyst is needed. The experiment performed by Bethune and colleagues became the first to deliver large quantities of SWNTs by arc-discharge in 1993 (Bethune et al. 1993). They experimented using a carbon anode that had a small portion of cobalt catalyst, and SWNTs were present in abundance in the soot product. Smalley and colleagues used a *laser ablation* process to develop high-quality SWNTs (Thess et al. 1996). The reactor used in the SWCNT synthesis is seen schematically in Fig. 6.10. The intense laser pulses were used to ablate a carbon target containing 0.5 at% of cobalt and nickel. The target was set to heat up to 1200°C in a tube furnace. During laser ablation, an inert gas flow was transmitted via the growth chamber to transfer the formed nanotubes horizontally to have been deposited on a cold finger. Journet and colleagues optimized SWNT growth by arc-discharge using a carbon anode treated with 1% of atomic yttrium and 4.2% of nickel as a catalyst (Journet et al. 1997). Fullerenes, graphitic polyhedrons with embedded metallic particles, and particle-shaped amorphous carbon or over-coating on the sidewalls of nanotubes are common by-products of SWNT growth by arc-discharge and laser ablation. Smalley and colleagues (Liu et al. 1998) developed a purification method for SWNT products, which is now commonly used by several researchers. This technique involved immersing SWNTs as developed in a long-term solution of nitric acid, oxidizing amorphous carbon species, and removing any metal catalyst species. The discovery of SWNT material created by arc-discharge and laser ablation is leading to the scientific understanding of the behavior of low-dimensional materials in general.

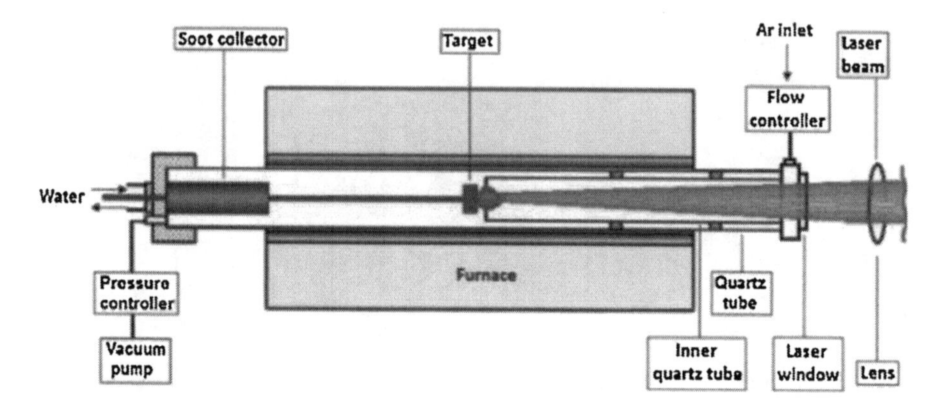

FIGURE 6.10 Reactor synthesis schematic diagram for single-walled CNTs using a laser ablation process. Reprinted with permission from Sensors and Actuators A: Physical Journal. Copyright, 2019, Elsevier (Han, Nag, Mukhopadhyay, and Xu 2019).

Currently, CVD is one of the most extensively investigated and still one of the most widely used CNT processing methods (Kumar and Ando 2010). Compared to two previously discussed processes, CVD synthesis doesn't need sophisticated equipment and less severe conditions, making it easier to produce on a wide scale and better for use in the synthesis of CNTs (Zhang et al. 2011). CVD growth is focused on the hydrocarbon decomposition to carbon atoms, followed by the carbon nanostructures synthesis on a variety of substrates containing catalysts. Size-correlated metal NPs are also used as catalysts, and that indicates the diameter of the synthesized nanotubes (for SWCNTs—0.5–5 nm, for MWCNTs—8–100 nm). The most widely used metals that act as catalysts in NP synthesis are nickel, cobalt, or iron NPs. CVD reactors are normally comprised of a reaction chamber and inert gas-filled tubes as well as hydrocarbon, shown in Fig. 6.11a. Ethylene and acetylene are frequently used for MWCNTs processing, while methane is used for SWCNTs. The process described is carried out in the 850–1000°C range for SWCNTs and 550–700°C for MWCNTs. Carbon is synthesized from hydrocarbons by thermal decomposition, which interacts with the metal NPs catalyst. It developed a semi-fullerene cap as the material for the creation of a nanotube cylindrical shell from the sources of hydrocarbon through the continuous flow of carbon to the catalyst particles after achieving a certain threshold carbon concentration, shown in Fig. 6.11b,c. The catalysts are also under progress and optimization for ultimate removal from nanotube tips and further purification to provide better-performing CNTs (Matsuzawa et al. 2014; Morsy, Helal, El-Okr, and Ibrahim 2014). The high defect densities in their architectures were a big pitfall for CVD-grown MWNTs.

However, the defeat of CVD-grown MWNTs remains thoroughly known but perhaps because they do not have enough thermal energy and low growth temperature for the annealing of nanotubes to fully crystalline structures. Even to this day, perfect MWNTs are still a challenge to develop.

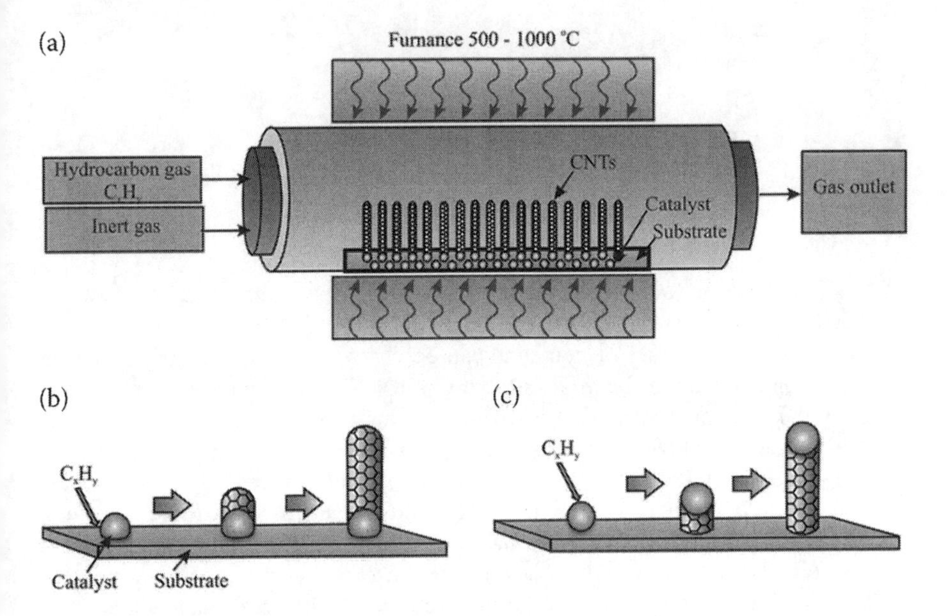

FIGURE 6.11 Schematic diagram of the mechanism of CVD deposition: a) The CNT-synthesized CVD reactor modified scheme, b) the CNT growth mechanism model of base growth, and c) tip-growth model of CNT growth mechanism (Zaytseva and Neumann 2016).

6.3 OPTICAL PROPERTIES OF CNT STRUCTURES

In this section, the optical properties and states of SWCNTs are summarily outlined. Additionally, the new optical properties of CNTs (time-resolved laser spectroscopy and single-nanotube) are discovered. A SWNT, around one nanometer in diameter and multiple nanometers long, is a 1D structure. SWNT consists of the 2D, single-layered graphene sheet rolled into a cylindrical shape. The SWNT is correlated with chirality or (n, m index) where n and m are integers. So, the chirality index defines the radius and angle, which explain the graphene sheet rolling up. Therefore, Fig. 6.5 shows the graphene sheet schematic diagram in a vector diagram. The chirality (C) of the graphene sheet can be defined as

$$C = na_1 + ma_2 \tag{6.1}$$

where a_1 and a_2 are basic vectors of a lattice. By linking the origins to points, the chiral index (n, m) structure is established for SWNTs. SWNTs with a chiral index can express their diameter (d) as

$$d = a\frac{\sqrt{n^2 + m^2 + nm}}{\pi} \tag{6.2}$$

where $a\,(=|a_1| = |a_2|)$ is calculated as 2.46 Å. So, the major significant aspect of

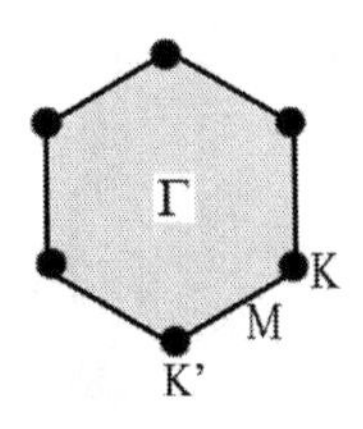

FIGURE 6.12 Schematic of graphene's reciprocal lattice and first-Brillouin zone. K, K', M, and Γ are the first-Brillouin zone's high symmetric points (Matsuda 2013).

SWCNTs is the electronic characteristics decided by the composition of the CNT itself, which is the chiral index (n, m).

The electronic structure of graphene can be clarified, as zone-folding graphene can be used to obtain the electronic structure of SWNTs. The graphene's reciprocal lattice in the dynamic space and first-Brillouin zone is presented in Fig. 6.12 in the reciprocal space, the first-Brillouin zone shape reveals a hexagonal type that reflects the graphene's periodic hexagon in the actual space. Therefore, the close-binding method is an advantageous step to understand graphene's electronic structure, where an electron's Coulomb interaction with another electron is excluded. Graphene is characterized by the basic lattice vectors a_1 and a_2.

Two carbon atoms are used in the unit cell, each with four valence electrons. Three of these electrons will be combined to form σ bonds, while the other will form a π bond. The delocalization of the π electrons normally dominates the electronic properties. Additionally, considering only the closest interactions, the graphene's energy-dispersion behavior is given by

$$E_g^{\pm}(k) = \pm\gamma_0 \frac{\sqrt{1 + \frac{4\cos(\sqrt{3}k_x a_0)\,\cos(k_y a_0)}{2}}}{2} + \frac{4\cos^2(k_y a_0)}{2} \tag{6.3}$$

where ± sign indicates the valence band (bonding) and conduction band (anti-bonding) within the k-dimensional space and γ_0 is the nearby transfer integral. The energy-dispersion of conduction and valence bands are symmetric w.r.t. the Fermi level. However, the Fermi-gap (E_g = 0) energy bands are degenerated at the K points, demonstrating that graphene is a zero-gap semiconductor.

According to the graphene's electronic structure from the above paragraph, the confinement approximation or the zone folding can deduce the SWNT electronic band structure. Wave vectors are confined along the circumference of the SWNT, but they are continuous over the nanotube axis. The zone-folding process divides the structure as in Fig. 6.13a,b (also known as the cutting line). Thus, there are a set of 1D dispersion correlations for the SWNT energy bands.

The fold approximation is enough to determine the properties of electronics; however, it is also important to realize that certain SWNTs have semiconducting states and others have metallic. This quality comes from the dispersion of graphene's zero-gap energy band structure at the K point. Since, the K point is permitted for SWNTs, i.e., is formed by a cutting line in Fig. 6.13a, in the lower panel band structure of Fig. 6.13a, the SWNTs will be metallic. According to this, the composition of SWNTs seems to have ($n - m = 3$ m) as a chiral index equation, where "m" is

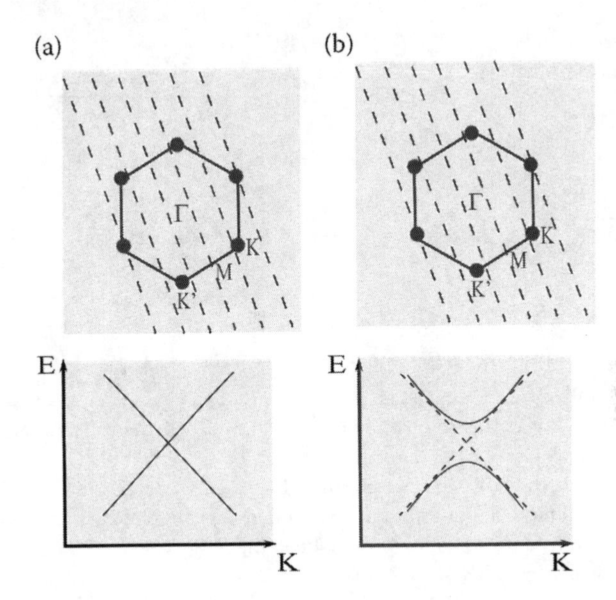

FIGURE 6.13 Schematic diagram of the first-Brillouin zone, including the energy band structure and cutting lines in a) metallic and b) semiconducting-CNTs (Matsuda 2013).

an integer. However, if the cutting line doesn't pass through the K point due to an unallowed SWCNT wave vector, then the single-walled nanotubes have a chirality of ($n - m \neq 3$ m), and will act as semiconductors. Fig. 6.13b shows the chiral indices of metallic SWNTs and semiconducting. One-third of single-walled nanotubes exhibit metallic states of electronics, whereas the remaining have semiconducting electronic states, according to this relationship. The metallic density of states, i.e., ($n - m = 3$ m) and semiconducting, i.e., ($n - m \neq 3$ m) SWNTs are depicted schematically in Fig. 6.14. The density of states is a valuable property for figuring out how SWNTs' electronic architectures function. The sharp singularity peak in von Hove's state density comes from the electronic confinement of a one-dimensional quantum in the SWNT. Semiconducting SWNTs' bandgap energy is inversely related to their diameter, as expected by a close estimation.

The optical properties of micelle-encapsulated single-walled nanotube have been extensively researched after O'Connell published them in 2002, finding strong photoluminescence, i.e., PL in the process (Bachilo et al. 2002; O'Connell et al. 2002). The interpretation of PL is brought on by the separation of SWNTs that do not associate with either metallic CNTs or semiconducting (Lefebvre, Homma, and Finnie 2003; O'Connell et al. 2002). The quenching of the electronic excitation and conversion of energy between the metallic nanotubes and semiconducting takes place without radiative recombination in the bundles (Lefebvre, Homma, and Finnie 2003; O'Connell et al. 2002). In other terms, it is also claimed that isolated air-suspended carbon tubes formed between the silicon pillars displayed clear PL signals (Lefebvre, Homma, and Finnie 2003). The standard 2D photoluminescence excitation, i.e., PLE map, demonstrates the contour plot between the PL-spectra of micelle-encapsulated single-walled nanotubes and the excitation wavelength is shown in Fig. 6.15

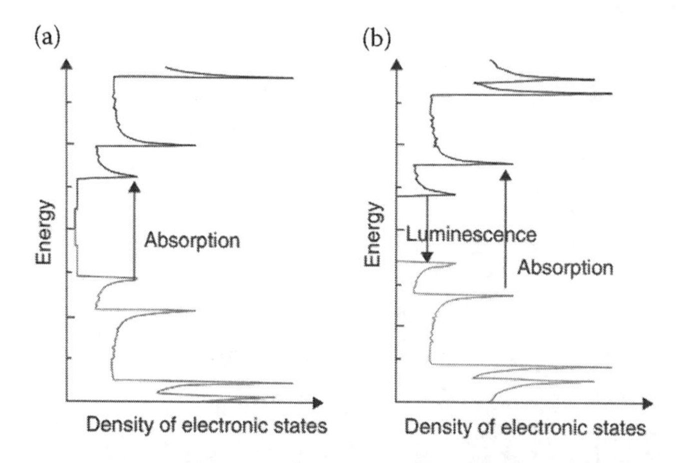

FIGURE 6.14 a) Metallic-CNT's density of state and b) semiconducting-CNT's density of state. This figure depicts the absorption and luminescence's optical transitions. Reprinted with permission from Journal of Carbon Nanotubes and Graphene for Photonic Applications. Copyright, 2013, Elsevier (Matsuda 2013).

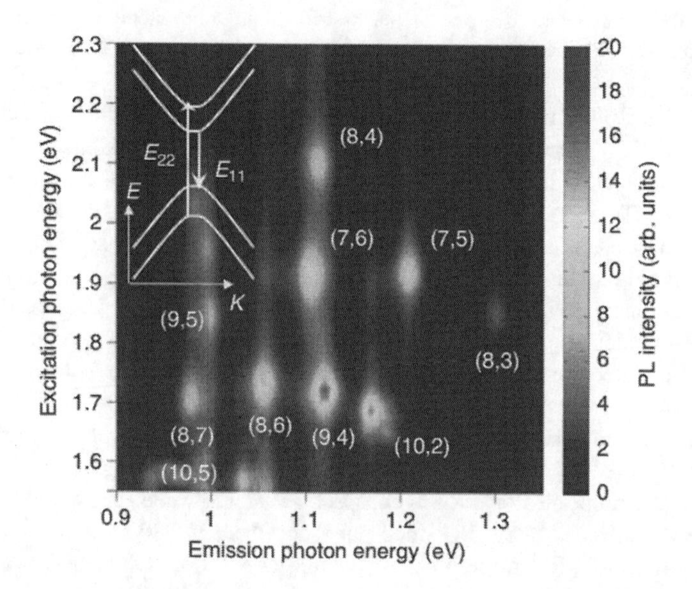

FIGURE 6.15 2D photoluminescence excitation (PLE) map of SWCNTs. Reprinted with permission from Journal of Carbon Nanotubes and Graphene for Photonic Applications. Copyright, 2013, Elsevier (Matsuda 2013).

(Lefebvre, Homma, and Finnie 2003). Other 2D PL signals, represented as spots, can be detected as sharp resonance characteristics in the single-walled nanotube electronic properties. The peak of absorption spectrum between the 2nd conduction sub-bands to valence state, as shown by E_{22} in the inset of Fig. 6.15, corresponds to the PLE signal's energy photon on the vertical axis. In addition, the maximum energy on the

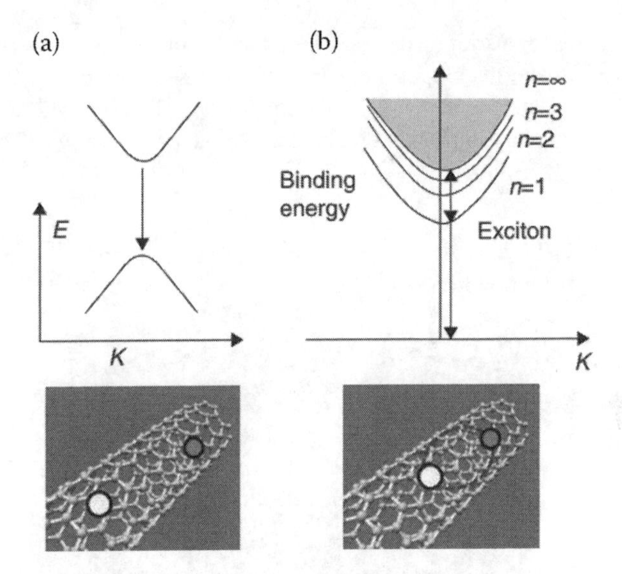

FIGURE 6.16 SWCNTs' electronic structures are shown in a schematic diagram in a) the one-electron image without interactions of coulomb and b) the exciton image with interactions of coulomb. Reprinted with permission from Journal of Carbon Nanotubes and Graphene for Photonic Applications. Copyright, 2013, Elsevier (Matsuda 2013).

horizontal axis is E_{11}, the PL energy corresponds to semiconducting SWNT. The energy sources E_{11} and E_{22} can be measured with the close binding technique in SWNTs (n, m). Any peak on the 2D PLE map can be attributed to (n, m) of SWNTs from projected E_{11} and E_{22} energy, as stated in Fig. 6.15 (Weisman and Bachilo 2003). The PL signal from semiconducting single-walled CNTs is thought to be caused by the optically excited recombination of free electrons in the conduction band and free holes in the valence band, according to most research groups in this field. The optical spectra's resonance characteristics are due to the sharp electronic states responding to the 1D von Hove singularity in this one-electron image, seen in Fig. 6.16a, without taking coulomb-interaction into account. As seen in Fig. 6.16b, the improved coulomb-interaction in one-dimensional SWNTs contributes to the creation of closely bound electrons and hole pairs, or "excitons," which are equivalent to a hydrogen-like phase in the solid. And, on the other hand, stated that in the one-nanometer cylindrical configuration, the interaction of a coulomb between the electron-hole pairs is largely improved and this interaction influences the optical spectrum of SWNTs (Ando 1997).

6.4 APPLICATIONS OF CNTS SUPPORTED BY SPR IN DIFFERENT FIELDS

6.4.1 Gas Sensor

Gas-sensing methods have been widely applied over the last two decades. At the start of a huge production of CNTs, scientists around the globe began to use CNT-based instruments for various studies. Since then, a multitude of scientific groups across the world have developed numerous CNT-based methods such as functionalization, doping, dielectric analysis, and deposition. In Fig. 6.17 and Fig. 6.18, two pathways of chemisorption and physisorption have been shown. The reversible

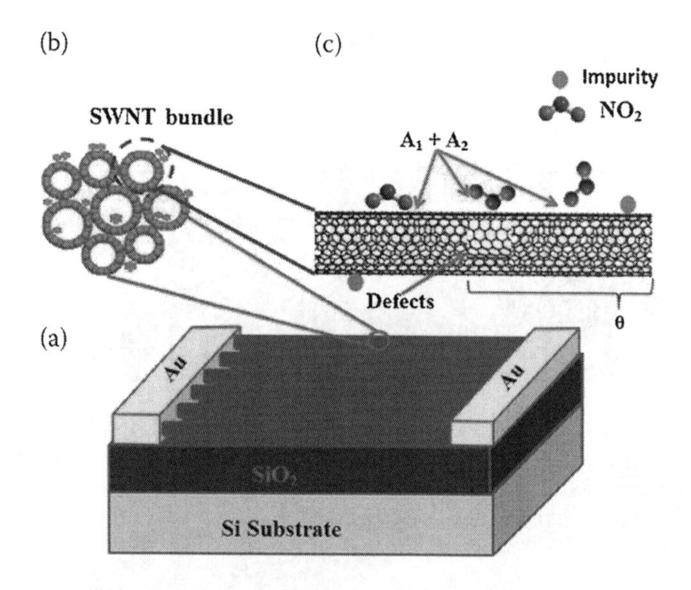

FIGURE 6.17 SWCNT-based sensor gas detection mechanism showing the chemisorption of the analyte molecules. Reprinted with permission from Sensors and Actuators A: Physical Journal. Copyright, 2019, Elsevier (Han, Nag, Mukhopadhyay, and Xu 2019).

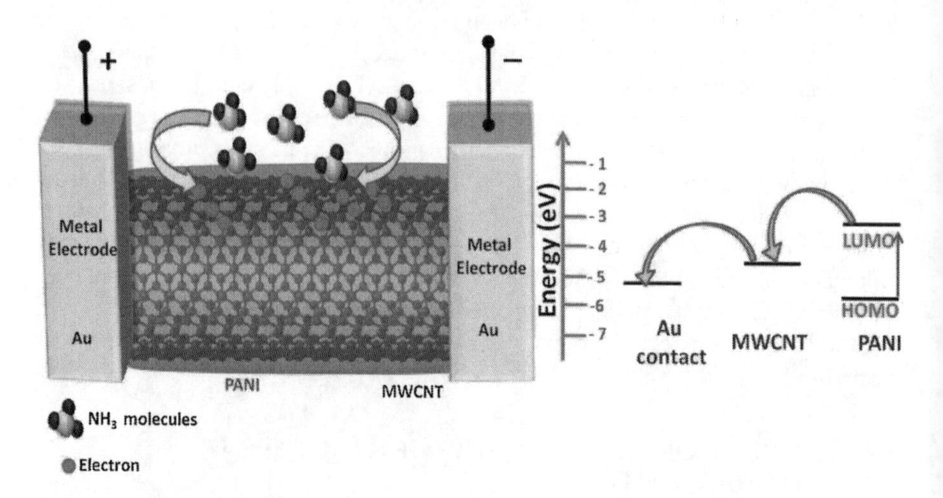

FIGURE 6.18 The gas-sensing system and the energy band diagram demonstrate that how ammonia gas adsorption contributing to the associated transfer of charge. Reprinted with permission from Sensors and Actuators A: Physical Journal. Copyright, 2019, Elsevier (Han 2019).

and permanent reaction concurrently occurs on the sensing surface during chemisorption. The different adsorption rate at different locations with associated activation energies is involved with the two types of reactions. When the gas is applied into the physisorption phase, the sensing rate is high, but the sensing rate drops gradually. This may be because the sensor's response becomes saturated until the later portion of its reaction, where reversible and irreversible sites are no longer present. The varying nature of SWCNT conductivity affects the activation energy (being made up of both metallic and semiconducting nanotubes). It was also proven from Raman spectroscopy and TEM that wall inspection where the presence of impurities in CNTs causes chemisorption. From Fig. 6.18, the energy difference shown in the band diagram of physisorption happens as a result of electron transfer between the measured gas and the nano-composite-based sensing area (He et al. 2009; Wu et al. 2013). During this scenario, a physisorption-chemisorption effect takes place for the gas molecules when they are exposed to it. Ion-dipole interaction between the gas molecules and nanotubes occurs during the reaction. In this case, the coplanar structure of the polyaniline composites leads to a simple electron transfer (NH_3). The location of polarons, because of net charge transfer between gas molecules and composites, was therefore created. This location then led to the resistance change. The degradation of the communication of nano-composites due to the gas molecule's adsorption may be another explanation for the improvement in the performance resistance. The weakening of nanocomposite connection makes it simpler without the exercise of outside energy to desorb the gas molecules. Although chemisorption and physisorption occur simultaneously, physisorption is predominantly than chemisorption, helping to respond quickly and to recover.

Angiola, Rutherglen, Galatsis, and Martucci (2016) demonstrated that transparent gas sensors can be developed with SWCNT, coupled with gold nanoparticles (AuNPs) to make the nanotubes suitable for exposure to potentially toxic gases, such as H_2, NO_2, and CO and with the localized surface plasmon resonance (LSPR) of AuNPs. The LSPR is prone to variation in the ambient medium's dielectric properties, which was applied in the development of the sensing systems. They used two distinct techniques: inkjet and droplet-casting, for depositing the CNTs on the plasmonic substrate. The gas-sensing experiments conducted at room temperature had no change in measurable absorbance. To prevent the alteration of the sensor's sensing material during testing, the operating temperature was placed at 150°C, which is significantly lower than the temperature where the sensitive surfaces were allowed to heat. Both the metallic and semi-conductive NPs were required for gas detection. When NPs were placed on the naked quartz layer, no difference was observed in optical properties. Fig. 6.19 illustrates the optical absorption spectra of m-SWCNTs and s-SWCNTs on Au NPs as seen in dry air and under hydrogen. There is a shift in the wavelength towards the left (blue-shift) of the LSPR for both experiments and the difference is more pronounced for m-SWCNTs. The inset of Fig. 6.19a display the optical absorbance change (OAC), which is calculated by exposure to gas absorbance and air absorbance, in order to further explain the variance of optical absorption. This inset from Fig. 6.19a shows how the variation in behavior to H_2 of the bare m-SWCNTs, the bare Au NPs, and Au NPs masked with m-SWCNTs. Both bare m-SWCNTs and bare Au NPs could not enable the

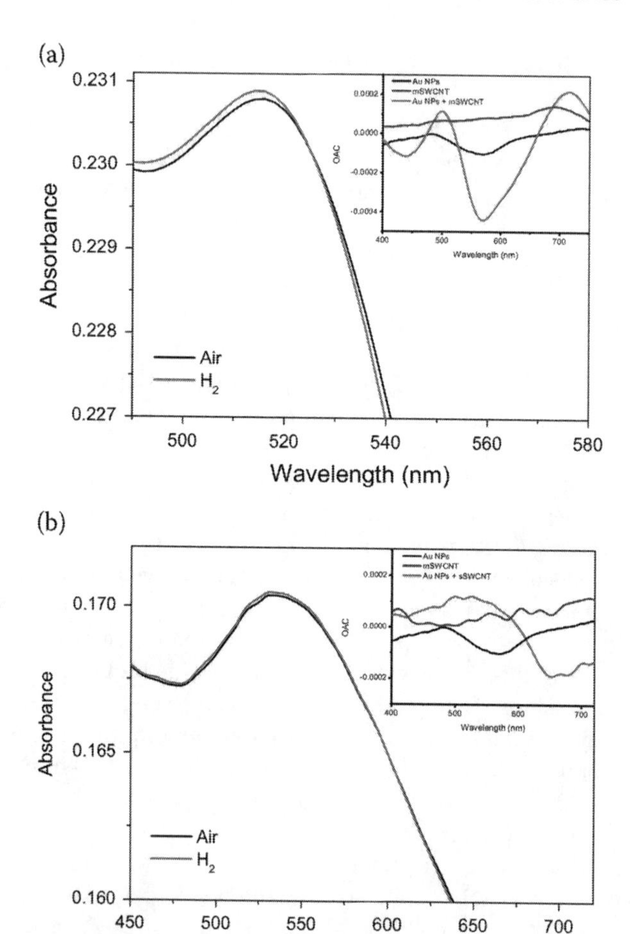

FIGURE 6.19 a) Spectrum of m-SWCNTs' optical absorption of Au-NPs collected on-air (black line) as well as H_2 (red) air at operating temperature = 150°C, with an inkjet printer. The inset shows the optical absorbance change of the bare m-SWCNTs, Au-NPs, and Au-NPs covered with m-SWCNTs. b) Spectrum of s-SWCNTs' optical absorption of Au-NPs collected on air (black line) as well as H_2 (red) air at operating temperature = 150°C, with an inkjet printer. The inset shows the optical absorbance change of the bare s-SWCNTs, Au-NPs, and Au-NPs covered with s-SWCNTs. Reprinted with permission from Sensors and Actuators B: Chemical Journal. Copyright, 2016, Elsevier (Angiola, Rutherglen, Galatsis, and Martucci 2016).

optical sensing properties of the material, whereas the m-SWCNTs with Au NPs coupling does so. Fig. 6.19b shows the same effects of experiments performed with s-SWCNTs. It is also important to obtain observable changes in SWCNT absorption in the presence of H_2 by using Au NPs, but s-SWCNTs have a poor response compared to m-SWNTS. The inset spectrum of OAC (see Fig. 6.19) yields the conclusion that the most intense absorption is reached at wavelengths of $\lambda = 570$ nm

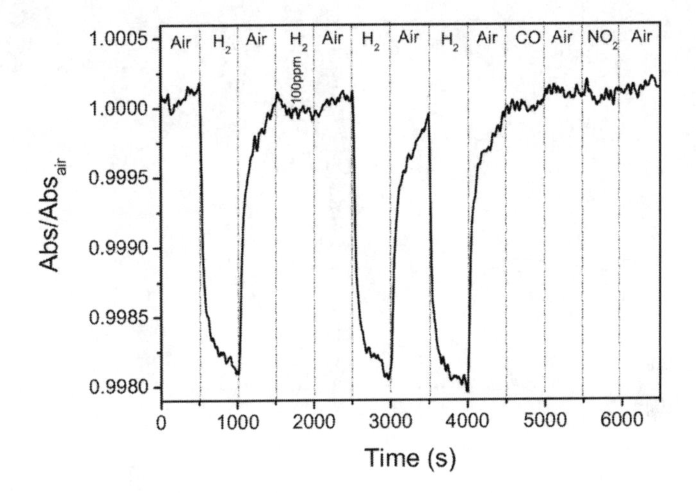

FIGURE 6.20 Metallic-SWCNTs' dynamic response deposited on Au-NPs at λ = 570 nm and operating temperature =150°C when subjected to various air or gas cycles. Reprinted with permission from Sensors and Actuators B: Chemical Journal. Copyright, 2016, Elsevier (Angiola, Rutherglen, Galatsis, and Martucci 2016).

and λ = 700 nm for m-SWCNTs and at λ = 540 nm for s-SWCNTs, which have been chosen for dynamic testing. As seen in Fig. 6.20, the variations in the SWCNTs' absorption and OAC spectrum illustrate sharper and quicker variations in absorption for the m-SWCNTs in relation to s-SWCNTs, as seen in Fig. 6.21. Additionally, the s-SWCNTs results show a significant drift, as seen in Fig. 6.21. It was not possible to discriminate between the different NO_2 and CO concentrations for s-SWCNTs as well as for m-SWCNTs, although for the 100 ppm hydrogen concentration in the case of m-SWCNTs, there was an appreciable difference in absorbance change seen in Fig. 6.20. Thus, s-SWNTs lacked oxidizing and reducing gas sensitivity, but the nanostructure consisting of Au-NPs and m-SWNTs worked well in the detection of hydrogen.

6.4.2 Chemical Sensor

Jiang and Wang (2019) presented a fiber-optic SPR sensor featured with MWCNT/PtNPs-based sensitive layers for the RI measurements. They built three distinct types of SPR sensors with distinct membranes, each using the self-layer-by-layer method: the SPR-coated Au film sensor, the SPR-coated Au film sensor with MWCNT mounted on it, and the SPR-coated Au film sensor with MWCNT/PtNPs mounted on it, respectively. Fig. 6.22 and Fig. 6.23 show the schematic SPR probe with MWCNT/PtNPs production and the measurement system of RI, respectively. Where the SMA adapter was clamped to both the Y-type 2 × 1 optical fiber-optic coupler and probe. The three types of optical sensor-based sensors were measured under the same test conditions. CNTs can enhance the optical fiber sensor's

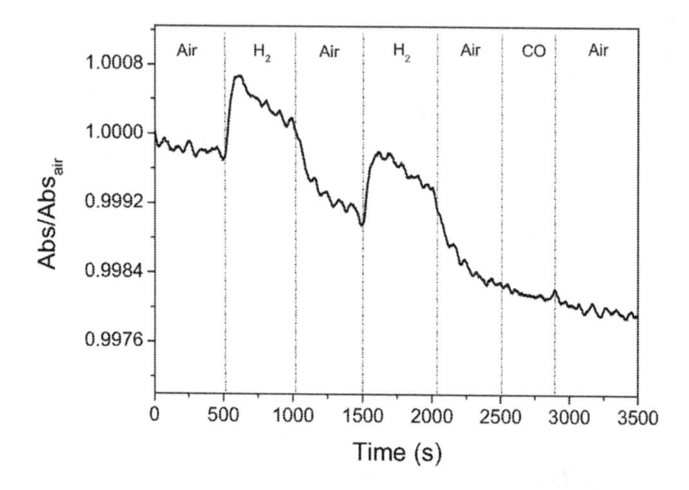

FIGURE 6.21 Semiconducting-SWCNTs' dynamic response deposited on Au-NPs at λ = 540 nm and operating temperature = 150°C when subjected to various air or gas cycles. Reprinted with permission from Sensors and Actuators B: Chemical Journal. Copyright, 2016, Elsevier (Angiola, Rutherglen, Galatsis, and Martucci 2016).

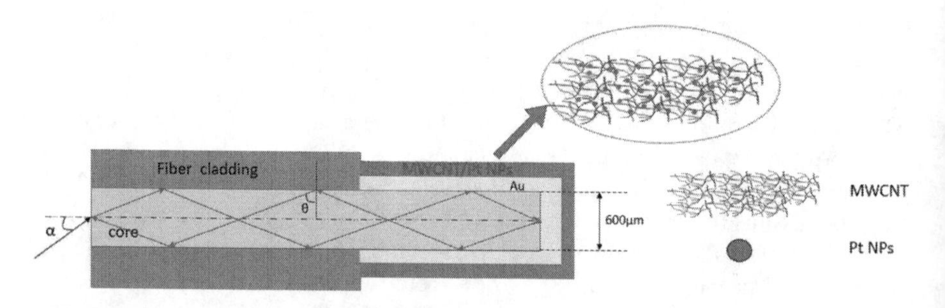

FIGURE 6.22 The SPR probe sensing model with MWCNT/PtNPs. Reprinted with permission from Optical Fiber Technology Journal. Copyright, 2019, Elsevier (Jiang and Wang 2019).

sensitivity, but in order to further enhance the sensor's capability as well as the sensor's FOM, the MWCNT/PtNPs composite is used.

However, Fig. 6.24 depicts the MWCNT/PtNPs-coated SPR spectrum of the experiment observed. This MWCNT to PtNPs volume ratio is 5:1, which is used for an optimized structure with the help of a large amount of platinum particles. The wavelength's resonant position moved towards the longer wavelengths, and further, the sensitivity is enhanced to 5,923 nm RIU^{-1}. Also, the inset of Fig. 6.24 illustrates a type of correlation between the resonant wavelength and the RI. From the graph, it can be concluded that the sensitivity has increased substantially and exhibits a linear correlation. Furthermore, the FOM of the sample is 29.32 RIU^{-1}, greater than the probe of an Au-coated thin-film.

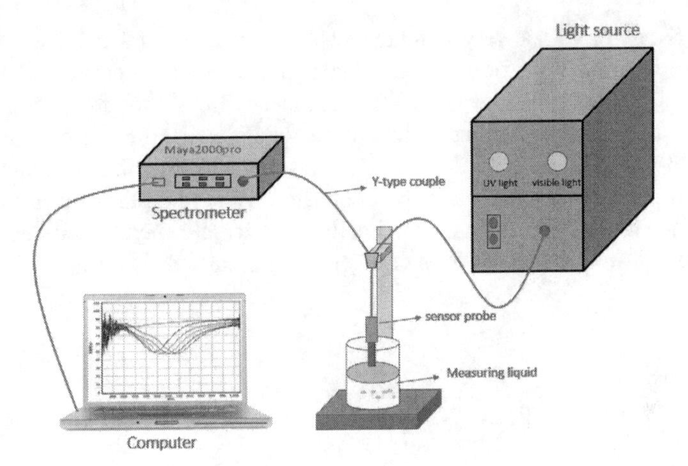

FIGURE 6.23 The experimental system's schematic diagram. Reprinted with permission from Optical Fiber Technology Journal. Copyright, 2019, Elsevier (Jiang and Wang 2019).

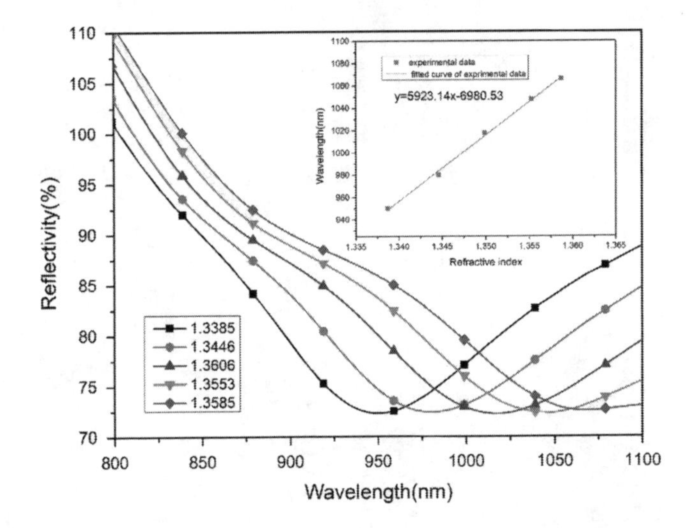

FIGURE 6.24 The reflection spectrum of SPR of the MWCNT/PtNPs film sensor probe as well as the inset of the figure indicates the relation between the RI and resonance wavelength. Reprinted with permission from Optical Fiber Technology Journal. Copyright, 2019, Elsevier (Jiang and Wang 2019).

Pathak and Gupta (2021) demonstrated the ability of palladium NPs, i.e., PdNPs, and polypyrrole, i.e., PPy, shell coated on CNTs in a plasmon-based sensor performed on a fiber optic. The synergistic modulation of PdNPs/PPy@CNTs-nanocomposite properties are made to communicate with hydrazine, which is investigated for SPR-sensing fiber-optic substrate with silver (Ag) as a plasmonic substance. The nanocomposite synthesis is achieved in-situ by the solvothermal method. Since PPy is a very conductive polymer, it is used in many chemical-sensing applications. It is observed that the alteration in the electronic configuration of PPy in the neutral state induces significant shifts in optical conductivity. Fig. 6.25a depicts the chemically interactive image concerned in nano-composite preparation. PdNPs/PPy with CNTs and PdNPs or CNTs, the final nano-composites, were deposited in deionized water before coated with Ag on the optical fiber layer. To execute the control experiment, PdNPs directly attached to surface-modified CNTs were also purified. SPR experiments were carried out using a laboratory-

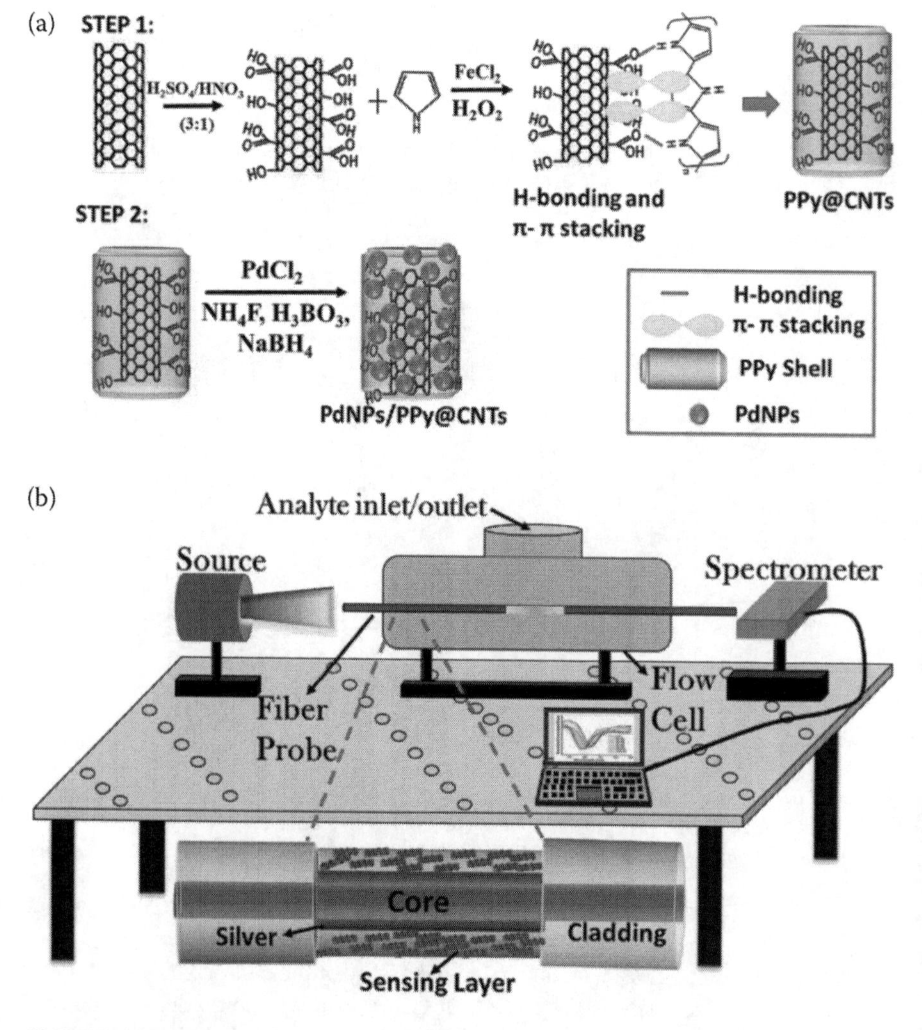

FIGURE 6.25 Schematic of a) involvement of chemical interactions in PdNP/PPy@CNT synthesis and b) experimental arrangement and fiber probe. Reprinted with permission from Sensors and Actuators B: Chemical Journal. Copyright, 2021, Elsevier (Pathak and Gupta 2021).

designed setup that operated in the mode of wavelength interrogation. The setup is represented schematically in Fig. 6.25b. The fiber-optic probe in a flow cell is paired with a polychromatic tungsten halogen lamp that has been used as a light source to enable the information transfer. For different parameters, including the nano-composite constitution, sensing layer thickness, and pH are measured, they are optimized to obtain the best sensing output. Fig. 6.26 displays different SEM images of the final nanocomposite. Fig. 6.26a indicates the PdNPs formation in a PPy and CNTs absence, with a method of direct reduction of palladium(II) to palladium(0). Fig. 6.26b shows the formation of a uniform shell around the CNTs using PPy. In Fig. 6.26c, the palladium nanoparticle's morphology contained in PPy with CNTs' embedded frameworks can be seen. The nano-composite was dip-coated on the Ag core, and the surface can be seen clearly in Fig. 6.26d. The Pd/CNT-mediated oxidation of hydrazine plays a critical role in the photocatalytic mechanism.

However, the interaction between hydrazine and PPy can also be a probable explanation of H-bonding. The degree of communication is determined and quantified by resonance wavelength and hydrazine concentration. In Fig. 6.27a, the SPR spectra are recorded for the concentration of hydrazine ranging from 0 to 1,500 nM. As the concentration of hydrazine increases, the resonance wavelength decreases.

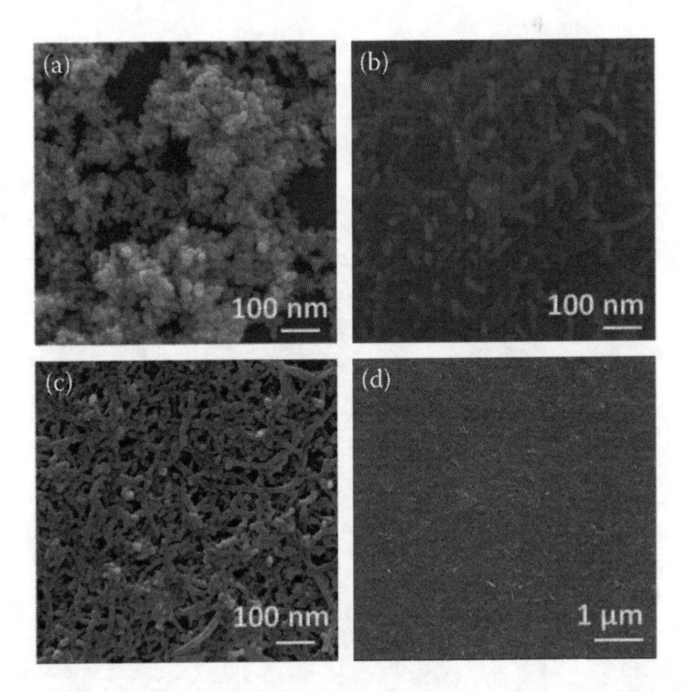

FIGURE 6.26 SEM images of a) palladium nanoparticles, b) CNTs coated with polypyrrole, c) polypyrrole-coated CNTs with a palladium coating, and d) resulting nanocomposite is coated on a fiber optic with an Ag coating. Reprinted with permission from Sensors and Actuators B: Chemical Journal. Copyright, 2021, Elsevier (Pathak and Gupta 2021).

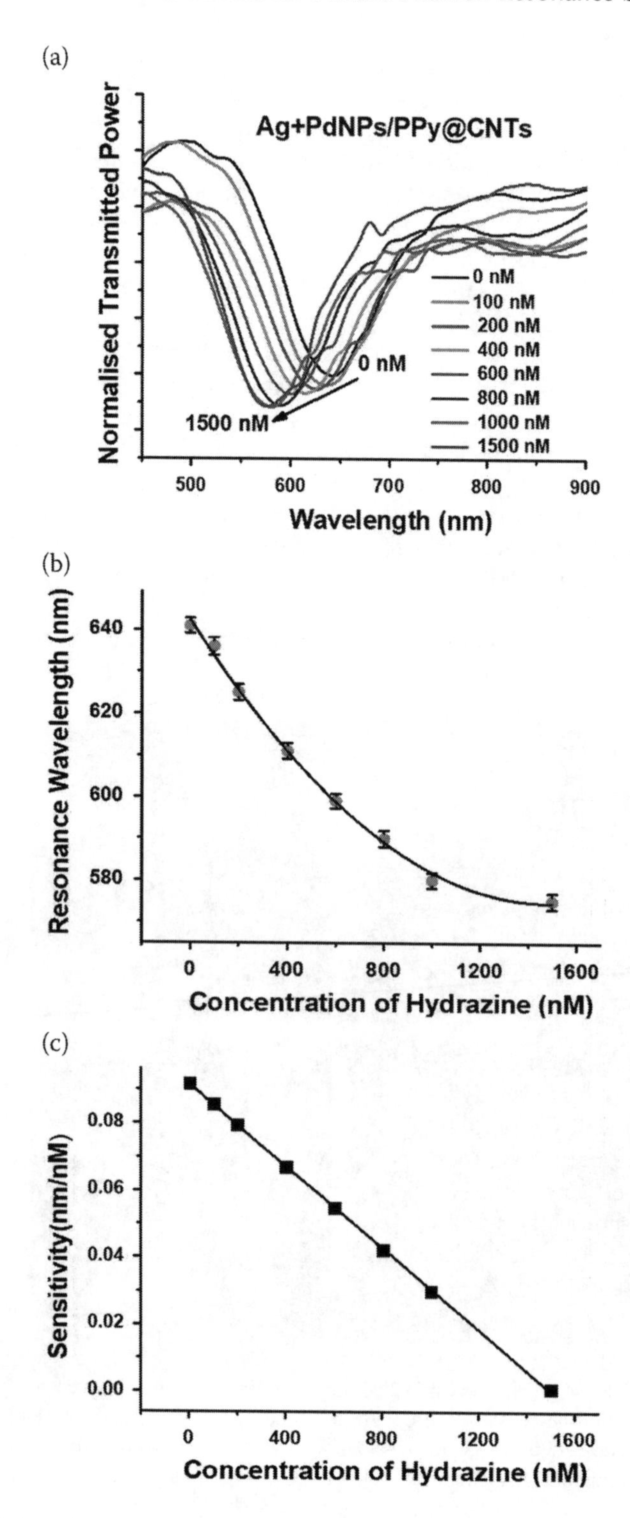

FIGURE 6.27 a) SPR plots, b) calibration curve, and c) sensor's sensitivity. Reprinted with permission from Sensors and Actuators B: Chemical Journal. Copyright, 2021, Elsevier (Pathak and Gupta 2021).

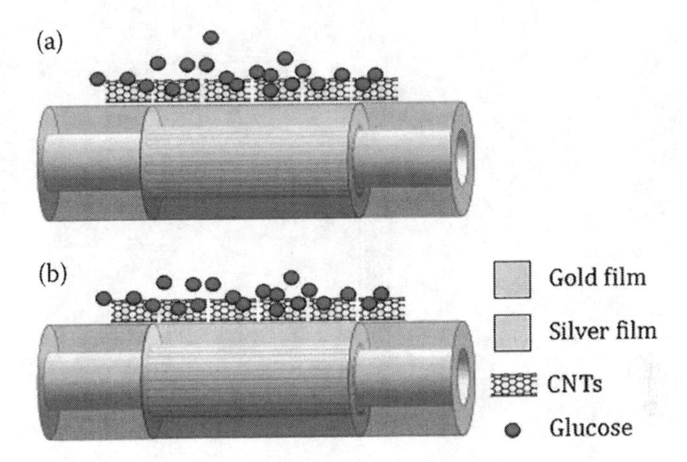

FIGURE 6.28 The sensing design images of a) the sensor of gold-PCF with CNT and b) the sensor of silver-PCF with CNT for the detection of glucose. Reprinted with permission from Optical Fiber Technology Journal. Copyright, 2018, Elsevier (Jing, Wang, and Wang 2018).

The wavelengths differ according to the hydrazine concentration and is highlighted in Fig. 6.27b. And in Fig. 6.27c, the derivative of this curve varies with hydrazine concentration and gives the sensor's sensitivity. The maximum sensitivity of the sensor is achieved at 0.09 nm nM^{-1}, with the LOD of 20 nM for the lowest hydrazine concentration.

Jing, Wang, and Wang (2018) proposed CNT deposited on a gold and silver film for a photonic crystal fiber (PCF) based SPR sensor to study the glucose detection in terms of RI-sensing characteristics. They introduced PCF in between two portions of multimode fiber (MMF), followed by a gold/silver coating and then further deposited with CNTs, as shown in Fig. 6.28. Introducing CNT in the developed model enhanced the confinement of electric field intensity around the sensing reason.

They observed that CNT-gold-PCF sensitivity is 1,016.09 nm RIU^{-1} higher than a standard gold-PCF sensor and CNT-silver-PCF sensitivity, compared to a standard silver-PCF-sensor, by 709.22 nm RIU^{-1}, respectively. The resolution and the penetration depth are also higher in the case of the CNT-deposited SPR sensor than the sensor without CNTs. Fig. 6.29 and Fig. 6.30 show the response analyses in both cases, respectively. Additionally, the aim was the bovine serum albumin (BSA) to investigate CNTs–gold–PCF sensors for the identification of biomolecules, which shows higher sensitivity, i.e., 8.18 nm $(mg\ mL^{-1})^{-1}$ and lower limits of detection, i.e., 2.5 μg mL^{-1} and has quicker reaction time, i.e., 8 s than conventional gold-PCF-based sensors.

6.5 ADVANTAGES OVER OTHER MATERIAL-BASED SENSORS

CNT-enabled nanocomposites have attracted a lot of interest compared to traditional composite materials because of their mechanical, optical, electrical, thermal, and chemical properties, including electrical conductivity and enhanced tensile strength.

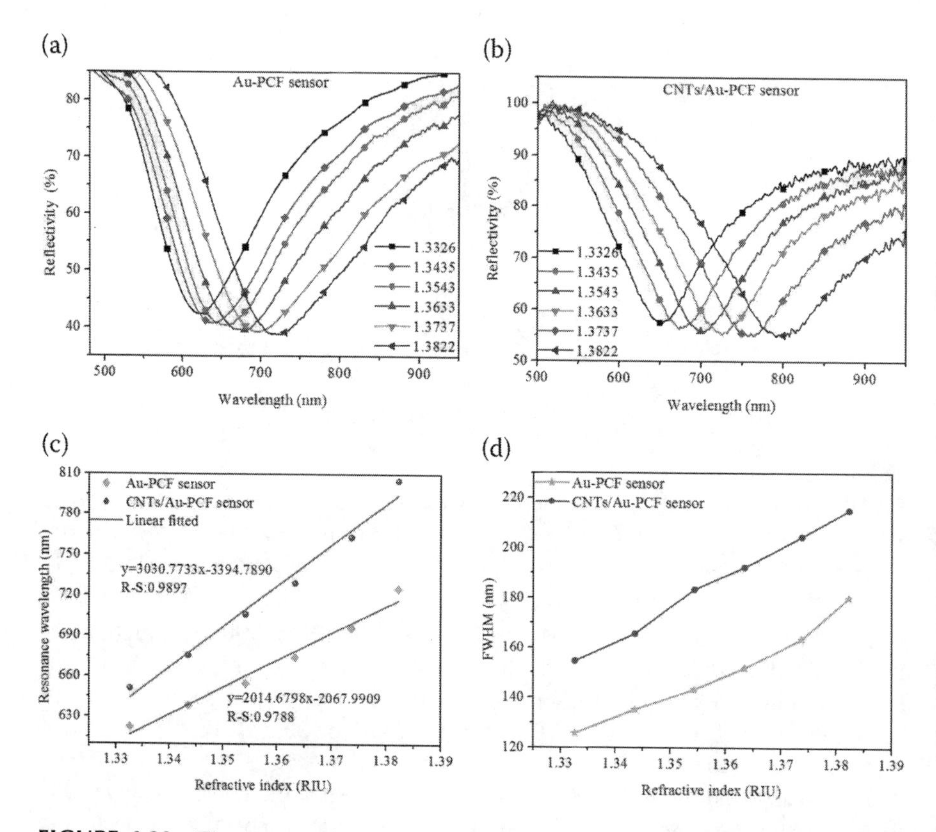

FIGURE 6.29 The resonance spectrum of a) gold-PCFbased sensor, b) gold-PCF with CNT-based sensor for different refractive indices of glucose solutions, c) the fitted profiles between the wavelength of SPR resonance and the refractive indices of the two types of sensors, and (d) the correlation between the full width at half maximum (FWHM) of the two types of sensors and their respective refractive indices. Reprinted with permission from Optical Fiber Technology Journal. Copyright, 2018, Elsevier (Jing, Wang, and Wang 2018).

Unlike other 2D materials, these materials promise improved wear resistance and breaking strength, antistatic properties, and weight reduction. Particularly, advanced CNT composites are thought to be capable of reducing the weight of spacecraft and aircraft by up to 30%. It is also extremely small and lightweight; one-sixth that of steel and 400 times more mechanical strength than steel and makes it a better replacement for the metallic wires.

They are often aimed for the development of polymers, medical, energy, chemical, and optical instruments. Multi-walled CNTs are known to have a higher growth and deployment rate among CNTs due to the enhanced similarity for shaping transparent electrodes, chemicals sensors, and conductive heating films in solar and thermal industries. There are plenty of resources available to produce CNT and can be made with a small amount of material at minimal cost. Regarding temperature fluctuations, it is resistant and improves the composites' conductive properties. In the field of gas sensing, CNT is highly sensitive

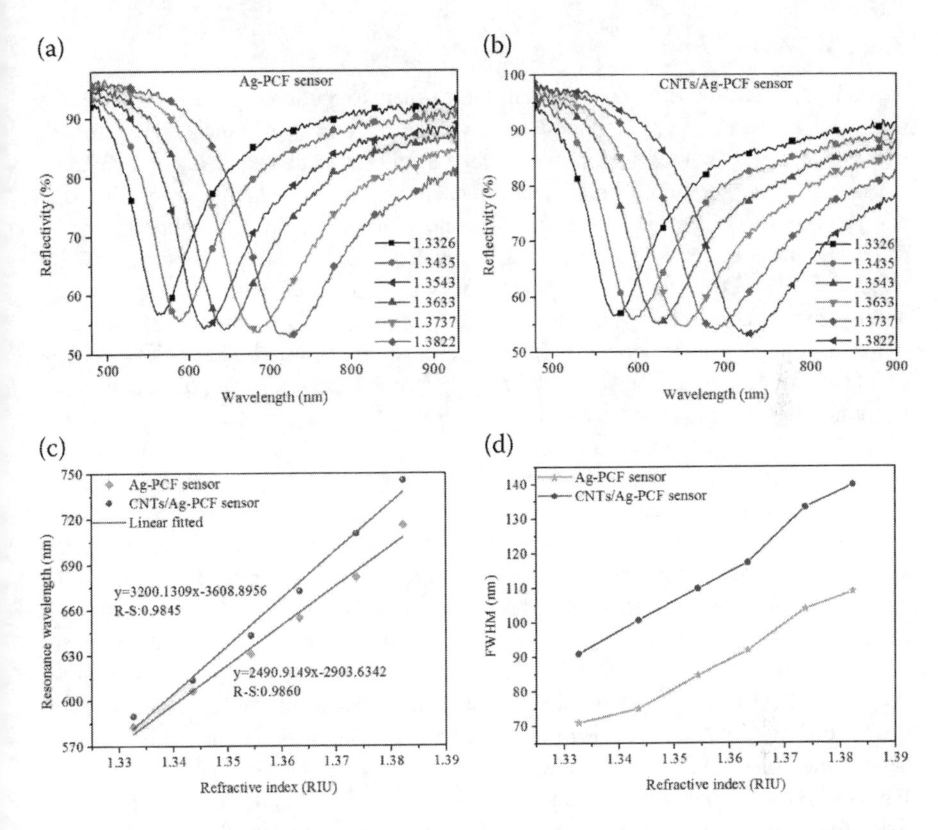

FIGURE 6.30 The resonance spectrum of a) silver-PCF-based sensor, b) silver-PCF with CNT-based sensor for different refractive indices of glucose solutions, c) the fitted profiles between the wavelength of SPR resonance and the refractive indices of the two types of sensors, and d) the correlation between the full width at half maximum (FWHM) of the two types of sensors and their respective refractive indices. Reprinted with permission from Optical Fiber Technology Journal. Copyright, 2018, Elsevier (Jing, Wang, and Wang 2018).

compared to other materials or metal oxides as it has a high surface area that makes the sensors respond faster to the gas molecule. The most remarkable features are for detecting NH_3 or NO_2 gases; CNT needs up to room temperature only, whereas metal oxides require the temperature above 200°C to detect the stated gas molecules. Due to the higher surface area, recovery time of the sensor is slow but it can be solved with the help of annealing the sensor at a high temperature using UV radiation. In the biomedical fields, an increasing number of laboratories are putting in significant effort and interest to investigate the use of carbon nanotubes in medicine. This is due to the variety of benefits that carbon nanotubes may have over other processes in diagnostic or therapeutic applications. Nevertheless, it is worth noting that the process of converting carbon nanotubes from an attractive nanomaterial to a successful medicinal component is still in its early stages.

6.6 SUMMARY

Nowadays, CNTs hold much potential due to their exceptional properties and large industrial applications. Their unique mechanical, electrical, chemical, thermal, and optical properties are reasons for this and are considered troublesome due to numerous reasons such as poor uniformity, lack of characterization techniques, and different compositions. A tremendous amount of research has been done to support the arc-discharge, laser ablation, and chemical vapor deposition techniques for synthesis to boost the CNT efficiency. Since CVD offers significant advantages, such as scalability, high-quality processing, accessibility, low-temperature synthesis, and economical, it gained more popularity in the chemical process market. The development of plasmonic-based sensors has drawn the attention of scientists to carbon-based nanomaterials. In this chapter, at a lower operative temperature, the Au-NPs and the m-SWCNTs' coupling have been shown to be required for the transparency of the design of a transparent sensor. The semiconductor-SWCNTs had low sensitivity to oxidizing or reducing agents, while the nano-composites' combination of Au NPs and m-SWCNTs showed better sensitivity to the detection of hydrogen. Another is the synthesis is carried out in-situ using a solvothermal process. Various parameters such as the content of nanophase, sensing layer thickness, and sample pH are measured to maximize their role in sensing output. A range of hydrazine concentrations is investigated, with a LOD of 20 nM, with the maximum sensitivity of 0.09 nm nM^{-1}. Also, the effect of CNTs on certain SPR sensors was investigated using a four-layer design that systematically performed the refractive index experiments with enhanced sensitivities of CNTs/gold-PCF having 1.50 times more than that of gold-PCF and CNTs/silver-PCF having 1.28 times more than that of silver-PCF, respectively. Therefore, carbon-based nanomaterials play various roles in plasmonic-based sensors, including sensitivity enhancement material, plasmonic layers, and sensing matrix material. The functions of graphene and its derivatives and carbon nanotubes in sensors have been addressed in this chapter.

REFERENCES

Abdulhalim, Ibrahim, Mohammad Zourob, and Akhlesh Lakhtakia. 2008. "Surface Plasmon Resonance for Biosensing: A Mini-Review." *Electromagnetics* 28 (3): 214–42.

Ando, Tsuneya. 1997. "Excitons in Carbon Nanotubes." *Journal of the Physical Society of Japan* 66 (4): 1066–73.

Angiola, Marco, Christopher Rutherglen, Kos Galatsis, and Alessandro Martucci. 2016. "Transparent Carbon Nanotube Film as Sensitive Material for Surface Plasmon Resonance Based Optical Sensors." *Sensors and Actuators B: Chemical* 236: 1098–103.

Aqel, Ahmad, Kholoud M. M. Abou El-Nour, Reda A. A. Ammar, and Abdulrahman Al-Warthan. 2012. "Carbon Nanotubes, Science and Technology Part (I) Structure, Synthesis and Characterisation." *Arabian Journal of Chemistry* 5 (1): 1–23.

Arora, Neha, and N. N. Sharma. 2014. "Arc Discharge Synthesis of Carbon Nanotubes: Comprehensive Review." *Diamond and Related Materials* 50: 135–50.

Bachilo, Sergei M., Michael S. Strano, Carter Kittrell, Robert H. Hauge, Richard E. Smalley, and R. Bruce Weisman. 2002. "Structure-Assigned Optical Spectra of Single-Walled Carbon Nanotubes." *Science* 298 (5602): 2361–6.

Bethune, D. S., Ch H. Kiang, M. S. De Vries, G. Gorman, R. Savoy, J. Vazquez, and R. Beyers. 1993. "Cobalt-Catalysed Growth of Carbon Nanotubes with Single-Atomic-Layer Walls." *Nature* 363 (6430): 605–7.

Bhattacharya, Mrinal. 2016. "Polymer Nanocomposites—A Comparison between Carbon Nanotubes, Graphene, and Clay as Nanofillers." *Materials* 9 (4): 262.

Chakoli, Ali Nabipour, Jin Mei He, and Yu Dong Huang. 2018. "Collagen/Aminated MWCNTs Nanocomposites for Biomedical Applications." *Materials Today Communications* 15: 128–33.

Chen, George Z., Xudong Fan, Angela Luget, Milo S. P. Shaffer, Derek J. Fray, and Alan H. Windle. 1998. "Electrolytic Conversion of Graphite to Carbon Nanotubes in Fused Salts." *Journal of Electroanalytical Chemistry* 446 (1–2): 1–6.

Chen, Chun-Ku, W. Lee Perry, Huifang Xu, Yingbing Jiang, and Jonathan Phillips. 2003. "Plasma Torch Production of Macroscopic Carbon Nanotube Structures." *Carbon* 41 (13): 2555–60.

Chowdhury, Zaira Zaman, Suresh Sagadevan, Rafie Bin Johan, Syed Tawab Shah, Abimola Adebesi, Sakinul Islam Md, and Rahman Faizur Rafique. 2018. "A Review on Electrochemically Modified Carbon Nanotubes (CNTs) Membrane for Desalination and Purification of Water." *Materials Research Express* 5 (10): 102001.

de Menezes, Beatriz Rossi Canuto, Karla Faquine Rodrigues, Beatriz Carvalho da Silva Fonseca, Renata Guimarães Ribas, Tha'is Larissa do Amaral Montanheiro, and Gilmar Patroc'inio Thim. 2019. "Recent Advances in the Use of Carbon Nanotubes as Smart Biomaterials." *Journal of Materials Chemistry B* 7 (9): 1343–60.

Eatemadi, Ali, Hadis Daraee, Hamzeh Karimkhanloo, Mohammad Kouhi, Nosratollah Zarghami, Abolfazl Akbarzadeh, Mozhgan Abasi, Younes Hanifehpour, and Sang Woo Joo. 2014. "Carbon Nanotubes: Properties, Synthesis, Purification, and Medical Applications." *Nanoscale Research Letters* 9 (1): 1–13.

Ebbesen, T. W., and P. M. Ajayan. 1992. "Large-Scale Synthesis of Carbon Nanotubes." *Nature* 358 (6383): 220–2.

Fenner, Ralph, and Edward Zdankiewicz. 2001. "Micromachined Water Vapor Sensors: A Review of Sensing Technologies." *IEEE Sensors Journal* 1 (4): 309–17.

Ferreira, Filipe V., Wesley Franceschi, Beatriz R. C. Menezes, Audrey F. Biagioni, Aparecido R. Coutinho, and Luciana S. Cividanes. 2019. "Synthesis, Characterization, and Applications of Carbon Nanotubes." In *Carbon-Based Nanofillers and Their Rubber Nanocomposites*, 1–45. Cambridge, United States: Elsevier.

Gupta, Banshi D, and Ravi Kant. 2018. "Recent Advances in Surface Plasmon Resonance Based Fiber Optic Chemical and Biosensors Utilizing Bulk and Nanostructures." *Optics & Laser Technology* 101: 144–61.

Han, Tao, Anindya Nag, Subhas Chandra Mukhopadhyay, and Yongzhao Xu. 2019. "Carbon Nanotubes and Its Gas-Sensing Applications: A Review." *Sensors and Actuators A: Physical* 291: 107–43.

He, Lifang, Yong Jia, Fanli Meng, Minqiang Li, and Jinhuai Liu. 2009. "Gas Sensors for Ammonia Detection Based on Polyaniline-Coated Multi-Wall Carbon Nanotubes." *Materials Science and Engineering: B* 163 (2): 76–81.

Hinman, Samuel S., Kristy S. McKeating, and Quan Cheng. 2018. "Surface Plasmon Resonance: Material and Interface Design for Universal Accessibility." *Analytical Chemistry* 90 (1): 19.

Homola, Jivrí, Sinclair S. Yee, and Günter Gauglitz. 1999. "Surface Plasmon Resonance Sensors." *Sensors and Actuators B: Chemical* 54 (1–2): 3–15.

Hong, Yong Cheol, and Han Sup Uhm. 2005. "Production of Carbon Nanotubes by Microwave Plasma Torch at Atmospheric Pressure." *Physics of Plasmas* 12 (5): 53504.

Hsu, W. K., J. P. Hare, M. Terrones, H. W. Kroto, D. R. M. Walton, and P. J. F. Harris. 1995. "Condensed-Phase Nanotubes." *Nature* 377 (6551): 687.

Iijima, Sumio. 1991. “Helical Microtubules of Graphitic Carbon.” *Nature* 354 (6348): 56–8.

Jašek, Ondvrej, Marek Eliáš, Lenka Zaj’ičková, Vit Kudrle, Martin Bublan, Jivrina Matějková, Antonin Rek, Jivr’i Burš’ik, and Magdaléna Kadleč’iková. 2006. “Carbon Nanotubes Synthesis in Microwave Plasma Torch at Atmospheric Pressure.” *Materials Science and Engineering: C* 26 (5–7): 1189–93.

Jiang, Xin, and Qi Wang. 2019. “Refractive Index Sensitivity Enhancement of Optical Fiber SPR Sensor Utilizing Layer of MWCNT/PtNPs Composite.” *Optical Fiber Technology* 51: 118–24.

Jing, Jian-Ying, Qi Wang, and Bo-Tao Wang. 2018. “Refractive Index Sensing Characteristics of Carbon Nanotube-Deposited Photonic Crystal Fiber SPR Sensor.” *Optical Fiber Technology* 43: 137–44.

Journet, Catherine, W. K. Maser, Patrick Bernier, Annick Loiseau, M. Lamy de La Chapelle, dl S. Lefrant, Philippe Deniard, R. Lee, and J. E. Fischer. 1997. “Large-Scale Production of Single-Walled Carbon Nanotubes by the Electric-Arc Technique.” *Nature* 388 (6644): 756–8.

Kim, Si-Jin, A-Young Lee, Han-Chul Park, So-Young Kim, Min-Cheol Kim, Jong-Min Lee, Seong-Bae Kim, Woo-Seong Kim, Youngjin Jeong, and Kyung-Won Park. 2015. “Carbon Nanotube Web-Based Current Collectors for High-Performance Lithium Ion Batteries.” *Materials Today Communications* 4: 149–55.

Koziol, Krzysztof, Bojan Obrad Boskovic, and Noorhana Yahya. 2010. “Synthesis of Carbon Nanostructures by CVD Method.” In *Carbon and Oxide Nanostructures*, 23–49. Berlin, Heidelberg: Springer.

Kumar, Mukul, and Yoshinori Ando. 2010. “Chemical Vapor Deposition of Carbon Nanotubes: A Review on Growth Mechanism and Mass Production.” *Journal of Nanoscience and Nanotechnology* 10 (6): 3739–58.

Lefebvre, J., Y. Homma, and Paul Finnie. 2003. “Bright Band Gap Photoluminescence from Unprocessed Single-Walled Carbon Nanotubes.” *Physical Review Letters* 90 (21): 217401.

Liu, Jie, Andrew G. Rinzler, Hongjie Dai, Jason H. Hafner, R. Kelley Bradley, Peter J. Boul, Adrian Lu, et al. 1998. “Fullerene Pipes.” *Science* 280 (5367): 1253–6.

Matsuda, K. 2013. “Fundamental Optical Properties of Carbon Nanotubes and Graphene.” In *Carbon Nanotubes and Graphene for Photonic Applications* (Woodhead Publishing Series in Electronic and Optical Materials) 3–25. Cambridge, England: Elsevier.

Matsuzawa, Yoko, Yuko Takada, Tetsuya Kodaira, Hideyuki Kihara, Hiromichi Kataura, and Masaru Yoshida. 2014. “Effective Nondestructive Purification of Single-Walled Carbon Nanotubes Based on High-Speed Centrifugation with a Photochemically Removable Dispersant.” *The Journal of Physical Chemistry C* 118 (9): 5013–9.

Meyyappan, Meyya. 2004. *Carbon Nanotubes: Science and Applications*. Boca Raton, FL: CRC Press.

Mishra, Akhilesh Kumar, and Satyendra Kumar Mishra. 2016. “Gas Sensing in Kretschmann Configuration Utilizing Bi-Metallic Layer of Rhodium-Silver in Visible Region.” *Sensors and Actuators B: Chemical* 237: 969–73.

Mishra, Satyendra K., Deepa Kumari, and Banshi D. Gupta. 2012. “Surface Plasmon Resonance Based Fiber Optic Ammonia Gas Sensor Using ITO and Polyaniline.” *Sensors and Actuators B: Chemical* 171: 976–983.

Morsy, Mohamed, Magdy Helal, Mohamed El-Okr, and Medhat Ibrahim. 2014. “Preparation, Purification and Characterization of High Purity Multi-Wall Carbon Nanotube.” *Spectrochimica Acta Part A: Molecular and Biomolecular Spectroscopy* 132: 594–8.

Mubarak, N. M., E. C. Abdullah, N. S. Jayakumar, and J. N. Sahu. 2014. “An Overview on Methods for the Production of Carbon Nanotubes.” *Journal of Industrial and Engineering Chemistry* 20 (4): 1186–97.

O'Connell, Michael J., Sergei M. Bachilo, Chad B. Huffman, Valerie C. Moore, Michael S. Strano, Erik H. Haroz, Kristy L. Rialon, et al. 2002. "Band Gap Fluorescence from Individual Single-Walled Carbon Nanotubes." *Science* 297 (5581): 593–6.

Pathak, Anisha, and Banshi D. Gupta. 2021. "Palladium Nanoparticles Embedded PPy Shell Coated CNTs towards a High Performance Hydrazine Detection through Optical Fiber Plasmonic Sensor." *Sensors and Actuators B: Chemical* 326: 128717.

Radhamani, A. V., Hon Chung Lau, and S. Ramakrishna. 2018. "CNT-Reinforced Metal and Steel Nanocomposites: A Comprehensive Assessment of Progress and Future Directions." *Composites Part A: Applied Science and Manufacturing* 114: 170–87.

Rauti, Rossana, Mattia Musto, Susanna Bosi, Maurizio Prato, and Laura Ballerini. 2019. "Properties and Behavior of Carbon Nanomaterials When Interfacing Neuronal Cells: How Far Have We Come?" *Carbon* 143: 430–46.

Rittersma, Z. M. 2002. "Recent Achievements in Miniaturised Humidity Sensors—A Review of Transduction Techniques." *Sensors and Actuators A: Physical* 96 (2–3): 196–210.

Saito, Riichiro, Mitsutaka Fujita, G. Dresselhaus, and U. M. S. Dresselhaus. 1992. "Electronic Structure of Chiral Graphene Tubules." *Applied Physics Letters* 60 (18): 2204–6.

Samuel Ratna Kumar P. S., D. S. Robinson Smart, and S. John Alexis. 2017. "Corrosion Behaviour of Aluminium Metal Matrix Reinforced with Multi-Wall Carbon Nanotube." *Journal of Asian Ceramic Societies* 5 (1): 71–5.

Shah, Khurshed A., and Bilal A. Tali. 2016. "Synthesis of Carbon Nanotubes by Catalytic Chemical Vapour Deposition: A Review on Carbon Sources, Catalysts and Substrates." *Materials Science in Semiconductor Processing* 41: 67–82.

Sharma, Anuj K., Rajan Jha, and B. D. Gupta. 2007. "Fiber-Optic Sensors Based on Surface Plasmon Resonance: A Comprehensive Review." *IEEE Sensors Journal* 7 (8): 1118–29.

Tabassum, Rana, Satyendra K. Mishra, and Banshi D. Gupta. 2013. "Surface Plasmon Resonance-Based Fiber Optic Hydrogen Sulphide Gas Sensor Utilizing Cu--ZnO Thin Films." *Physical Chemistry Chemical Physics* 15 (28): 11868–74.

Thess, Andreas, Roland Lee, Pavel Nikolaev, Hongjie Dai, Pierre Petit, Jerome Robert, Chunhui Xu, et al. 1996. "Crystalline Ropes of Metallic Carbon Nanotubes." *Science* 273 (5274): 483–7.

Traversa, Enrico. 1995. "Ceramic Sensors for Humidity Detection: The State-of-the-Art and Future Developments." *Sensors and Actuators B: Chemical* 23 (2–3): 135–56.

Wang, Ruiqian, Lijuan Xie, Saima Hameed, Chen Wang, and Yibin Ying. 2018. "Mechanisms and Applications of Carbon Nanotubes in Terahertz Devices: A Review." *Carbon* 132: 42–58.

Weisman, R. Bruce, and Sergei M. Bachilo. 2003. "Dependence of Optical Transition Energies on Structure for Single-Walled Carbon Nanotubes in Aqueous Suspension: An Empirical Kataura Plot." *Nano Letters* 3 (9): 1235–8.

Wu, Zuquan, Xiangdong Chen, Shibu Zhu, Zuowan Zhou, Yao Yao, Wei Quan, and Bin Liu. 2013. "Enhanced Sensitivity of Ammonia Sensor Using Graphene/Polyaniline Nanocomposite." *Sensors and Actuators B: Chemical* 178: 485–93.

Xie, Sishen, Wenzhi Li, Zhengwei Pan, Baohe Chang, and Lianfeng Sun. 2000. "Mechanical and Physical Properties on Carbon Nanotube." *Journal of Physics and Chemistry of Solids* 61 (7): 1153–8.

Yin, Ming-jie, Bobo Gu, Quan-Fu An, Chengbin Yang, Yong Liang Guan, and Ken-Tye Yong. 2018. "Recent Development of Fiber-Optic Chemical Sensors and Biosensors: Mechanisms, Materials, Micro/Nano-Fabrications and Applications." *Coordination Chemistry Reviews* 376: 348–92.

Zahid, Muhammad Umer, Erum Pervaiz, Arshad Hussain, Muhammad Imran Shahzad, and Muhammad Bilal Khan Niazi. 2018. "Synthesis of Carbon Nanomaterials from Different Pyrolysis Techniques: A Review." *Materials Research Express* 5 (5): 52002.

Zaytseva, Olga, and Günter Neumann. 2016. "Carbon Nanomaterials: Production, Impact on Plant Development, Agricultural and Environmental Applications." *Chemical and Biological Technologies in Agriculture* 3 (1): 1–26.

Zhang, Y. F., M. N. Gamo, C. Y. Xiao, and T. Ando. 2002. "Liquid Phase Synthesis of Carbon Nanotubes." *Physica B: Condensed Matter* 323 (1–4): 293–5.

Zhang, Qiang, Jia-Qi Huang, Meng-Qiang Zhao, Wei-Zhong Qian, and Fei Wei. 2011. "Carbon Nanotube Mass Production: Principles and Processes." *ChemSusChem* 4 (7): 864–89.

7 Recent Trends of Transition-Metal Dichalcogenides (TMDC) Material for SPR Sensors

7.1 INTRODUCTION

The TMDC material is one of the 2D materials, and is an MX_2-type semiconductor material that is comprised of a transition metal atom and a chalcogen atom, so it is named a transition metal dichalcogenide (TMDC). The significant advancements in the development of atomically thin TMDC material became popular in various research areas including high-performance electronics, spintronics, flexible electronics, nanoelectronics, optoelectronics, energy harvesting, energy storage, superconductor, non-linear optics, photonics, sensor, DNA sequencing, and the biomedical field because of the unique features including direct bandgap, strong spin-orbit coupling, atomic-scale thickness, large specific surface area, and van der Waals (vdW) gap (Zhou, Sun, and Bai 2020). Fig. 7.1 shows the complete family of TMDC materials and the application area of TMDCs.

Since TMDC materials have a wide range of properties, including mechanical, thermal conductivity, thermoelectric, optical, and electrical transport, they can be used in numerous applications, briefly discussed in this section.

7.1.1 Properties of TMDCs

7.1.1.1 Mechanical Properties

The mechanical properties such as elasticity, braking, and stretching of a freely suspended ultra-thin MoS_2 layer were studied (Bertolazzi, Brivio, and Kis 2011; Castellanos-Gomez et al. 2012) in which the tip of AFM was mounted on top of a thin film coated over a pre-patterned small hole on a SiO_2 substrate for the deflection measurement of an MoS_2 membrane with force applied. Fig. 7.2 shows the bending test experiment setup and loading curves with the sample thickness of 5–20 layers where the MoS_2 membrane pretension, σ_0^{2D}, and elastic modulus, E^{2D}, are determined by a least-squares fit of the experimental curves; however, it was found its numerical value in both studies slightly differs.

The mechanical behavior of several monolayer TMDCs concerning large elastic deformation was investigated (Li, Medhekar, and Shenoy 2013) by calculation of the

DOI: 10.1201/9781003190738-7

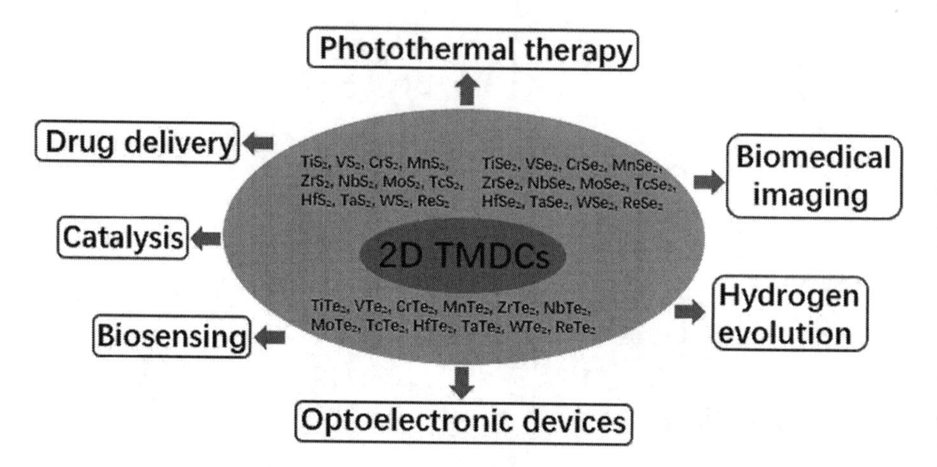

FIGURE 7.1 Summary of the types and applications of 2D TMDCs (Zhou, Sun, and Bai 2020).

first-principles' density function. Fig. 7.3a–c shows the computed tensile stress, σ, versus uniaxial strain, ε curve, for the armchair (x), zigzag (y), and biaxial tensions and found that for all monolayer TMDCs with small strain, the stress manifests a linear response for the strain applied in all loading directions. However, for monolayer TMDCs with a larger strain, the stress-strain response no longer holds the linear characteristics, and also, the stress-induced armchair direction is much larger than the zigzag direction.

It was found that the anisotropy stress response is inversely proportional to the ultimate strength of the monolayer TMDCs' sheet, i.e., the anisotropy factor is more significant for the lower young modulus and ultimate strength of TMDCs. The elastic properties of TMDCs' material were characterized by an amount of charge transfer from a transition metal to chalcogens. Also, Fig. 7.4a,b shows that the young modulus, E, and ultimate strength, σ^*, are functions of charge transfer, ΔQ, which infers that the TMDCs' material mechanical properties linearly increase with charge transfer from a transition metal to chalcogen atoms.

7.1.1.2 Thermal Conductivity

In a monolayer MoS_2, the edge roughness and size are influenced by thermal conductivity, and is described by the Boltzmann transport equation of phonon coupled with relaxation time approximation of an acoustic phonon (such as transverse, longitudinal, and out-of-plane) (Wei et al. 2014). Although thermal conductivity of samples of various sizes was calculated by first-principles' simulations and inferred that the thermal conductivity of standard samples with 1 μm size should be greater than 83 W m K^{-1} at room temperature (Li, Carrete, and Mingo 2013), this contradicts prior results (Fig. 7.5).

The anharmonic behavior of phonons and umklapp scattering was due to the intrinsic thermal conductivity in a monolayer MoS_2 that was studied (Cai, Lan, Zhang, and Zhang 2014) and that found a 23.2 W m K^{-1} thermal conductivity at room temperature. However, monolayer MoS_2 thermal conductivity using DFT methods was found at about 1.35 W m K^{-1} (Liu, Zhang, Pei, and Zhang 2013). The thermal conductivity was anisotropic and showed (Jiang, Zhuang, and Rabczuk 2013) its value

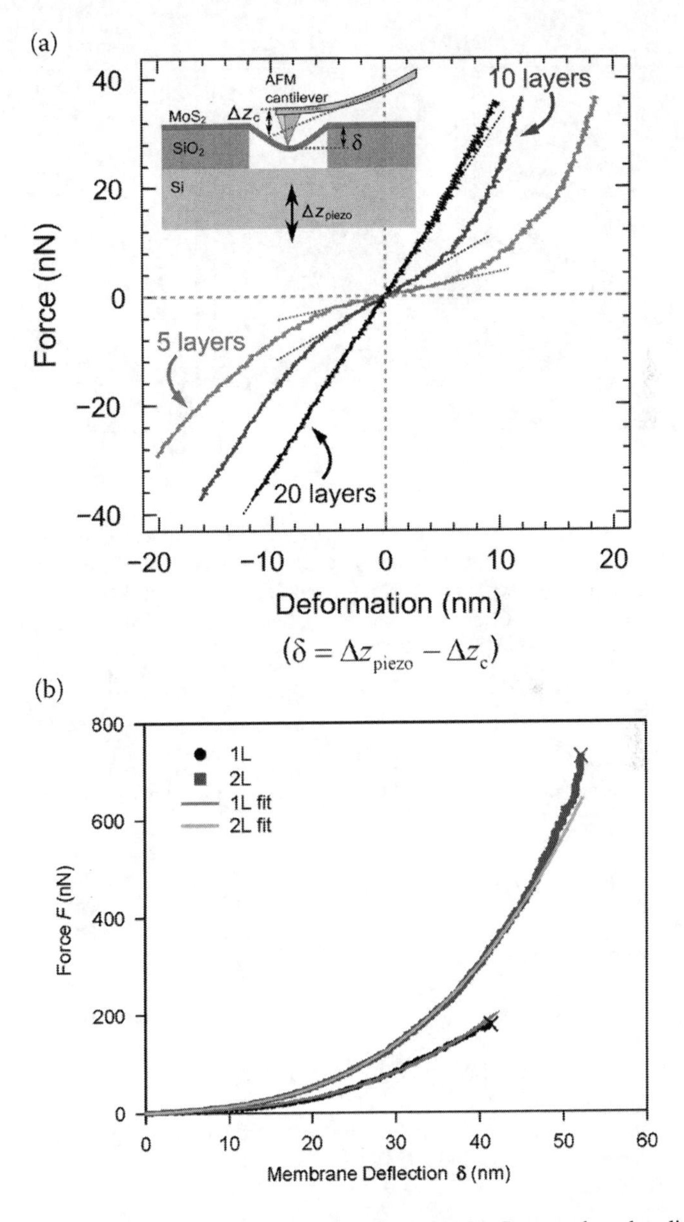

FIGURE 7.2 a) Schematic diagram of freely suspended MoS_2 nanosheet bending test experiment and its measured force vs. deformation curve for 5, 10, and 20 layers of MoS_2 nanosheets (Castellanos-Gomez et al. 2012). b) Loading curves for single and bilayer MoS_2. Reprinted with permission from ACS Nano, Copyright, 2011, American Chemical Society (Bertolazzi, Brivio, and Kis 2011).

for the armchair MoS_2 nanoribbon was about 673.6 W m K^{-1} under room temperature, while the zigzag nanoribbon had a value of 841.1 W m K^{-1}. It was also observed that thermal conductivity was inversely proportional to temperature.

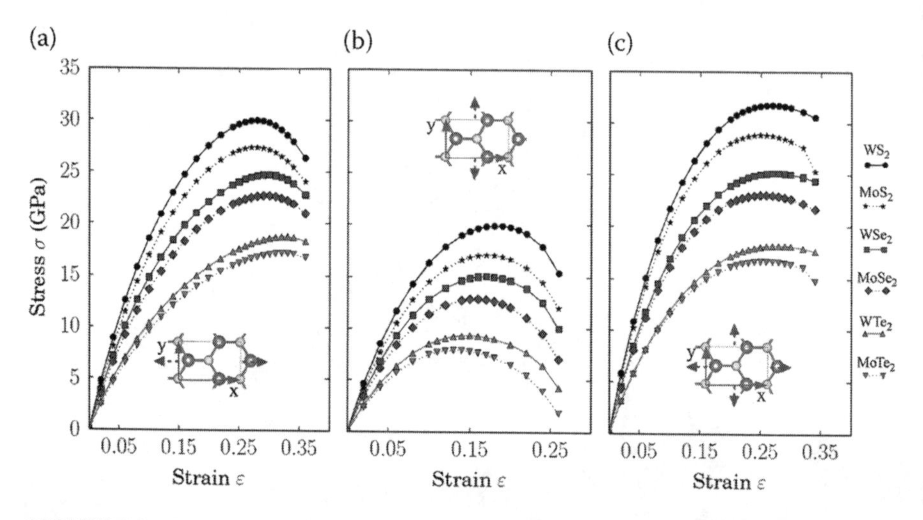

FIGURE 7.3 Response of tensile stress, σ, versus uniaxial strain, ε, for monolayer TMDCs, MX_2 (M = Mo, W; X = S, Se, Te) in the a) armchair direction, b) zigzag direction, and c) biaxial tension. Reprinted with permission from The Journal of Physical Chemistry C, Copyright, 2013, American Chemical Society (Li, Medhekar, and Shenoy 2013).

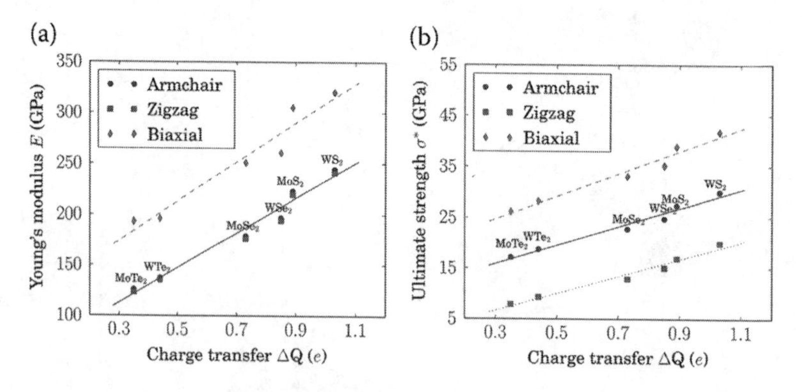

FIGURE 7.4 a) Young's modulus and (b) the ultimate strength variation with the charge transfer, ΔQ, in monolayer TMDCs, MX_2 (M = Mo, W; X = S, Se, Te). Reprinted with permission from The Journal of Physical Chemistry C, Copyright, 2013, American Chemical Society (Li, Medhekar, and Shenoy 2013).

7.1.1.3 Thermoelectric Properties

The thermoelectric properties of monolayer TMDCs were reported (Huang, Da, and Liang 2013; Huang et al. 2014) were described by a 2D ballistic transport method. It governed from complete electronic band structures with dispersion relation of phonon energy evaluated using first-principles' calculations, depending on the orientation of crystal and temperature of p-type and n-type materials. However, such materials usually have a low figure of merit. First-principles' calculations and semiclassical Boltzmann transport theory described the thermoelectric properties of

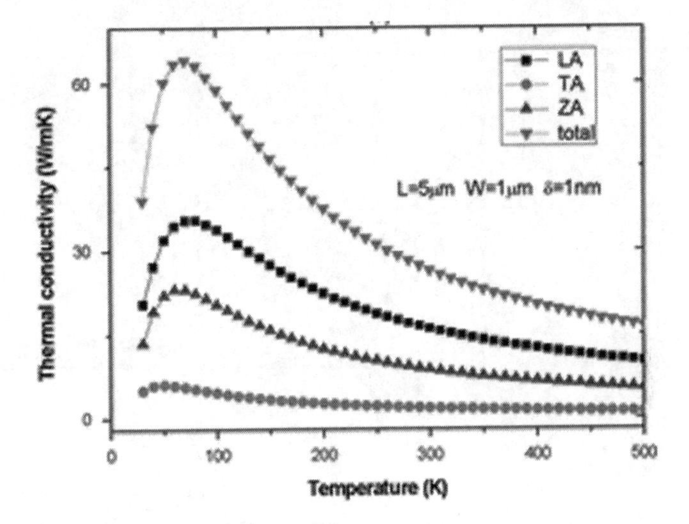

FIGURE 7.5 Variation of thermal conductivity of a MoS_2 monolayer flake with temperature. Reprinted with permission from Applied Physics Letters, Copyright 2014, the American Institute of Physics (Wei et al. 2014).

bulk and monolayer $MoSe_2$ and WSe_2 and found that WSe_2 is better than $MoSe_2$ (Kumar and Schwingenschlögl 2015). In the presence of off-resonant light, thermoelectric transport of the MoS_2 monolayer and relevant group-VI dichalcogenides was analyzed analytically (Tahir and Schwingenschlögl 2014) and showed as light intensity increased, the direct bandgap decreased, which resulted in a strong spin splitting in the conduction band and a drastic increase in thermoelectric transport.

7.1.1.4 Optical Properties

The study of optical properties in TMDC materials has significant contributions in electronics, optoelectronics, spintronics, and valley electronics. The optical band structure of materials is characterized by a tool called photoluminescence.

The reflectance measurements of monolayers of TMDCs at room temperature were used to determine their dielectric function. Fig. 7.6a–d shows the absolute reflectance spectra of TMDC monolayers on fused silica. In the reflectance spectra of all four monolayers of TMDCs, the two lowest energy peaks describe the excitonic features associated with interband transitions at the K (K^1) point in the Brillouin zone (Li, Chernikov, et al. 2014). Here, A and B characterize the valance band splitting with spin-orbit coupling characterized. Higher-lying interband transitions showed a spectrally broad response at higher photon energies.

The constrained analysis of the Kramers–Kronig relation yielded dielectric functions, and the resulting real and imaginary parts for MoS_2, $MoSe_2$, WS_2, and WSe_2 for the 1.5–3.0 eV spectral range are shown in Fig. 7.6e–l.

Fig. 7.7 compares the measured dielectric function of monolayer TMDC crystals and their corresponding bulk materials. It shows similarities between both data sets and variations in the spectral response, such as resonance broadening in the bulk

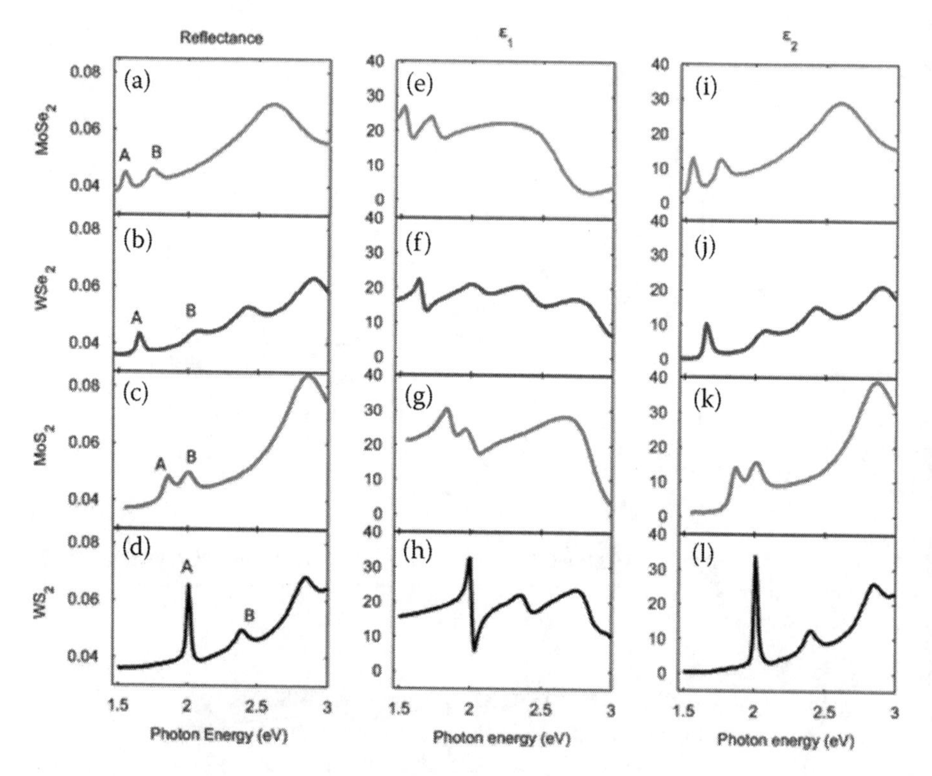

FIGURE 7.6 Optical response of monolayers of TMDC, such as $MoSe_2$, WSe_2, MoS_2, and WS_2 exfoliated on fused silica: (a)–(d) Measured reflectance spectra; (e)–(h) real part of dielectric function, ε_1; and (i)–(l) imaginary part of dielectric function, ε_2. The peaks labeled *A* and *B* in (a)–(d) correspond to excitons from the two spin-orbit split transitions at the *K* point of the Brillouin zone. Reprinted with permission from Physical Review B, Copyright, 2014, American Physical Society (Li, Chernikov, et al. 2014).

material relative to monolayers that occur due to the addition of an optical transition and a carrier relaxation channel caused by interlayer coupling. Furthermore, it observes a modest shift in the resonance energies of the dielectric function of a monolayer and corresponding bulk material.

The investigation of optical characteristics of monolayer TMDCs, such as MoS_2, $MoSe_2$, WS_2, and WSe_2 by spectroscopic ellipsometry, was reported (Liu et al. 2014), and the measurement of refractive index (RI) and extinction coefficient spectra of these thin films are shown in Fig. 7.8.

In a monolayer TMDC, optical functions have a direct correlation with bandgap magnitude and the binding energy of exciton. The RI increases with increasing wavelength over the 193–550-nm spectral range, and then approaches maxima and decreases with a wavelength until 1,700 nm. It revealed that the RI dispersive response exhibits many anomalous dispersions features below 800 nm and reaches a constant value of 3.5–4.0 in the range of near-infrared frequency range, and the MoS_2 monolayer has an extremely high RI of about 6.5 at a wavelength of 450 nm.

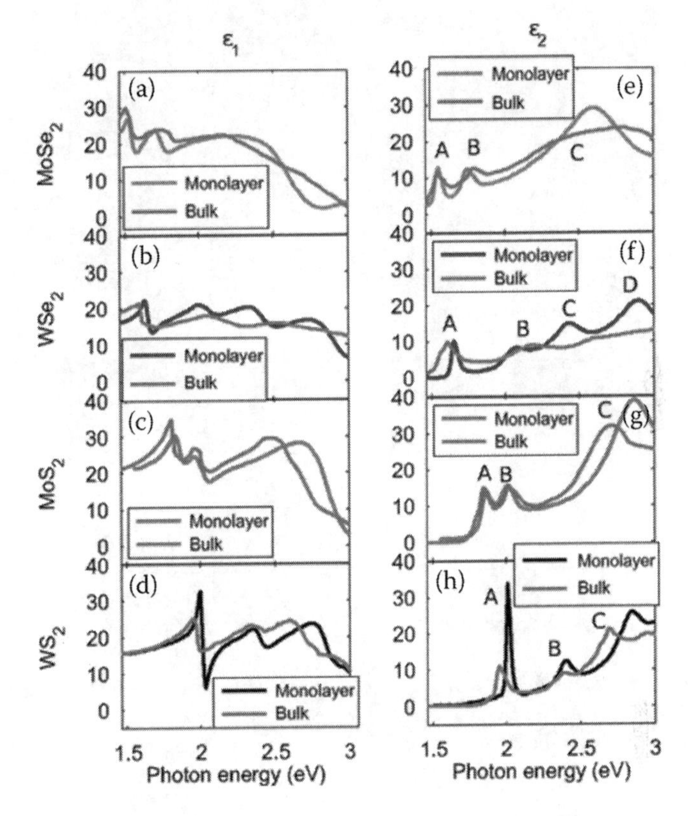

FIGURE 7.7 Dielectric function comparison between monolayer TMDCs (color) and bulk TMDCs (gray). Reprinted with permission from Physical Review B, Copyright, 2014, American Physical Society (Li, Chernikov, et al. 2014).

Fig. 7.9 shows the absorption spectra of monolayer TMDCs films such as MoS_2, $MoSe_2$, WS_2, and WSe_2 that can divide into the low-energy region and high-energy region. The low-energy region signifies the low absorption due to excitonic transitions, while high energy signifies a strong absorption. However, the discrete exciton states are modeled by a broadened Lorentzian line shape in the thin-film monolayer TMDCs.

The optical constants n and k in the visible range for ultrathin MoS_2 and NbS_2 crystals have been calculated (Castellanos-Gomez, Agrat, and Rubio-Bollinger 2010). Nonlinear optical characteristics of TMDC nanosheets were investigated (Dong et al. 2015), in which nonlinear optical characteristics like SHG occur due to a lack of center of symmetry or center inversion present in the crystal lattice of a monolayer TMDC material. The purpose of SHG is to study the material thickness and different crystal structures.

7.1.1.5 Electric Transport

The carrier scattering and transport in the 2D TMDC layer is confined in the material plane. The in-plane carrier mobility is principally expressed by momentum scattering time, τ_D

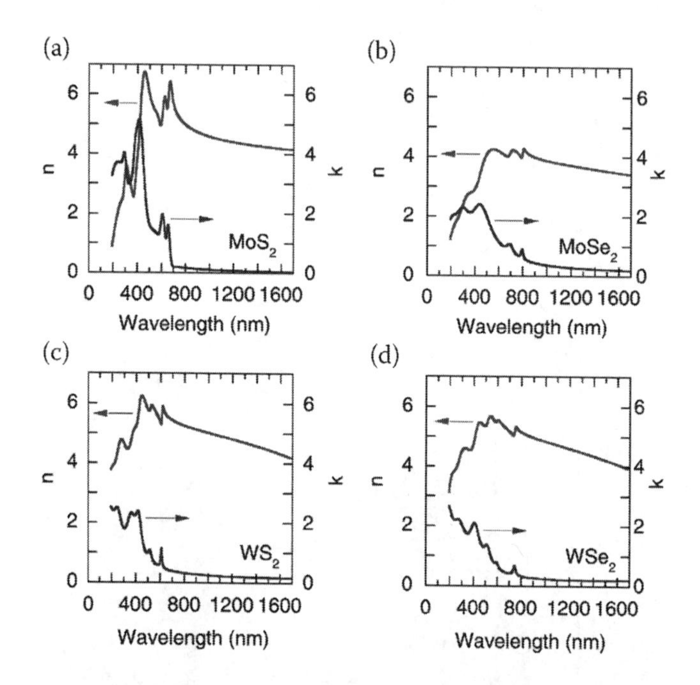

FIGURE 7.8 Refractive index, *n*, and extinction coefficient, *k*, of thin-film monolayer TMDCs: a) MoS_2, b) $MoSe_2$, c) WS_2, and d) WSe_2. Reprinted with permission from Applied Physics Letters, Copyright 2014, the American Institute of Physics (Liu et al. 2014).

$$\mu = \frac{e\tau_D}{m*}$$

where $m*$ is the in-plane effective mass.

Scattering influences the carrier mobility in a material due to the optical and acoustic phonon, charged impurities, scattering due to roughness, and surface interface phonon. The layer thickness, temperature, carrier concentration, effective mass of the carrier, band structure of electrons, and phonon influence these scattering mechanisms, further affecting carrier mobility and transport (Kolobov and Tominaga 2016). The acoustic component dominates carrier mobility at low temperatures, i.e., $T < 100$ K, while the optical component dominates at higher temperatures. However, Coulomb scattering has a dominant effect in 2D TMDCs at low temperatures and is caused due to randomly charged impurities inside the 2D TMDC layer or on its surfaces.

However, the scattering effect due to surface phonon and roughness is very significant in fragile 2D materials. MoS_2 has a limited phonon at room temperature, and its mobility is about 410 cm^2 $(V\ s)^{-1}$; comparable values are predicted for other monolayer TMDCs (Kaasbjerg, Thygesen, and Jacobsen 2012).

The electron's mobility and hole's mobility for the various structural modifications of MoS_2 were computed by first-principal calculations coupled to the Boltzmann transport equation (Kan et al. 2014). It shows that 2H-MoS_2 was found isotropic and its electron and hole mobility were found to be 1.2×10^2 and 3.8×10^2 cm^2 $(V\ s)^{-1}$;

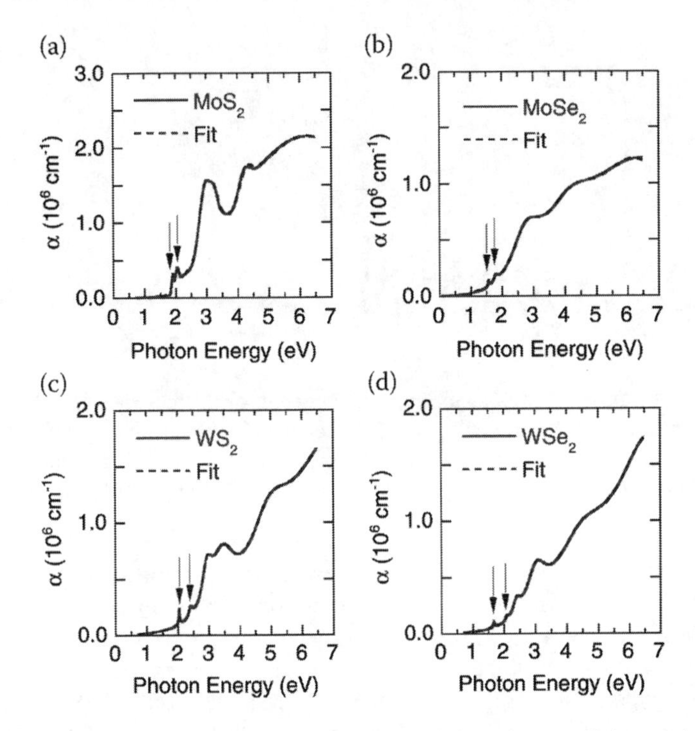

FIGURE 7.9 Thin-film monolayer TMDCs' optical absorption coefficients: a) MoS_2, b) $MoSe_2$, c) WS_2, and d) WSe_2. Reprinted with permission from Applied Physics Letters, Copyright 2014, the American Institute of Physics (Liu et al. 2014).

however, ZT-MoS_2 was anisotropic, and its electron and hole mobility was found to be 4.1 (2.1) × 10^3 and 6.4 (5.7) × 10^4 cm^2 $(V\ s)^{-1}$ in the x and y directions, respectively; that is one or two orders of magnitude higher compared to 2H-MoS_2. The effective mass of electrons and holes reduced from 0.49 m_e to 0.12 m_e and 0.60 m_e to 0.05 m_e, respectively, as 2H-MoS_2 converted to ZT-MoS_2, which increased mobility.

DC probe techniques measure the electrical conductivity of a monolayer $MoSe_2$ and are found at approximately 10^{-4} Ω^{-1} cm^{-1} at room temperature. The value of electrical conductivity for $MoSe_2$ compared to other materials is significantly less due to the presence of the surface state, nano-crystallinity, and less thickness of the film.

7.2 HISTORY OF TMDCS

Today, the great success in the development and realization of 2D materials has become a research trend due to their extraordinary properties. Every year, the library of 2D materials is growing, and more than 150 researches were reported on exotic layered materials. Fig. 7.10a (Choi et al. 2017) shows the publication list of 2D materials where the research trend is increasing toward the TMDC materials because it is thin, transparent, and flexible. Fig. 7.10b (Zappa 2017) is based on SCOPUS data from Elsevier B.V., and indicates the number of publications per year on TMDC and its applications in sensor development.

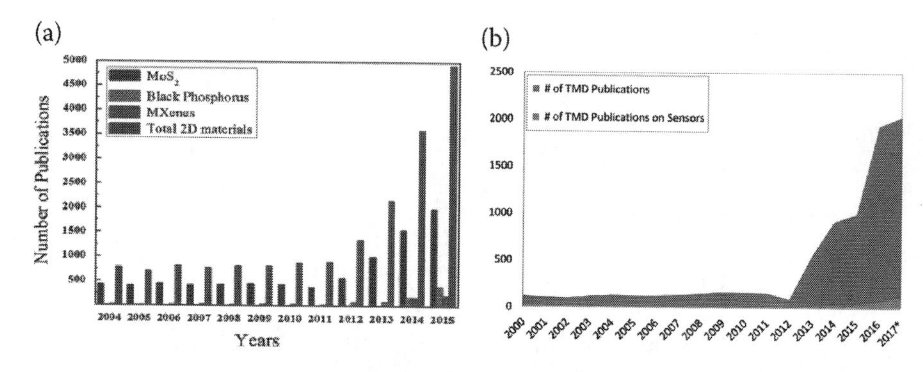

FIGURE 7.10 a) Bar chart of annual publication report on 2D materials including TMDCs, black phosphorous, MXenes, and total 2D materials. Reprinted with permission from Materials Today. Copyright, 2017, Elsevier (Choi et al. 2017). b) Annual publication report on TMDCs (in blue) and on TMDC-based sensors (in red) (Zappa 2017).

The history of TMDC materials is not new; however, MoS_2 is one of the oldest samples in the TMDC material group known for over 2.9 billion years (Golden et al. 2013). The brief review on TMDC materials and their properties is first reported by Wilson and Yoffe (1969). The TMDC family of layered materials has characteristics that range from insulator, semiconductor, metal, and superconductor and further depend on the elemental configuration and electronic properties (Chhowalla et al. 2013). The families of transition metals and chalcogen elements is highlighted in the periodic table shown in Fig. 7.11 (Meng et al. 2021), which are the essential building blocks of about 40 different TMDC materials. There is a wide range of elements as well as three distinct phases: rhombohedral symmetry (3R), hexagonal symmetry (2H), and trigonal symmetry (1T). The bandgap of monolayer and bulk TMDC materials are summarized in Fig. 7.12 (Duan et al. 2015). Since TMDCs have a wide range of properties, including mechanical, thermal conductivity, thermoelectric, optical, and electrical transport, they can be used in various applications, briefly discussed in this section.

7.3 CHALLENGES OF TMDC

The major challenges in 2D TMDC materials are their synthesis and layer transfer on the other substrate for the various applications. However, several synthesis processes of the 2D TMDC layer are reported in literature. There are mainly two methods for the growth of 2D TMDCs: the top-down method and bottom-up method. In the top-down method, the TMDCs' bulk form is exfoliated into a monolayer or multilayer structure, while the bottom-up method uses chemical vapor deposition (CVD) or molecular epitaxy techniques for the synthesis of TMDC. However, both approaches for the synthesis of 2D TMDCs are described in this section.

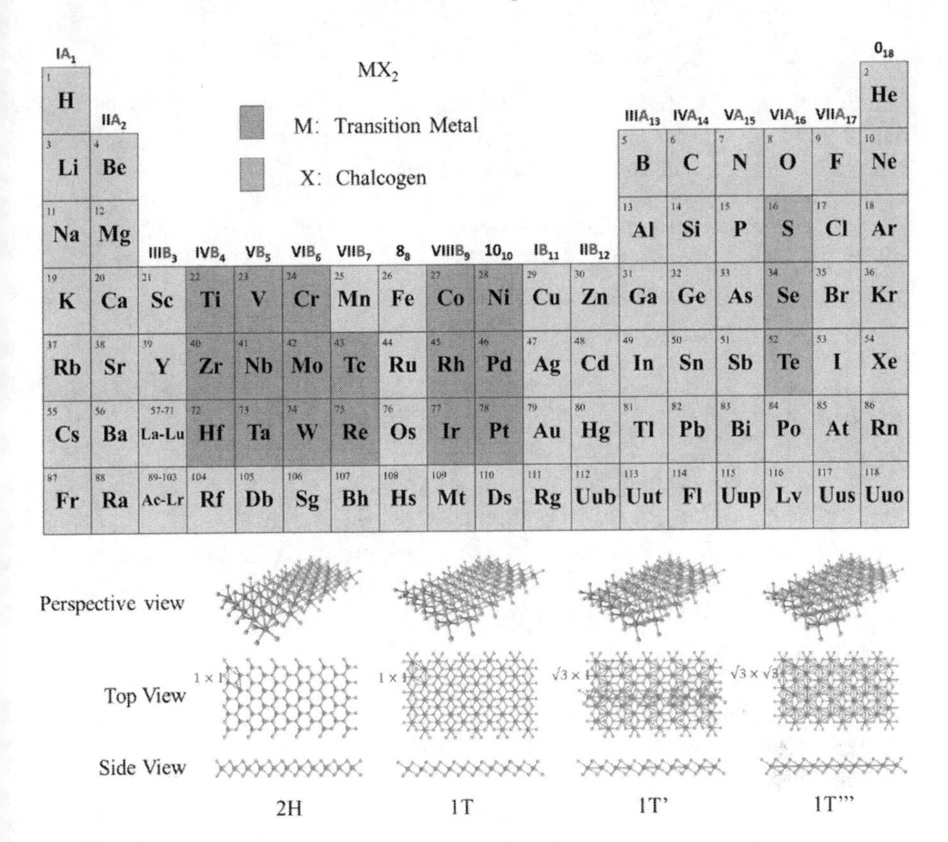

FIGURE 7.11 The family of the transition metals and chalcogen elements in the periodic table for the layered TMDC materials (top). Structural polytypes (2H, 1 T, 1 T′ and 1 T''') of TMDC materials (bottom). Reprinted with permission from Biomaterial. Copyright, 2021, Elsevier (Meng et al. 2021).

Bandgap (eV)		Mo	W	Ti	Zr	Hf	V	Nb	Ta	Ni	Pd	Pt
S	monolayer	1.8-2.1	1.8-2.1	~0.65	~1.2	~1.3	~1.1	metal	metal	~0.6	~1.2	~1.9
	Bulk	1.0-1.3	1.3-1.4	~0.3	~1.6	~1.6	metal	metal	metal	~0.3	~1.1	~1.8
Se	monolayer	1.4-1.7	1.5-1.7	~0.51	~0.7	~0.7	metal	metal	metal	~0.12	~1.1	~1.5
	Bulk	1.1-1.4	1.2-1.5	metal	~0.8	~0.6	metal	metal	metal	metal	~1.3	~1.4
Te	monolayer	1.1-1.3	~1.03	~0.1	~0.4	~0.3	metal	metal	metal	metal	~0.3	~0.8
	Bulk	1.0-1.2	metal	metal	metal	metal	metal	metal	metal	metal	~0.2	~0.8

FIGURE 7.12 Summary of common TMDCs and their bandgaps. Reprinted with permission from Chemical Society Reviews, Copyright 2015 Royal Society of Chemistry (Duan et al. 2015).

7.3.1 Top-Down Method

The growth of 2D TMDC by the top-down method is stripping of 2D TMDCs by bulk TMDCs with a layered structure by the processes of mechanical exfoliation, liquid exfoliation, and electrochemical exfoliation.

7.3.1.1 Mechanical Exfoliation Process

The mechanical exfoliation process is a top-down method and the most effective process to develop the cleanest layered TMDC material in atomically thin nanosheets and crystal form. It creates either a single layer or multiple layers of crystals or crystalline flake materials from the bulk crystals of TMDC materials. But, the growth of thin-film TMDCs with the desired geometry and their performance are limited by this process method, which is a reliable area of research to overcome these growth challenges.

The mechanical exfoliation process for the growth of monolayer or multilayer TMDCs, MoS_2, and WSe_2 in crystalline form is reported (Li et al. 2012; Li, Wu, Yin, and Zhang 2014; Yuan et al. 2016). Typically, the mechanical exfoliation process stripping method happens when the required TMDC layer is stripped away from the bulk crystal by an adhesive tape. These newly cleaved, thin TMDC crystals on the tape are attached to the target substrate's contact and rubbed with plastic tweezers. Monolayer and multilayer TMDC nanosheets are formed on the substrate after removal of the tape. The monolayer to quadruple-layer formation of MoS_2 on a SiO_2 with an Si substrate is shown in Fig. 7.13.

AFM also reveals the layers' thicknesses of MoS_2 nanosheets, ranging from 0.8 nm, 1.5 nm, 2.1 nm, and 2.9 nm for one layer to four layers. The optical contrast difference is often used to distinguish the number of layers of TMDC nanosheets by mechanical exfoliation.

7.3.1.2 Liquid Exfoliation Process

The liquid exfoliation process is a top-down method and the most effective exfoliation process for developing single or multilayer exfoliated TMDC nanosheets on a large scale. This method exploits the mixing and dispersing of various materials to develop the layered TMDC nanocomposites and their hybrids (Coleman et al. 2011; Smith et al. 2011). The intercalation method was reported (Dines 1975), in which the bulk TMDC powder is dispersed with solution based on a lithium compound for more than a day to enable lithium ions to intercalate to water. The lithium intercalation method (Gordon et al. 2002) was demonstrated for the monolayers of WS_2, MoS_2, and $MoSe_2$.

However, the demonstration of electrochemical lithiation for the layered TMDCs MoS_2, WS_2, TiS_2, TaS_2, and ZrS_2 was reported (Zeng et al. 2011; Zhang, Zhang, Su, and Wei 2015), shown in Fig. 7.14 as a three-step process. Step 1, the TMDC bulk materials are used as a cathode of the electrochemical bath, while lithium foil is used as an anode to supply lithium ions. After that, the lithium intercalation process was carried out in the galvanostatic discharge mode with a current density of 0.05 mA. Step 2, following lithium insertion, the intercalated compound (e.g., Lix(TMDC)) was extracted using a sonication process and a

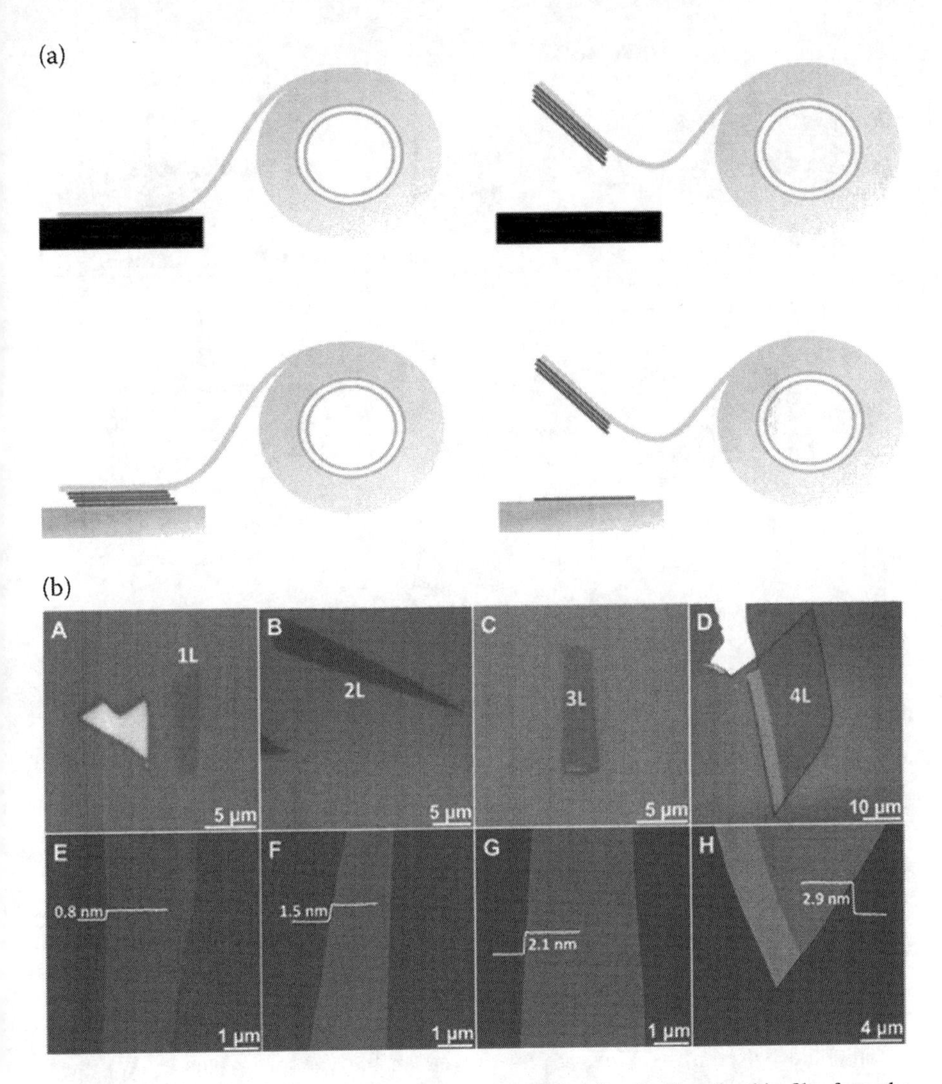

FIGURE 7.13 a) Schematic of the adhesive tape method for cleaving the thin film from the TMDC crystal. Reprinted with permission from Reviews of Modern Physics, Copyright 2011, American Physical Society (Novoselov 2011). b) Mechanically exfoliated single and few-layer MoS_2 nanosheets on 300 nm SiO_2/Si and the optical microscopic (A–D) and AFM (E–H) images of single-layer (1L, thickness 0.8 nm; A and E), double-layer (2L, thickness 1.5 nm; B and F), triple-layer (3L, thickness 2.1 nm; C and G), and quadruple-layer (4L, thickness 2.9 nm; D and H) MoS_2 nanosheets. Reprinted with permission from Accounts of Chemical Research, Copyright 2014 American Chemical Society (Li, Wu, Yin, and Zhang 2014).

series of acetone rinses. Finally, in step 3, corresponding TMDCs were collected as 2D nanosheets.

It is inferred that lithium plays a dual role in this study, supplying Li^+ ions for intercalation with bulk materials and reducing Li(OH) and H_2 gas after reaction with water, which helps exfoliated nanosheets separate from water.

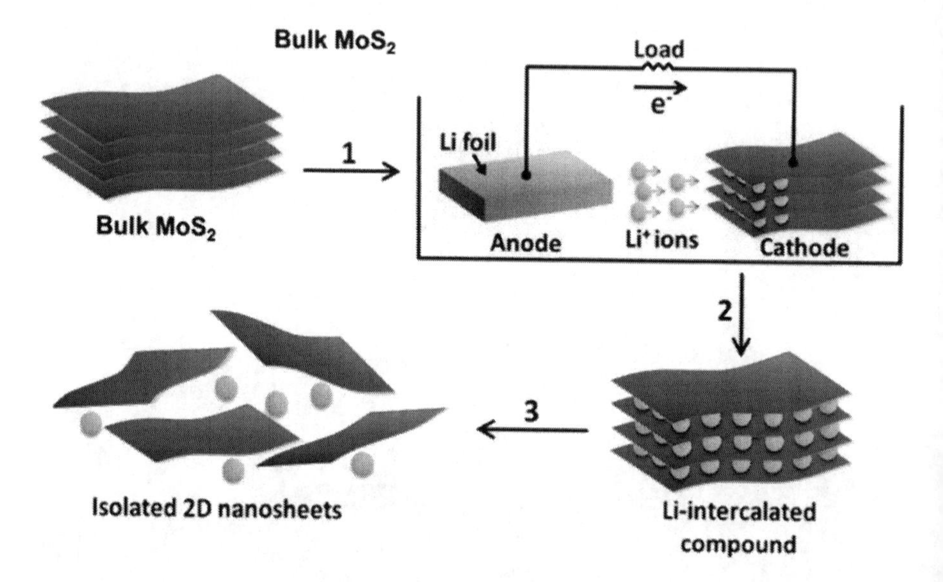

FIGURE 7.14 Electrochemical lithiation process for 2D TMDC nanosheet fabrication. Reprinted with permission from Nanoscale, Copyright 2015 Royal Society of Chemistry (Zhang, Zhang, Su, and Wei 2015).

7.3.1.3 Electrochemical Exfoliation Process

Liu et al. (2004) demonstrated the electrochemical exfoliation experimental setup for bulk MoS_2 crystals, shown in Fig. 7.15. In this process, DC bias is connected between the Pt wire and MoS_2, where low bias voltage wet the bulk MoS_2 while the increase in bias voltage results crystal exfoliates. Fig. 7.15b shows the dissociated flakes of MoS_2 from its bulk crystal and in Fig. 7.15c it is dissolved in solution. Fig. 7.16e shows the electrochemical exfoliation mechanism of bulk MoS_2 crystals.

The positive bias was first applied to the electrode to oxidate water that releases –O and –OH radicals around the bulk MoS_2 crystals. The radicals and/or SO_2^{4-} anions bind between the MoS_2 layers to weaken the van der Waals (vdW) interactions. Second, when radicals and/or anions are oxidized, O_2 and/or SO_2 are released, causing the MoS_2 interlayers to expand significantly. Finally, the erupting gas separates the MoS_2 flakes from the bulk crystal and suspends them in the solution. The fact that bulk MoS_2 should be oxidized during electrochemical exfoliation may affect the exfoliated MoS_2 nanosheets, unless the conditions are optimized, is a significant issue.

7.3.2 Bottom-Up Methods

The bottom-up method is one of the material preparation techniques for 2D TMDC material synthesis, including chemical vapor deposition (CVD) and hydrothermal synthesis.

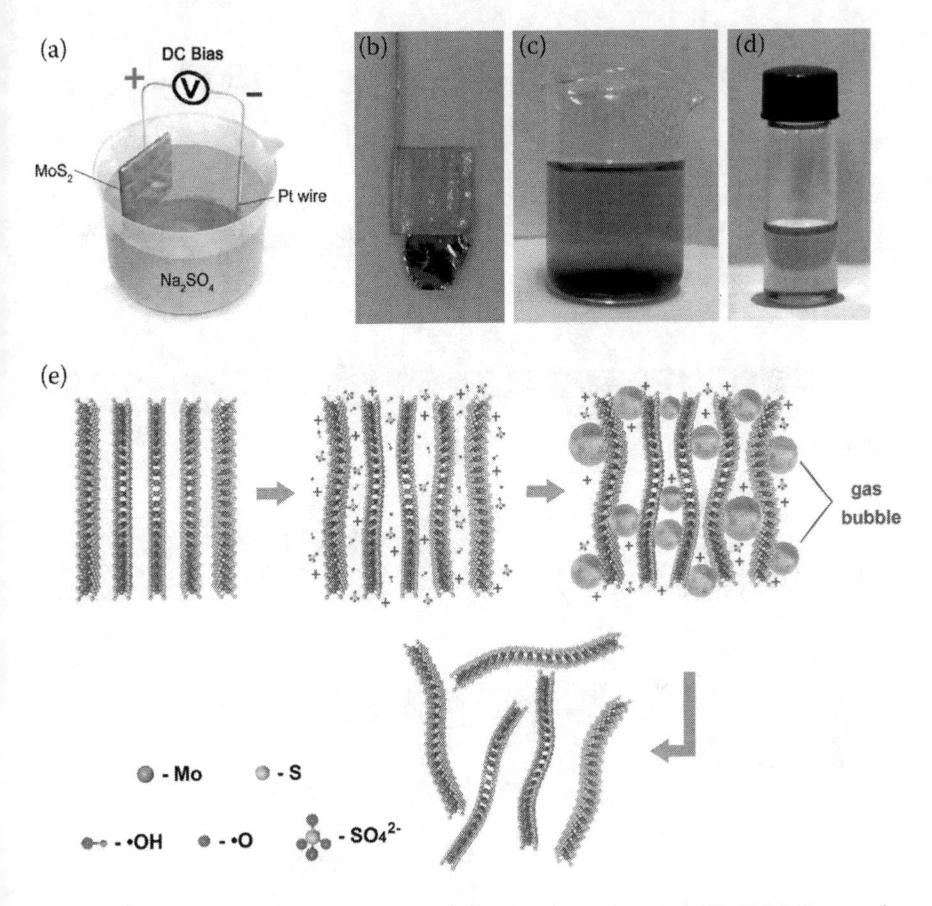

FIGURE 7.15 a) Experimental setup for electrochemical exfoliation of bulk MoS_2 crystals. b) Photograph of a bulk MoS_2 crystal held by a Pt clamp before exfoliation. c) Exfoliated MoS_2 flakes suspended in an Na_2SO_4 solution. d) MoS_2 nanosheets dispersed in an NMP solution. e) Schematic illustration for a mechanism of electrochemical exfoliation of bulk MoS_2 crystals. Reprinted with permission from ACS Nano, Copyright 2014, American Chemical Society (Liu et al. 2014).

7.3.2.1 Chemical Vapor Deposition

The most extensively used bottom-up method is to grow an atomically thin TMDC layer with a large area on the target substrate. There are several methods for TMDC layer growth using CVD techniques reported in the literature, including metal chalcogenisation synthesis (Wang, Feng, Wu, and Jiao 2013), thermolysis of thiosalts (Liu et al. 2012), vapor pressure reaction between transition metal and chalcogen precursors (Lin et al. 2012; Shi, Li, and Li 2015; Su et al. 2014), growth of TMDC alloys (Li, Duan, et al. 2014), and van der Waals epitaxy (Saito, Fons, Kolobov, and Tominaga 2015). However, sulfurization and selenization are two methods for the TMDC layers' growth by the CVD process where the vapor phase reaction conducts by transition metals, along with sulfur or selenium, sometimes

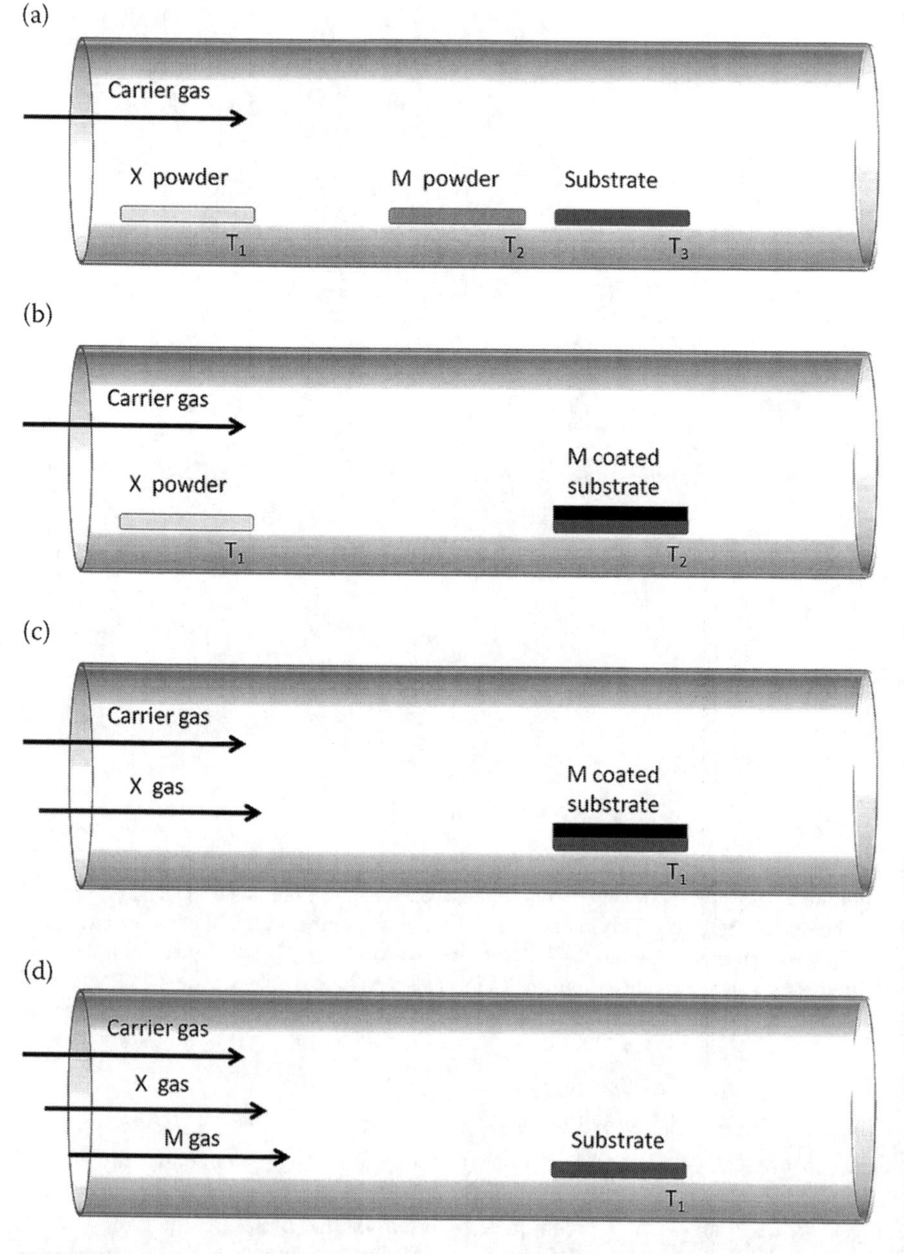

FIGURE 7.16 Schematics of CVD methods for the deposition of TMDCs using the vapor phase: a) Metal (M) and chalcogen (X) powders; b) metal or metal oxides are deposited on a substrate and chalcogen powders; c) metal or metal oxides deposited on substrate and chalcogen supplied as gaseous precursors; and d) metal and chalcogen compounds supplied by gaseous precursors. Reprinted with permission from RSC Advances, Copyright 2015 Royal Society of Chemistry (Bosi 2015).

called synthesis via metal chalcogenization (Li, Duan, et al. 2014). The TMDC layers' growth in the CVD method mainly depends on its process parameters, such as vacuum pressure, carrier gas flow rate, deposition temperature, and temperature ramp that further affect its crystal structure, orientation, surface morphology, and electronic properties.

Mainly the CVD system generates the vapor precursor of source material and delivers those vapor reactants through the reactor at a particular pressure and temperature. In the TMDC layers' growth, the vapor phase reaction is conducted for the transition metal-coated substrate in the presence of a chalcogen atmosphere, but it can also be carried out by a reaction under the vapor phase of both transition metals and chalcogen.

Fig. 7.16 shows a schematic of a three-zone CVD furnace, which is comprised of three zones in a reaction chamber that is patterned for substrates, placement of subsequent precursors, and a heating system with a temperature controller. There are two methods for the synthesis of a monolayer MoS_2 in CVD. In the first method, Mo is deposited on the precursor, and then gaseous sulfur sulphurizes it into the MoS_2 precursor. However, in the second method, gaseous Mo and sulfur are fed in the CVD system to form the MoS_2 layer on the substrate due to the vapor reaction. Fig. 7.17 depicts the two different methods of MoS_2 layers by the CVD method (Bosi 2015).

7.3.2.2 Hydrothermal Synthesis

Hydrothermal synthesis is one of the bottom-up techniques for the 2D TMDC materials synthesis. Guo et al. (2015) demonstrated the hydrothermal synthesis process for the layered MoS_2 shown in Fig. 7.17, and the detailed steps are as follows. First, dissolve the 0.25-g $Na_2MoO_4{\cdot}2H_2O$ in 25 mL of water, and then add 0.1-M HCL to the solution to maintain a pH value of 6.5. Second, the solution prepared by the addition of 0.5 g l-cysteine and 75 mL water is kept for 10 min ultra-sonication. Third, this mixture was put into the 100-ml Teflon-lined stainless-steel autoclave and heated at 200°C temperature up to 36 h. Fourth, after cooling, the mixture was centrifuged at 12,000 rpm for 30 min to extract black precipitate. Lastly, the black precipitate was washed adequately in water and kept in a vacuum dried at 80°C for 24 h. This process deals with the pressurized autoclave-based aqueous solutions reaction for the development of the TMDC layer.

In an autoclave, the solution's temperature and pressure can be raised to the water boiling point to reach the vapor saturation pressure. However, the TMDC-layered material is produced with the addition of Mo, W, and S, Se into the autoclave and heated at 773 K up to 3 h in the presence of a noble gas.

This synthesis process can generate high-quality and uniform-size TMDC materials ranging from nanometers to microns, but the layer thickness is not controllable.

7.3.3 Layer Transfer Method

The synthesis of thin-layer TMDC material on any substrate is very important in many fundamental and applied research applications (Gurarslan 2014; Lee et al. 2013). Since the layer growth temperature of TMDC materials is very high, the

(a)

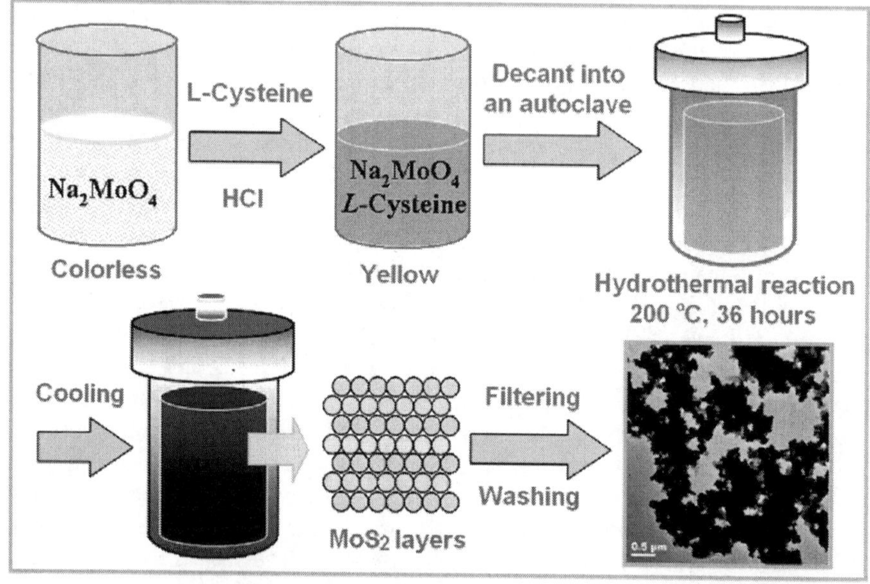

(b)

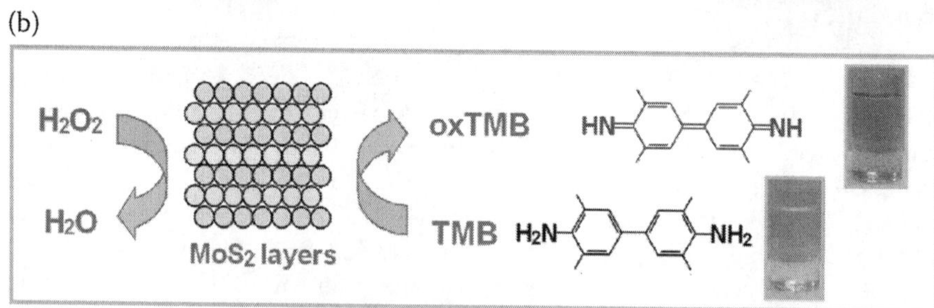

FIGURE 7.17 Hydrothermal method for the preparation of MoS_2. Reprinted with permission from Analyst, Copyright 2015, Royal Society of Chemistry (Guo et al. 2015).

synthesis process is not appropriate for the temperature-sensitive substrates like polymers; however, the use of TMDC material in such a substrate is essential in many applications. So, the layer transfer techniques are developed to transfer the TMDC layer on any substrate, and one of the techniques that maintains the quality of grown TMDC monolayers was demonstrated by Lee et al. (2013).

The grown sample of MoS_2 was chopped into three parts and treated for 30 s in DI water, isopropyl alcohol, and acetone. The surface of the grown monolayer is hydrophobic, so acetone and isopropyl alcohol are spread out on the MoS_2 surface, but the water remained as a droplet. The grown MoS_2 monolayer began to break up into small chunks and float on the water droplets after 30 s, indicating that the as-grown MoS_2 monolayer can be easily extracted from the substrate using DI water.

Gurarslan et al. (2014) reported surface energy-assisted transfer techniques for transferring substrate growth of centimeter-scale monolayer or multilayer TMDC

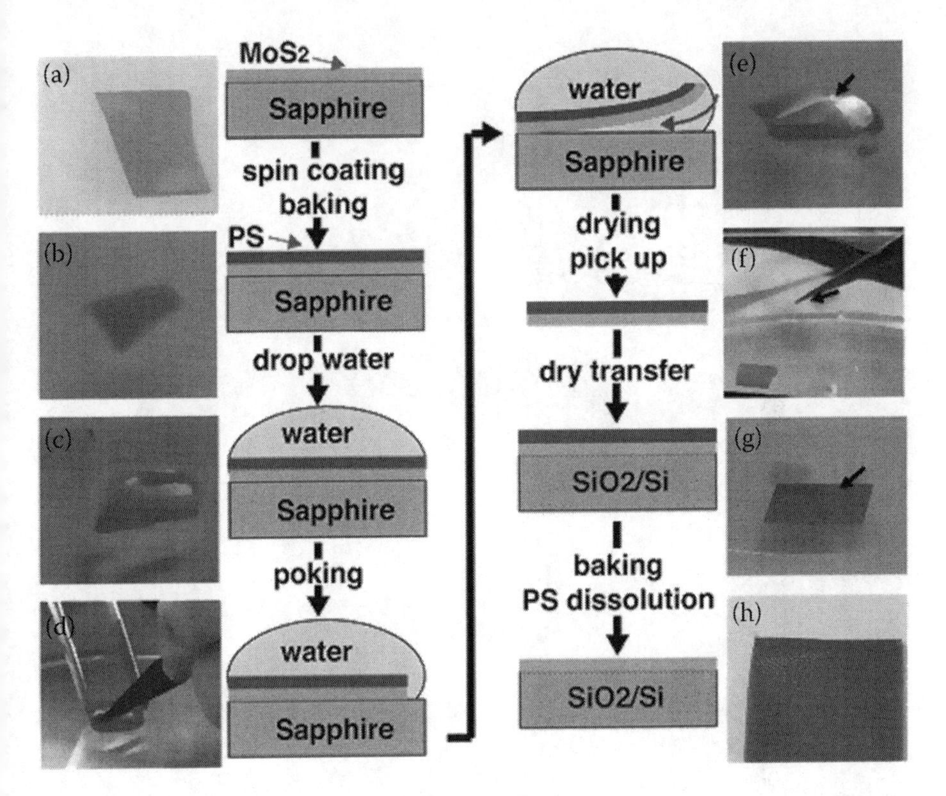

FIGURE 7.18 Flow diagram schematic of surface-energy-assisted transfer process: (a–h) Typical images of the transfer process. The arrows in (e–g) point toward the MoS_2 film for visual convenience. Reprinted with permission from ACS nano, Copyright 2014, American Chemical Society (Gurarslan et al. 2014).

material films to any specific substrates without cracks, wrinkles, or polymer residues; the systematic procedure is shown in Fig. 7.18.

7.4 CLASSIFICATION OF TMDC MATERIALS

The classification of TMDC materials mainly depends on the transition metal family, chalcogen family, atomic arrangement, bandgap, and number of layers, shown in Fig. 7.19. The classification of TMDC materials mainly depends on the transition metal and chalcogen family, atomic arrangement, bandgap, and several layers. In the periodic table, the transition metals are classified as Group IVB, VB, VIB, VIIB, and VIII, but VB and VIIB transition metals with chalcogens, such as S, Se, and Te, are mainly reported in the literature and research because of their feasibility in the synthesis process. In TMDC materials, the structural polymorph and crystal geometry depend on the atomic arrangement of transition metals and chalcogen elements, which were already discussed in section 7.2. It is also inferred that these atomic arrangements also alter the electronic properties of the TMDC materials that result in different bandgaps, which further classify them into the metal to

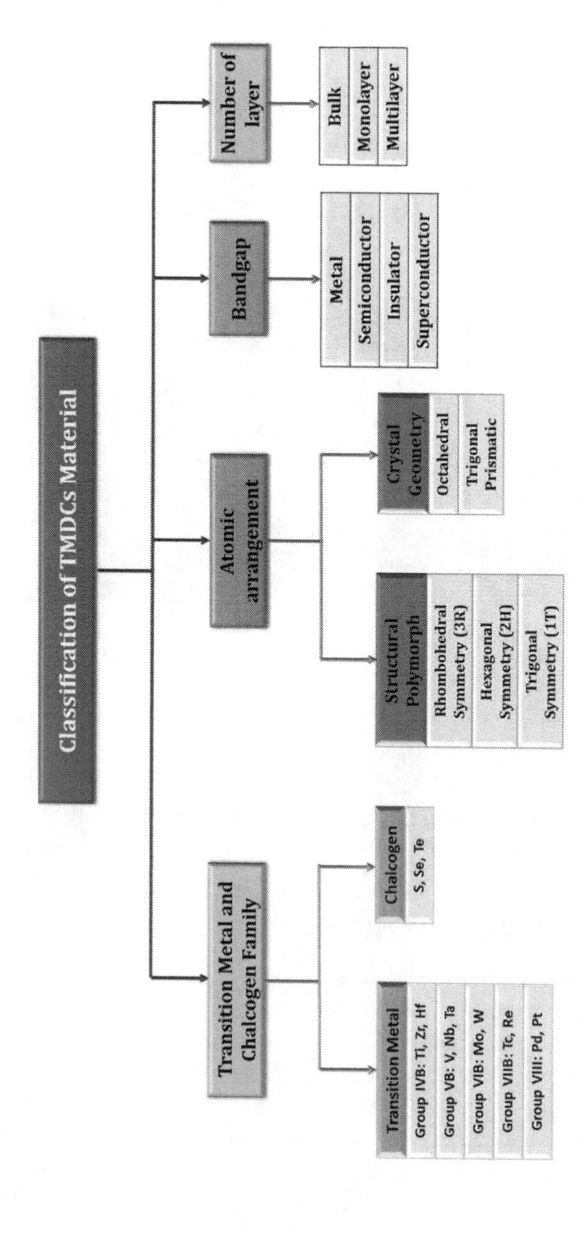

FIGURE 7.19 Brief classification of TMDC materials.

superconductor range. Finally, the TMDC materials are also categorized as bulk or layered TMDCs that, again, have enormous applications, including semiconductor devices, nonlinear optics, catalysts, and biorecognization layers.

7.5 TMDC MATERIAL WITH SPR PHENOMENA AND APPLICATION

Surface plasmon resonance (SPR) phenomena is seen mainly in the metal-dielectric interface, which is further utilized in the optical-based sensor for sensing purposes such as chemical, gas, biomolecules, and their adulteration. However, the SPR phenomenon are realized in the prism and optical fiber including SMF and MMF. In SPR-assisted optical sensors, the performance matrix includes minimum reflectivity, sensitivity, full-width half maxima (FWHM), figure of merit (FOM), and detection accuracy, which play a crucial role in the detection of a sensing medium. This section presents a brief review of conventional SPRs with TMDC material and their performance matrix measurement.

Xu, Wu, and Ang (2018) proposed a near-infrared SPR RI sensor with a Kretschmann configuration, where MoS_2-coated Al thin film is deposited over the chalcogenide (2S2G) glass prism. However, a sensor performance in the near-infrared regime was investigated based on Al thickness's alteration, the number of MoS_2 layers, other 2D TMDC materials, and analyte RI. It also suggests that the WSe_2 sensor performs best at wavelengths of 785 nm, while the MoS_2 sensor performs better at wavelengths of 1,150 nm and 1,540 nm. Furthermore, the sensor's sensitivity at the operating wavelength of 1,540 nm is reported to be 970 RIU^{-1}.

Similarly, Jia et al. (2019) theoretically investigated the SPR sensor using the newly developed TMDC material, $PtSe_2$, which has an exceptional optoelectronic property and a structure similar to graphene and phosphorous. The sensor in the Kretschmann structure is shown in Fig. 7.20.

A comparative study of the performance matrix based on the number of $PtSe_2$ layers, thickness variation in $PtSe_2$, and metal for the SPR biochemical sensors are summarized in Table 1 of Jia et al. (2019) for the sensing medium RI range of

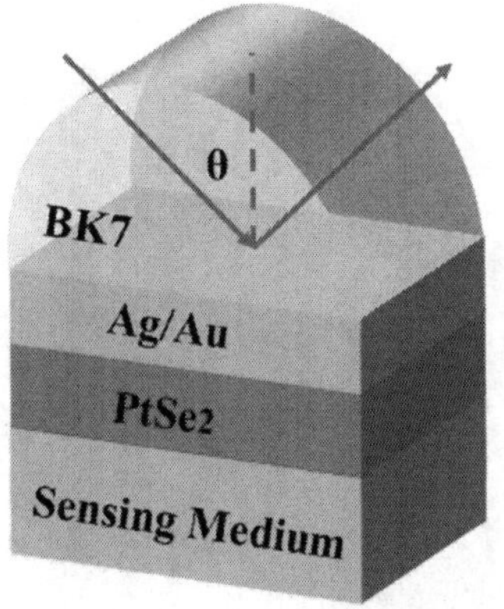

FIGURE 7.20 Schematic of $PtSe_2$-based SPR biochemical sensor for sensitivity improvement (Jia et al. 2019).

1.330–1.335. Furthermore, the simulation result for the sensor configuration with Ag/$PtSe_2$ has a sensitivity of 162 RIU^{-1}, while Au/$PtSe_2$ has a sensitivity of 165 RIU^{-1}, indicating that the sensing mechanism could be used in real-time applications.

The WS_2 nanosheet overlayer-based SPR sensor with metal film in the Kretschmann configuration was demonstrated experimentally by Wang et al. (2018) in Fig. 7.21. Here, the sensitivity of SPR is determined by the WS_2 layer thickness, and it can be changed with different coating concentrations of a WS_2-based ethanol suspension on it.

The details about the fabrication of the proposed SPR sensor and its characterization are discussed. Glass slides were first cleaned for 10 min in an ultrasonic bath. Since the gold film has a weak adhesion bond over the glass slide's surface, and it can be enhanced by depositing a thin chromium layer of 5 nm and then a 50-nm gold film over the glass slide using a vacuum evaporating system in the subsequent metal deposition process. However, the WS_2-modified SPR chip was made by coating the gold film with the prepared WS_2 alcohol suspension. MKNANO Tec. Co., Ltd. provides alcohol suspension of WS_2 used in the experiment (WS_2 nanosheet concentration: 1 mg mL^{-1}; average nanosheet size: 20–200 nm). This alcohol suspension of WS_2 was first decanted in a centrifuge tube for 30 min of ultrasonication to prevent agglomeration. After that, the suspension was lowered directly on the surface of the gold film and left to evaporate the alcohol naturally for 10 h at room temperature. As a result, the WS_2 film with a certain thickness was able to adhere to the gold layer firmly.

Fig. 7.22 shows the Raman spectrum of the WS_2 layer over the gold film that characterizes its structure and thickness quality measured by a Raman microscope excited by a laser with a 514.5 nm wavelength.

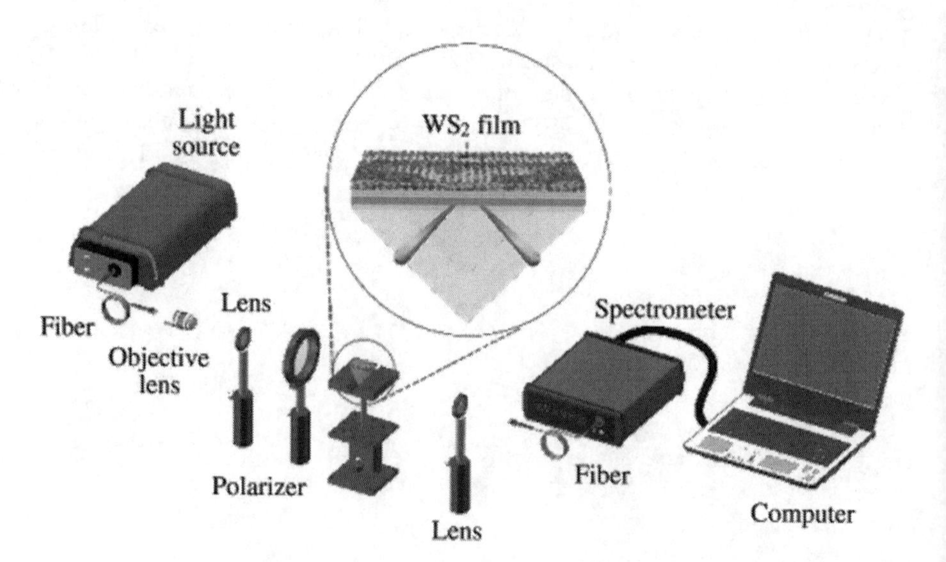

FIGURE 7.21 Test setup of RI-sensing measurement. Reprinted with permission from Photonics Research, Copyright, 2018, OSA Publishing (Wang et al. 2018).

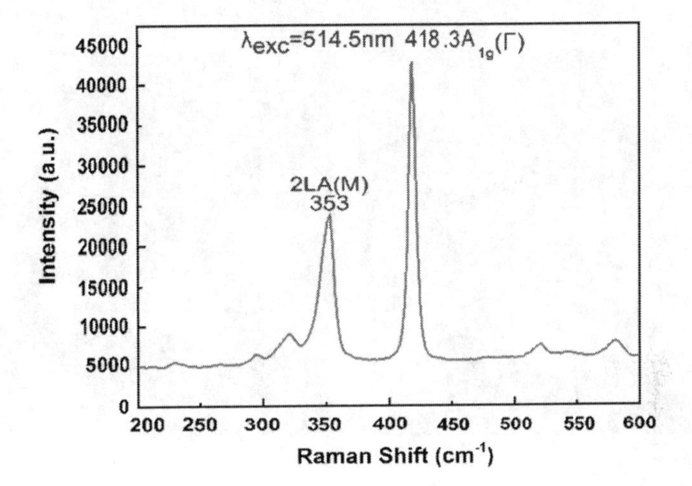

FIGURE 7.22 Raman spectrum of the coated WS_2 layers on the SPR sensor structure. Reprinted with permission from Photonics Research, Copyright, 2018, OSA Publishing (Wang et al. 2018).

SEM images of the thickness and surface morphology of the WS_2 layer coated over the gold film-based sensor are shown in Fig. 7.23.

However, the sensor's sensitivity was reported to be up to 2,459.3 nm RIU for a single coating of the WS_2 suspension in contrast to the sensor without a WS_2 layer, resulting in a 26.6% of sensitivity improvement. Furthermore, the WS_2 layer has unique benefits of metal oxidation prevention, resonance wavelength tuning, bio-capability, and is a promising candidate for vapor and gas sensing.

So far, we have discussed the TMDC layer-based conventional SPR sensor with a Kretschmann structure, but various literature have reported on the optical fiber-based SPR sensor using TMDC materials.

Odacı and Aydemir (2021) presented a comparative study of an SPR-assisted fiber-optic sensor where an Au/Ag bimetallic layer is coated with a 2D TMDC material. Fig. 7.24 shows the schematic of the SPR configuration in an optical fiber.

TMDC layers MoS_2, WS_2, $MoSe_2$, and WSe_2 are resonant sensor layers that affect the sensor's performance matrix. Even the Matlab simulation analysis shows that the $MoSe_2$ has a sensitivity value of 8,096 nm RIU^{-1} with a change in the RI of 0.0025 of the sensing mediums. On the other hand, the figure of merit and accuracy for monolayer WS_2 is 136.89 RIU^{-1} and 0.34, respectively. Although the performance matrix of SPR-based fiber-optic sensors with varying numbers of TMDC layers and TMDC material-based heterostructures are summarized in Tables 2 and 3 of Odacı and Aydemir (2021) and it is deduced that as the number of TMDC layers increases, the sensitivity decreases, which holds true for all TMDC materials. Therefore, this comparative study shows that the number of TMDC layers and their heterostructures characterize the SPR-assisted fiber-optic sensor and their performance matrix.

An SPR in a side-polished optical fiber coated with titanium, followed by TMDC materials such as MoS_2 and WS_2 covering the visible to mid-infrared region, was

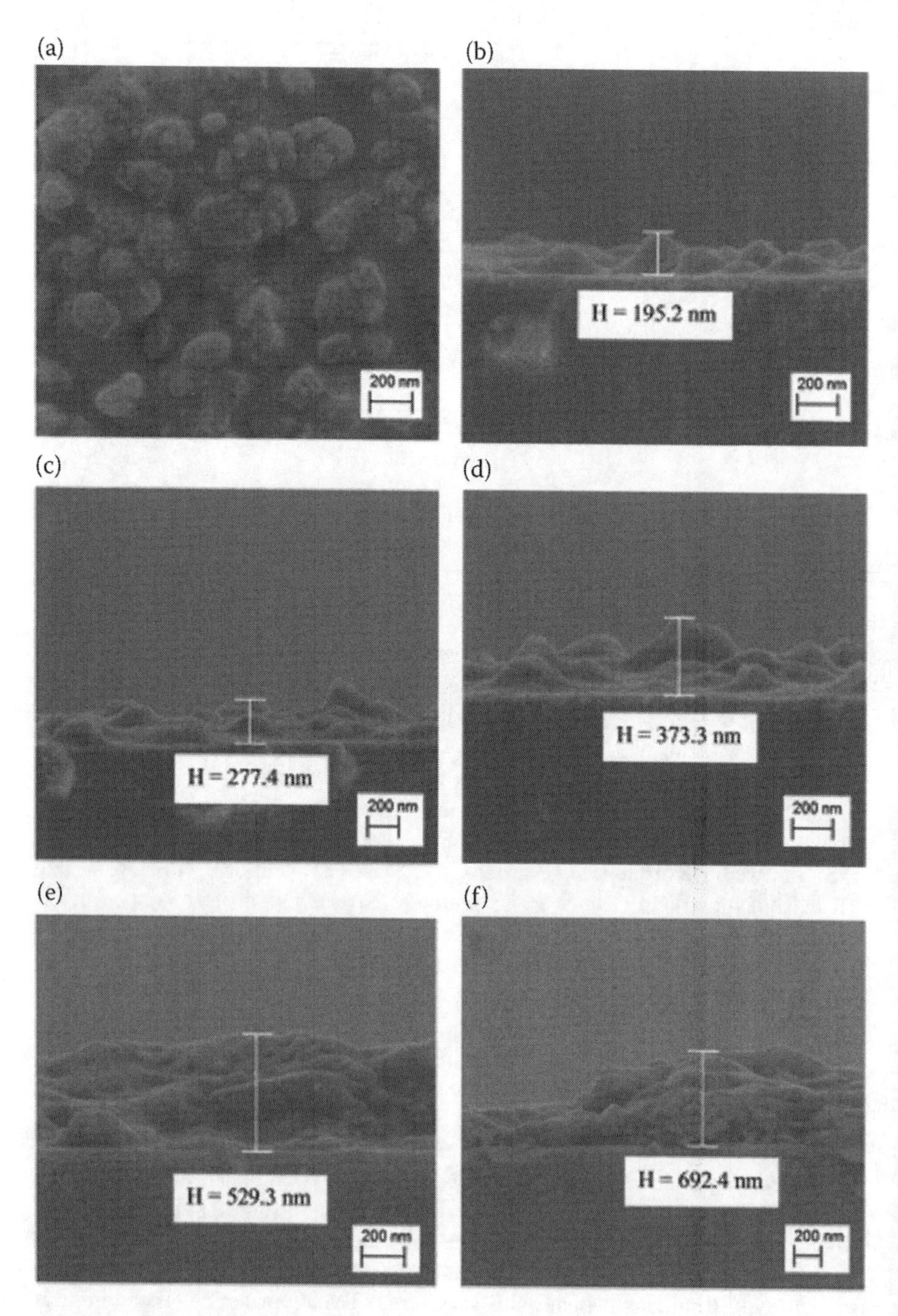

FIGURE 7.23 a) Surface morphology of WS_2-based SPR sensor. SEM image of WS_2 thickness after a number of times of repeated post-coating: b) one; c) two; d) three; e) four; and f) five. Reprinted with permission from Photonics Research, Copyright, 2018, OSA publishing (Wang et al. 2018).

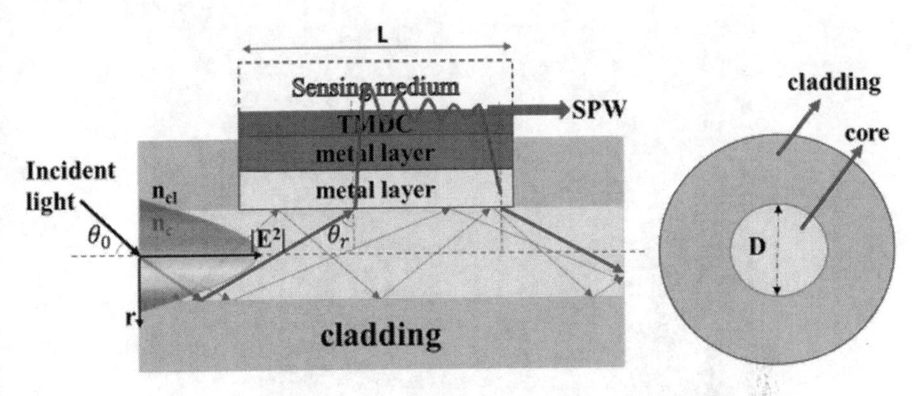

FIGURE 7.24 Schematic of an SPR-based fiber-optic sensor. Reprinted with permission from Results in Optics, Copyright, 2021, Elsevier (Odacı and Aydemir 2021).

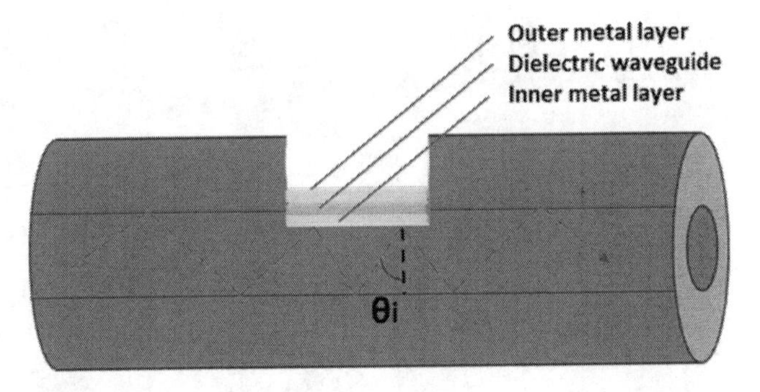

FIGURE 7.25 Schematic of propagation of light in the optical fiber sensor (Zakaria et al. 2019).

demonstrated experimentally by Zakaria et al. (2019) as a humidity sensor. The schematic of this sensor and propagation of light in such a sensor is shown in Fig. 7.25. The thickness of the Ti layer on the side-polished fiber has a significant impact on the evanescent field and optical absorption in the interaction with MoS_2 or WS_2, which improves sensitivity.

The fabrication process details of the side-polished optical-fiber-based SPR sensor are as follows. The SMF with 9 μm core and 125 μm cladding diameters is polished with an abrasive polishing wheel, and the process necessitates the use of a pair of fiber holders to fix the SMF in place, and a small portion of the SMF's coating is stripped away. The smoothness of the polished region is controlled by abrasive paper fastened to the mechanical wheel. This process has a near-100% success rate, significantly reduces polishing time, and results in a uniformly polished active region.

In this case, the 3 mm flat transition region of SMF is considered for the actual sensing region, as shown in Fig. 7.26a, which is obtained using field emission scanning electron microscopy (FESEM). During the polishing process, the fiber loss

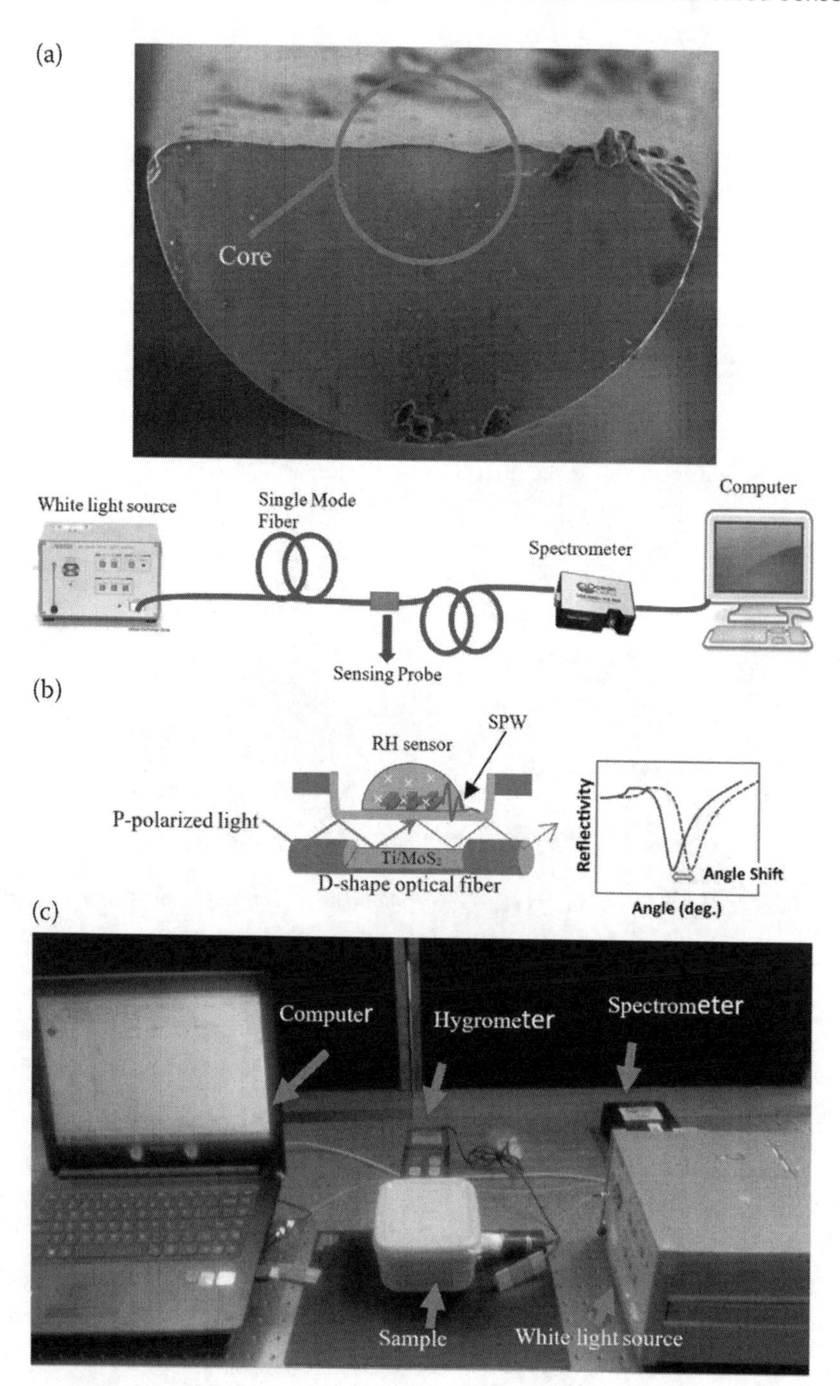

FIGURE 7.26 a) Field emission scanning electron microscopy (FESEM) image of side-polished single-mode fiber (SMF). b) Schematic diagram of experimental setup of Ti/MoS_2/WS_2 RH sensor and surface plasmon wave (SPW). c) Test setup of humidity sensor (Zakaria et al. 2019).

is measured as −2 dB after removing the SMF cladding. In the experiment, the transmission attenuation of side-polished fiber (SPF) is measured using a spectrometer and broadband light source over the range of 900–1,500 nm wavelengths. An electron beam evaporation machine is used to coat the polished section of the flat surface with Ti layers of different thicknesses, including 5 nm, 13 nm, and 36 nm, followed by a 0.2-mL drop-cast MoS_2 and WS_2.

Fig. 7.26b and c show the schematic diagram and its experimental setup for the characterization of the resonance peak of TiO_2/MoS_2- and WS_2-based side-polished fiber sensor in which a white light source is connected to one end of the fiber and the output end of the fiber is connected to an optical spectrum analyzer with a 2 nm spectral resolution and 700–1,800 nm spectral range.

As a result, it is deduced that Ti's optimized thickness, 36 nm with MoS_2, causes a weak redshift in the localized surface plasmon resonance (LSPR) mode but a large blueshift in WS_2. Sensitivity is determined in this sensor configuration by shifting the transmission dip in response to relative humidity changes, and Ti/MoS_2 is more sensitive than Ti/Ws_2, based on the transmission dip.

The reflectivity and resonance angles for the Kretschmann configuration of an SPR biosensor is comprised of a SF10 prism, an Si layer, and a thin-film gold layer with an enhanced MoS_2 nanosheet is investigated using the transfer matrix method by Ouyang, Zeng, Dinh, et al. (2016).

It is also deduced that the resonance angle and FWHM of reflectivity curves for the sample solutions of fixed RI could be improved by optimizing the thickness of gold, Si, and MoS_2 layers. Also, the optimum configuration of SPR based biosensor has a prism with a 50-nm thin film of gold, and further coated by a 7-nm silicon layer and a monolayer of MoS_2 with an excitation wavelength 633 nm light source. The reported sensitivity is 10% for Silicon and MoS_2 monolayer while 8% for only MoS_2 monolayer.

Similarly, Ouyang, Zeng, Jiang, et al. (2016) also present a theoretical and comparative study on Silicon nanosheet with different 2D TMDC material-based SPR biosensors for sensitivity enhancement. The SPR sensor in Kretschmann configuration is shown in Fig. 7.27 consists of an SF10 triangular prism coated with a thin gold film that further has a layer of silicon nanosheet, TMDC material, and biomolecular analyte layer for sensing medium.

It investigates the sensor's sensitivity enhancement due to TMDC material by optimizing parameters for the silicon-MX_2 model where MX_2 represents MoS_2, $MoSe_2$, WS_2, and WSe_2. Also, the performance matrices such as reflectivity, sensitivity, and FWHM are addressed in the visible and near-infrared regions using the Fresnel equation and transfer matrix method.

The analysis shows that Si and the number of TMDC layers change the reflectivity and resonance angles because of their effective light absorption. The sensitivity decreases as the number of TMDC layers increases due to the energy loss increases within the layer. So, the SPR sensor with the optimized value of 35 nm gold film, 7 nm Si nanosheet, and a monolayer WS_2 had 155.68° RIU^{-1}, sensitivity under the illumination of the source at the excitation wavelength, 600 nm.

Nur et al. (2019) proposed a Kretschmann configuration-based SPR biosensor based on WS_2 as a 2D TMDC material on top of the Al_2O_3 layer. The sensor

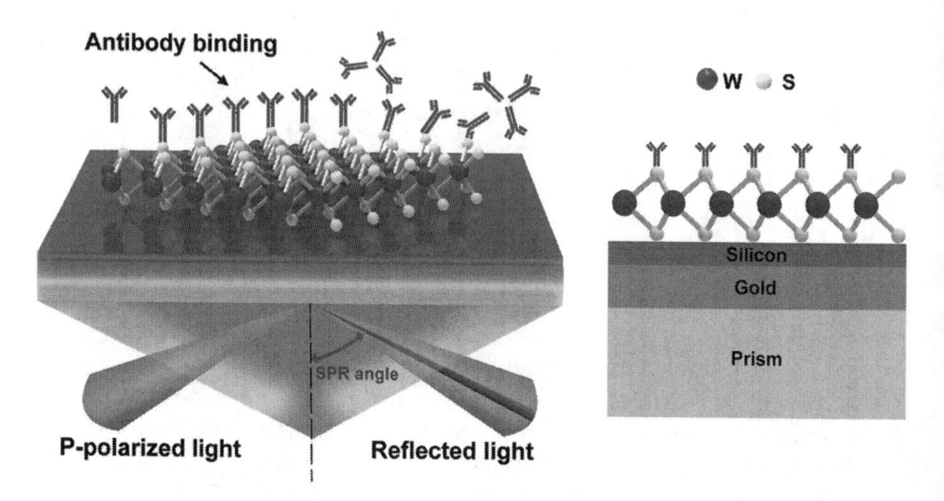

FIGURE 7.27 Silicon-WS_2/nanosheets-enhanced SPR biosensor (Ouyang, Zeng, Jiang, et al. 2016).

comprises five layers: prism, metal, Al_2O_3 layer, TMDC WS_2 layer, and sensing region. The prism material CaF_2 was chosen because of its low RI, which impacts the sensor's performance matrix. The sensor configuration of the WS_2 monolayer and six layers of Al_2O_3 has 227.5° RIU^{-1} sensitivity, 1.1123 detection accuracy, and 28.26 RIU^{-1} quality factor at the 633 nm wavelength.

Feng, Liu, and Teng (2018) reported a new configuration of a Kretschmann configuration-based SPR biosensor with the hybrid structure of graphene and MoS_2, as shown in Fig. 7.28. This configuration adopts the MgF_2 prism for the low RI to improve sensitivity and figure of merit.

Here, the absentee layer is introduced between the prism and the metal for sensitivity improvement; however, the absentee layers chosen for sensor optimization are air, SiO_2, KCl, Si_3N_4, TiO_2, and PbS. The configuration has a graphene layer over the metal that can prevent oxidation and characterize the visible range's complex RI. Therefore, the hybrid layer of the graphene and MoS_2 layer over the MgF_2 prism has significant effects on the sensing region's field distribution that enhances the biosensor's performances. However, the performance matrix of this SPR-based sensor with the consideration of different absentee layers is summarized in Table 1 of Feng, Liu, and Teng (2018). This work also reported the 540.8° RIU^{-1} sensitivity and 145 RIU^{-1} FOM by optimizing the configuration parameter's modulation and different absentee material.

Similarly, Lin et al. (2016) studied the effect of the graphene/MoS_2 layer on the metal surface to improve the sensitivity of a conventional SPR biosensor, which has a low sensitivity in angular interrogation. Here, the four layers of MoS_2 sheets and monolayer graphene sheets were introduced between two gold films in the Kretschmann configuration; a sensitivity of 182° RIU^{-1} was reported. Moreover, one can further tune the biosensor's sensitivity by changing the thickness of graphene and MoS_2.

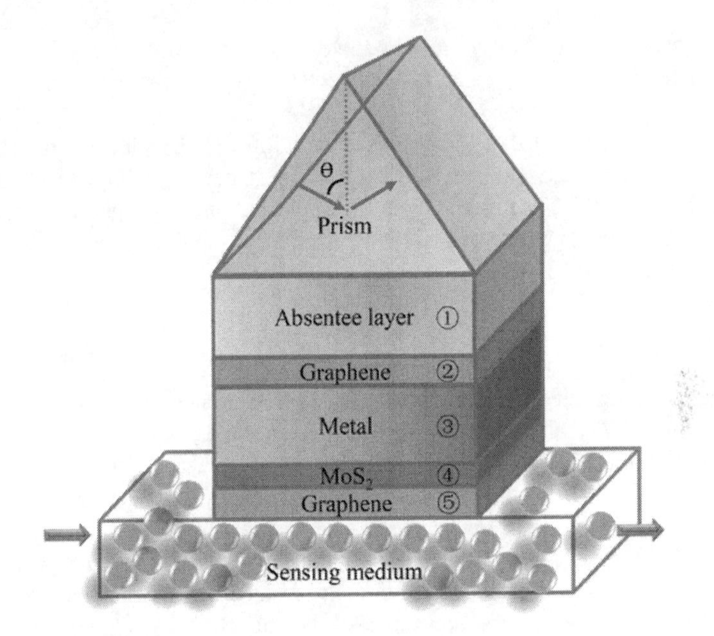

FIGURE 7.28 Schematic of MgF_2 prism with a graphene-MoS_2 hybrid structure-based biosensor. Reprinted with permission from Applied Optics, Copyright, 2018, OSA Publishing.

Hasib, Nur, Rizal, and Shushama (2019) reported an SPR biosensor with five layers using a Kretschmann configuration that constitutes the prism (CaF_2, BK7, SF10, 2S2G), Ag or Au layer with black phosphorus followed by TMDC materials such as MoS_2, $MoSe_2$, WS_2, and WSe_2, as shown in Fig. 7.29.

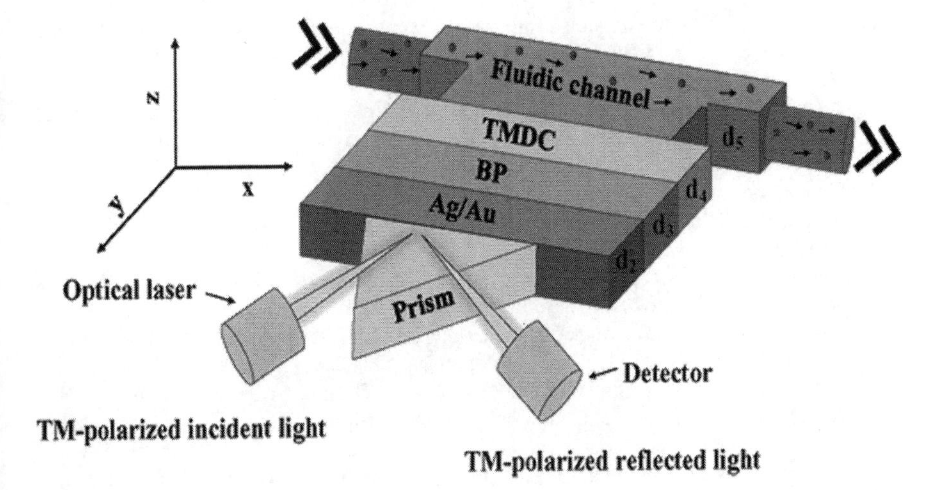

FIGURE 7.29 Schematics of the BP TMDC-based SPR biosensor (Hasib, Nur, Rizal, and Shushama 2019).

The sensor configuration with WS_2 material reported a 375° RIU^{-1} sensitivity, 0.9210 detection accuracy, and 65.78 1 RIU^{-1} quality factor by using a TM-polarized light at 633 nm.

Here, it is addressed that the SPR sensor with TMDC material has many advantages over the conventional SPR. However, it has incredible results when it incorporates other 2D materials such as graphene, blue phosphorous, and oxide layers to further enhance the performance matrix and the detection of many sensing mediums.

7.6 SUMMARY

This chapter presented details about the TMDC materials, including research publications on TMDC, properties of TMDC, brief classifications, and processing methods. It found that the TMDC materials have versatile applications in many research areas and, based on their application, several processing methods were proposed in the literature. Finally, a brief review was shown on the TMDC material application in the SPR-assisted optical sensor both theoretically and experimentally, which showed its feasibility potential in many areas of sensing, including chemical, gas, and biomolecule. It was also reported that the SPR phenomena with TMDC materials in the optical sensor significantly affect their performance matrices, such as reflectivity, sensitivity, the figure of merit, FWHM, and detection accuracy. However, one can utilize TMDC materials with many other materials, such as oxide, graphene, and blue phosphorous, to improve sensor design and sensitivity performance.

REFERENCES

Bertolazzi, Simone, Jacopo Brivio, and Andras Kis. 2011. Stretching and Breaking of Ultrathin MoS 2." *ACS Nano* 5 (12): 9703–9. 10.1021/nn203879f.

Bosi, Matteo. 2015. "Growth and Synthesis of Mono and Few-Layers Transition Metal Dichalcogenides by Vapour Techniques: A Review." *RSC Advances* 5 (92): 75500–18. 10.1039/c5ra09356b.

Cai, Yongqing, Jinghua Lan, Gang Zhang, and Yong Wei Zhang. 2014. "Lattice Vibrational Modes and Phonon Thermal Conductivity of Monolayer MoS_2." *Physical Review B – Condensed Matter and Materials Physics* 89 (3): 035438. 10.1103/PhysRevB.89.035438.

Castellanos-Gomez, A., N. Agrat, and G. Rubio-Bollinger. 2010. "Optical Identification of Atomically Thin Dichalcogenide Crystals." *Applied Physics Letters* 96 (21): 213116. 10.1063/1.3442495.

Castellanos-Gomez, Andres, Menno Poot, Gary A. Steele, Herre S.J. van der Zant, Nicolás Agraït, and Gabino Rubio-Bollinger. 2012. "Mechanical Properties of Freely Suspended Semiconducting Graphene-like Layers Based on MoS2." *Nanoscale Research Letters* 7 (1): 233. 10.1186/1556-276X-7-233.

Chhowalla, Manish, Hyeon Suk Shin, Goki Eda, Lain Jong Li, Kian Ping Loh, and Hua Zhang. 2013. "The Chemistry of Two-Dimensional Layered Transition Metal Dichalcogenide Nanosheets." *Nature Chemistry* 5: 263–75. 10.1038/nchem.1589.

Choi, Wonbong, Nitin Choudhary, Gang Hee Han, Juhong Park, Deji Akinwande, and Young Hee Lee. 2017. "Recent Development of Two-Dimensional Transition Metal

Dichalcogenides and Their Applications." *Materials Today*. 20 (3): 116–30. 10.1016/j.mattod.2016.10.002.

Coleman, Jonathan N., Mustafa Lotya, Arlene O'Neill, Shane D. Bergin, Paul J. King, Umar Khan, Karen Young, et al. 2011. "Two-Dimensional Nanosheets Produced by Liquid Exfoliation of Layered Materials." *Science* 331 (6017): 568–71. 10.1126/science.1194975.

Dines, Martin B. 1975. "Lithium Intercalation via N-Butyllithium of the Layered Transition Metal Dichalcogenides." *Materials Research Bulletin* 10 (4): 287–91. 10.1016/0025-5408(75)90115-4.

Dong, Ningning, Yuanxin Li, Yanyan Feng, Saifeng Zhang, Xiaoyan Zhang, Chunxia Chang, Jintai Fan, Long Zhang, and Jun Wang. 2015. "Optical Limiting and Theoretical Modelling of Layered Transition Metal Dichalcogenide Nanosheets." *Scientific Reports* 5 (1): 14646. 10.1038/srep14646.

Duan, Xidong, Chen Wang, Anlian Pan, Ruqin Yu, and Xiangfeng Duan. 2015. "Two-Dimensional Transition Metal Dichalcogenides as Atomically Thin Semiconductors: Opportunities and Challenges." *Chemical Society Reviews*. 44: 8859–76. 10.1039/c5cs00507h.

Feng, Yuncai, Youwen Liu, and Jinghua Teng. 2018. "Design of an Ultrasensitive SPR Biosensor Based on a Graphene-MoS_2 Hybrid Structure with a MgF_2 Prism." *Applied Optics* 57 (14): 3639. 10.1364/ao.57.003639.

Golden, Joshua, Melissa McMillan, Robert T. Downs, Grethe Hystad, Ian Goldstein, Holly J. Stein, Aaron Zimmerman, Dimitri A. Sverjensky, John T. Armstrong, and Robert M. Hazen. 2013. "Rhenium Variations in Molybdenite (MoS_2): Evidence for Progressive Subsurface Oxidation." *Earth and Planetary Science Letters* 366 (March): 1–5. 10.1016/j.epsl.2013.01.034.

Gordon, R. A., D. Yang, E. D. Crozier, D. T. Jiang, and R. F. Frindt. 2002. "Structures of Exfoliated Single Layers of WS_2, MoS_2, and $MoSe_2$ in Aqueous Suspension." *Physical Review B – Condensed Matter and Materials Physics* 65 (12): 1254071–9. 10.1103/PhysRevB.65.125407.

Guo, Xinrong, Yong Wang, Fangying Wu, Yongnian Ni, and Serge Kokot. 2015. "A Colorimetric Method of Analysis for Trace Amounts of Hydrogen Peroxide with the Use of the Nano-Properties of Molybdenum Disulfide." *Analyst* 140 (4): 1119–26. 10.1039/c4an01950d.

Gurarslan, Alper, Yifei Yu, Liqin Su, Yiling Yu, Francisco Suarez, Shanshan Yao, Yong Zhu, Mehmet Ozturk, Yong Zhang, and Linyou Cao. 2014. "Surface-Energy-Assisted Perfect Transfer of Centimeter-Scale Monolayer and Few-Layer MoS_2 Films onto Arbitrary Substrates." *ACS Nano* 8 (11): 11522–8. 10.1021/nn5057673.

Hasib, Mohammad Hasibul Hasan, Jannati Nabiha Nur, Conrad Rizal, and Kamrun Nahar Shushama. 2019. "Improved Transition Metal Dichalcogenides-Based Surface Plasmon Resonance Biosensors." *Condensed Matter* 4 (2): 49. 10.3390/condmat4020049.

Huang, Wen, Haixia Da, and Gengchiau Liang. 2013. "Thermoelectric Performance of MX2 (M Mo,W; X S,Se) Monolayers." *Journal of Applied Physics* 113 (10): 104304. 10.1063/1.4794363.

Huang, Wen, Xin Luo, Chee Kwan Gan, Su Ying Quek, and Gengchiau Liang. 2014. "Theoretical Study of Thermoelectric Properties of Few-Layer MoS_2 and WSe_2." *Physical Chemistry Chemical Physics* 16 (22): 10866–74. 10.1039/c4cp00487f.

Jia, Yue, Zhongfu Li, Haiqi Wang, Muhammad Saeed, and Houzhi Cai. 2019. "Sensitivity Enhancement of a Surface Plasmon Resonance Sensor with Platinum Diselenide." *Sensors* 20 (1): 131. 10.3390/s20010131.

Jiang, Jin Wu, Xiaoying Zhuang, and Timon Rabczuk. 2013. "Orientation Dependent Thermal Conductance in Single-Layer MoS_2." *Scientific Reports* 3 (1): 1–4. 10.1038/srep02209.

Kaasbjerg, Kristen, Kristian S. Thygesen, and Karsten W. Jacobsen. 2012. "Phonon-Limited Mobility in n-Type Single-Layer MoS_2 from First Principles." *Physical Review B – Condensed Matter and Materials Physics* 85 (11): 115317. 10.1103/PhysRevB.85.115317.

Kan, M., J. Y. Wang, X. W. Li, S. H. Zhang, Y. W. Li, Y. Kawazoe, Q. Sun, and P. Jena. 2014. "Structures and Phase Transition of a MoS2 Monolayer." *Journal of Physical Chemistry C* 118 (3): 1515–22. 10.1021/jp4076355.

Kolobov, Alexander V., and Junji Tominaga. 2016. "Structure and Physico-Chemical Properties of Single Layer and Few-Layer TMDCs." *Springer Series in Materials Science* 239: 109–63. 10.1007/978-3-319-31450-1_5.

Kumar, S., and U. Schwingenschlögl. 2015. "Thermoelectric Response of Bulk and Monolayer $MoSe_2$ and WSe_2." *Chemistry of Materials* 27 (4): 1278–84. 10.1021/cm504244b.

Lee, Yi Hsien, Lili Yu, Han Wang, Wenjing Fang, Xi Ling, Yumeng Shi, Cheng te Lin, et al. 2013. "Synthesis and Transfer of Single-Layer Transition Metal Disulfides on Diverse Surfaces." *Nano Letters* 13 (4): 1852–7. 10.1021/nl400687n.

Li, Hai, Jumiati Wu, Zongyou Yin, and Hua Zhang. 2014. "Preparation and Applications of Mechanically Exfoliated Single-Layer and Multilayer MoS_2 and WSe_2 Nanosheets." *Accounts of Chemical Research* 47 (4): 1067–75. 10.1021/ar4002312.

Li, Hai, Zongyou Yin, Qiyuan He, Hong Li, Xiao Huang, Gang Lu, Derrick Wen Hui Fam, Alfred Iing Yoong Tok, Qing Zhang, and Hua Zhang. 2012. "Fabrication of Single- and Multilayer MoS_2 Film-Based Field-Effect Transistors for Sensing NO at Room Temperature." *Small* 8 (1): 63–67. 10.1002/smll.201101016.

Li, Wu, J. Carrete, and Natalio Mingo. 2013. "Thermal Conductivity and Phonon Linewidths of Monolayer MoS_2 from First Principles." *Applied Physics Letters* 103 (25): 253103. 10.1063/1.4850995.

Li, Junwen, Nikhil V. Medhekar, and Vivek B. Shenoy. 2013. "Bonding Charge Density and Ultimate Strength of Monolayer Transition Metal Dichalcogenides." *Journal of Physical Chemistry C* 117 (30): 15842–8. 10.1021/jp403986v.

Li, Yilei, Alexey Chernikov, Xian Zhang, Albert Rigosi, Heather M. Hill, Arend M. van der Zande, Daniel A. Chenet, En Min Shih, James Hone, and Tony F. Heinz. 2014. "Measurement of the Optical Dielectric Function of Monolayer Transition-Metal Dichalcogenides: MoS2, $MoSE_2$, WS_2, and WS E_2." *Physical Review B – Condensed Matter and Materials Physics* 90 (20): 205422. 10.1103/PhysRevB.90.205422.

Li, Honglai, Xidong Duan, Xueping Wu, Xiujuan Zhuang, Hong Zhou, Qinglin Zhang, Xiaoli Zhu, et al. 2014. "Growth of Alloy MoS_2xSe_2(1-x) Nanosheets with Fully Tunable Chemical Compositions and Optical Properties." *Journal of the American Chemical Society* 136 (10): 3756–9. 10.1021/ja500069b.

Lin, Yu Chuan, Wenjing Zhang, Jing Kai Huang, Keng Ku Liu, Yi Hsien Lee, Chi te Liang, Chih Wei Chu, and Lain Jong Li. 2012. "Wafer-Scale MoS2 Thin Layers Prepared by MoO3 Sulfurization." *Nanoscale* 4 (20): 6637–41. 10.1039/c2nr31833d.

Liu, Hsiang Lin, Chih Chiang Shen, Sheng Han Su, Chang Lung Hsu, Ming Yang Li, and Lain Jong Li. 2014. "Optical Properties of Monolayer Transition Metal Dichalcogenides Probed by Spectroscopic Ellipsometry." *Applied Physics Letters* 105 (20): 201905. 10.1063/1.4901836.

Lin, Zhitao, Leyong Jiang, Leiming Wu, Jun Guo, Xiaoyu Dai, Yuanjiang Xiang, and Dianyuan Fan. 2016. "Tuning and Sensitivity Enhancement of Surface Plasmon Resonance Biosensor with Graphene Covered Au-MoS_2-Au Films." *IEEE Photonics Journal* 8 (6): 1–8. 10.1109/JPHOT.2016.2631407.

Liu, Keng Ku, Wenjing Zhang, Yi Hsien Lee, Yu Chuan Lin, Mu Tung Chang, Ching Yuan Su, Chia Seng Chang, et al. 2012. "Growth of Large-Area and Highly Crystalline

MoS_2 Thin Layers on Insulating Substrates." *Nano Letters* 12 (3): 1538–44. 10.1021/nl2043612.

Liu, Xiangjun, Gang Zhang, Qing Xiang Pei, and Yong Wei Zhang. 2013. "Phonon Thermal Conductivity of Monolayer MoS_2 Sheet and Nanoribbons." *Applied Physics Letters* 103 (13): 133113. 10.1063/1.4823509.

Meng, Si, Yuyan Zhang, Huide Wang, Lude Wang, Tiantian Kong, Han Zhang, and S. Meng. 2021. "Recent Advances on TMDCs for Medical Diagnosis." *Biomaterials* 269: 120471. 10.1016/j.biomaterials.2020.120471.

Novoselov, K. S. 2011. "Nobel Lecture: Graphene: Materials in the Flatland." *Reviews of Modern Physics* 83 (3): 837–49. 10.1103/RevModPhys.83.837.

Nur, Jannati Nabiha, Mohammad Hasibul Hasan Hasib, Fairuj Asrafy, Kamrun Nahar Shushama, Reefat Inum, and Md Masud Rana. 2019. "Improvement of the Performance Parameters of the Surface Plasmon Resonance Biosensor Using $Al2O_3$ and WS_2." *Optical and Quantum Electronics* 51 (6): 174. 10.1007/s11082-019-1886-9.

Odacı, Cem, and Umut Aydemir. 2021. "The Surface Plasmon Resonance-Based Fiber Optic Sensors: A Theoretical Comparative Study with 2D TMDC Materials." *Results in Optics* 3 (May): 100063. 10.1016/j.rio.2021.100063.

Ouyang, Qingling, Shuwen Zeng, Xuan Quyen Dinh, Philippe Coquet, and Ken Tye Yong. 2016. "Sensitivity Enhancement of MoS_2 Nanosheet Based Surface Plasmon Resonance Biosensor." In *Procedia Engineering*, 140: 134–9. 10.1016/j.proeng.2015.08.1114.

Ouyang, Qingling, Shuwen Zeng, Li Jiang, Liying Hong, Gaixia Xu, Xuan Quyen Dinh, Jun Qian, et al. 2016. "Sensitivity Enhancement of Transition Metal Dichalcogenides/Silicon Nanostructure-Based Surface Plasmon Resonance Biosensor." *Scientific Reports* 6 (1): 1–13. 10.1038/srep28190.

Saito, Yuta, Paul Fons, Alexander V. Kolobov, and Junji Tominaga. 2015. "Self-organized van Der Waals Epitaxy of Layered Chalcogenide Structures." *Physica Status Solidi (b)* 252 (10): 2151–8. 10.1002/pssb.201552335.

Shi, Yumeng, Henan Li, and Lain Jong Li. 2015. "Recent Advances in Controlled Synthesis of Two-Dimensional Transition Metal Dichalcogenides via Vapour Deposition Techniques." *Chemical Society Reviews* 44: 2744–56. 10.1039/c4cs00256c.

Smith, Ronan J., Paul J. King, Mustafa Lotya, Christian Wirtz, Umar Khan, Sukanta De, Arlene O'Neill, et al. 2011. "Large-Scale Exfoliation of Inorganic Layered Compounds in Aqueous Surfactant Solutions." *Advanced Materials* 23 (34): 3944–8. 10.1002/adma.201102584.

Su, Sheng-Han, Wei-Ting Hsu, Chang-Lung Hsu, Chang-Hsiao Chen, Ming-Hui Chiu, Yung-Chang Lin, Wen-Hao Chang, Kazu Suenaga, Jr-Hau He, and Lain-Jong Li. 2014. "Controllable Synthesis of Band-Gap-Tunable and Monolayer Transition-Metal Dichalcogenide Alloys." *Frontiers in Energy Research* 2 (July): 27. 10.3389/fenrg.2014.00027.

Tahir, M., and U. Schwingenschlögl. 2014. "Tunable Thermoelectricity in Monolayers of MoS2 and Other Group-VI Dichalcogenides." *New Journal of Physics* 16 (11): 115003. 10.1088/1367-2630/16/11/115003.

Wang, Xinsheng, Hongbin Feng, Yongmin Wu, and Liying Jiao. 2013. "Controlled Synthesis of Highly Crystalline MoS_2 Flakes by Chemical Vapor Deposition." *Journal of the American Chemical Society* 135 (14): 5304–7. 10.1021/ja4013485.

Wang, Zhan Yu, Yan Li Zhou, Xue Qing Wang, Fei Wang, Qiang Sun, Zheng Xiao Guo, and Yu Jia. 2015. "Effects of In-Plane Stiffness and Charge Transfer on Thermal Expansion of Monolayer Transition Metal Dichalcogenide." *Chinese Physics B* 24 (2): 026501. 10.1088/1674-1056/24/2/026501.

Wang, Hao, Hui Zhang, Jiangli Dong, Shiqi Hu, Wenguo Zhu, Wentao Qiu, Huihui Lu, et al. 2018. "Sensitivity-Enhanced Surface Plasmon Resonance Sensor Utilizing a Tungsten

Disulfide (WS_2) Nanosheets Overlayer." *Photonics Research* 6 (6): 485. 10.1364/prj.6.000485.

Wei, Xiaolin, Yongchun Wang, Yulu Shen, Guofeng Xie, Huaping Xiao, Jianxin Zhong, and Gang Zhang. 2014. "Phonon Thermal Conductivity of Monolayer MoS_2: A Comparison with Single Layer Graphene." *Applied Physics Letters* 105 (10): 103902. 10.1063/1.4895344.

Wilson, J. A., and A. D. Yoffe. 1969. "The Transition Metal Dichalcogenides Discussion and Interpretation of the Observed Optical, Electrical and Structural Properties." *Advances in Physics* 18 (73): 193–335. 10.1080/00018736900101307.

Xu, Yi, Lin Wu, and Lay Kee Ang. 2018. "MoS_2 Based Highly Sensitive Near-Infrared Surface Plasmon Resonance Refractive Index Sensor."*IEEE Journal of Selected Topics in Quantum Electronics* 25(2): 1–7. 10.1109/JSTQE.2018.2868795.

Yuan, Lin, Jun Ge, Xianglin Peng, Qian Zhang, Zefei Wu, Yu Jian, Xiaolu Xiong, Hongxing Yin, and Junfeng Han. 2016. "A Reliable Way of Mechanical Exfoliation of Large Scale Two Dimensional Materials with High Quality." *AIP Advances* 6 (12): 125201. 10.1063/1.4967967.

Zakaria, Rozalina, Nur Zainuddin, Tan Leong, Rosnadiya Rosli, Muhammad Rusdi, Sulaiman Harun, and Iraj Sadegh Amiri. 2019. "Investigation of Surface Plasmon Resonance (SPR) in MoS_2- and WS_2-Protected Titanium Side-Polished Optical Fiber as a Humidity Sensor." *Micromachines* 10 (7): 465. 10.3390/mi10070465.

Zappa, Dario. 2017. "Molybdenum Dichalcogenides for Environmental Chemical Sensing." *Materials* 10 (12): 1418. 10.3390/ma10121418.

Zeng, Zhiyuan, Zongyou Yin, Xiao Huang, Hai Li, Qiyuan He, Gang Lu, Freddy Boey, and Hua Zhang. 2011. "Single-Layer Semiconducting Nanosheets: High-Yield Preparation and Device Fabrication." *Angewandte Chemie International Edition* 50 (47): 11093–7. 10.1002/anie.201106004.

Zhang, Wensi, Panpan Zhang, Zhiqiang Su, and Gang Wei. 2015. "Synthesis and Sensor Applications of MoS2-Based Nanocomposites." *Nanoscale* 7 (15): 18364–78. 10.1039/c5nr06121k.

Zhou, Xiaofei, Hainan Sun, and Xue Bai. 2020. "Two-Dimensional Transition Metal Dichalcogenides: Synthesis, Biomedical Applications and Biosafety Evaluation." *Frontiers in Bioengineering and Biotechnology* 8: 236. 10.3389/fbioe.2020.00236.

8 Application of SPR Sensors for Clinical Diagnosis

8.1 INTRODUCTION

Nowadays, two-dimensional (2D) materials/nanomaterials are used in various applications, especially in biosensing, due to their superior chemical and physical properties. Biosensors are mainly used for detection and measurement of various biomolecules found in the human body, as well as bacteria, viruses, and cells. Various techniques such as fluorescence, field-effect transistors, surface-enhanced Raman scattering, electrochemical, photoelectrochemical, colorimetric, and surface plasmon resonance (SPR)/localized surface plasmon resonance (LSPR) are used in biosensing (Su et al. 2019). To enhance the performance of these techniques, 2D nanomaterials such as graphene oxide (GO), boron nanosheets (B NSs), graphite carbon nitride (g-C_3N_4), transition-metal oxides (such as MnO_2, MoO_3, and WO_3), silicate clays, transition-metal dichalcogenides (TMD, e.g. WS_2, MoS_2), boron nitride (BN), black phosphorus (BP), layered double hydroxides (LDHs), antimonene, tin-telluride-nanosheets (SnTe NSs), and MXenes play a vital role (Hu et al. 2019; Su et al. 2019). A schematic illustration of 2D nanomaterials is shown in Fig. 8.1. These are the types of layered nanomaterials based on the composition and crystal structures. In each layer, chemical bonds exist to connect all atoms on the plane and a weak van der Waals interlinkage helps to connect the layer stack to form the bulk crystals. The excellent chemical, physical, electronic, biological, and optical properties of these 2D materials are due to high surface charge, sustained shape, and an outrageous surface-to-volume ratio, used in various applications such as catalysis, energy, sensing, antibacterials, and drug delivery. Most of the commonly used 2D nanomaterials belong to the graphene family, such as graphene, GO, and reduced graphene oxide (rGO). Graphene is a type of thin nanomaterial due to its layered structure and monoatomic thickness. Graphene has an aloft-specific surface area, compared with a carbon nanotube, that helps in proper functionalization. Then, with the help of partial oxidation of graphene, graphene produces GO that increases the property of hydrophilicity and enhances the functionalization. Thereafter, most of the oxygenated groups can be removed from GO and produce the rGO. It increases absorption due to more sp^2 carbon in the near-infrared region. Thus, due to excellent electrical/thermal conductivities and biocompatibility properties, graphene and its families have various applications in day-to-day life such as bioengineering, biosensors, bioimaging, tissue engineering, gene/drug delivery, and antifungal phenomena.

DOI: 10.1201/9781003190738-8

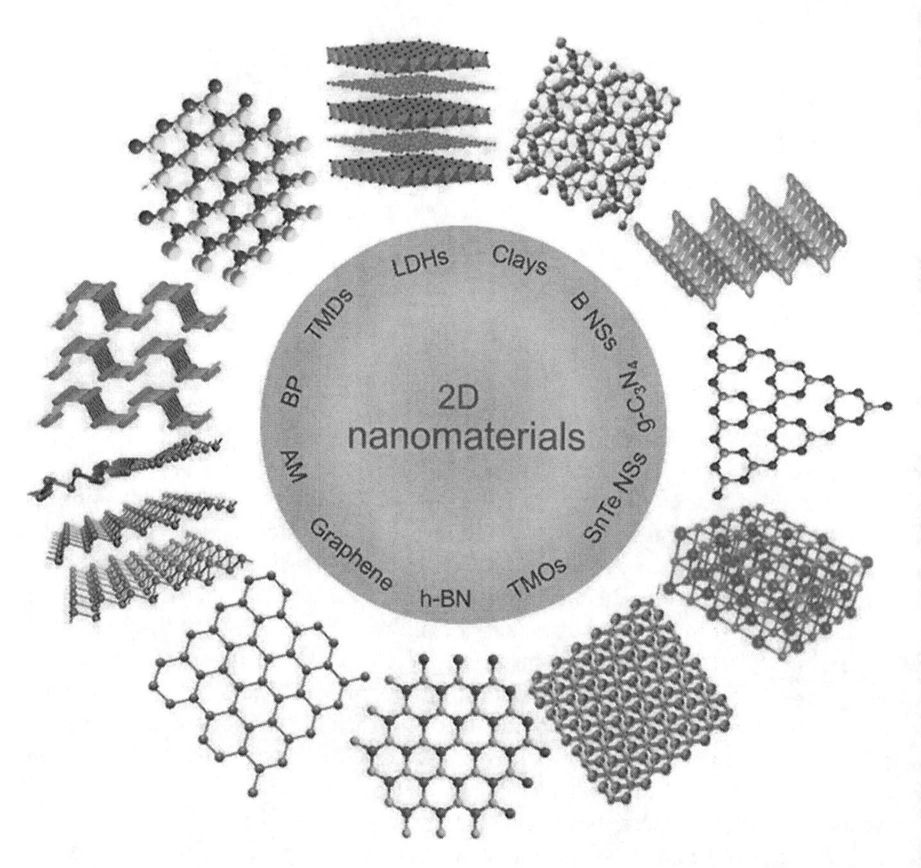

FIGURE 8.1 Various types of 2D nanomaterials. Reprinted with permission from Science Bulletin Journal. Copyright, 2019, Elsevier (Hu et al. 2019).

Similarly, silicate clays have an aqueous stability, shear-thinning, and high drug loading properties. Due to these properties, these are commonly preferred in the pharmaceutical industry for biosensing, tissue engineering, and drug delivery. Apart from these materials, MXenes also emerged as novel 2D nanomaterials due to their peculiar sensing abilities, such as graphene-like morphology, unique surface chemistry, metallic conductivity, redox capability, good hydrophilicity, and strong mechanical properties. Due to these properties, it has huge applications in several research areas, especially in biosensing and gas sensing due to adsorption characteristics and unique compositions (Deshmukh, Kovářík, and Khadheer Pasha 2020). Similarly, hexagonal boron nitride (HBN) nanosheets are a kind of 2D material used for catalysis, bioimaging, drug delivery, and sensor applications. HBN works in visible and IR regions due to its large bandgap and they have high surface area and conductivity that also helps in sensor applications. Özkan et al. (Özkan, Atar, and Yola 2019) developed the HBN nanosheets immobilized core-shell nanoparticles (Ag@AuNPs)-based SPR sensor for etoposide (ETO) detection. ETO is a kind of drug used to cure different types of cancers. It is also a type of podophyllotoxin that is used for the treatment of neuroblastoma, ovarian cancer,

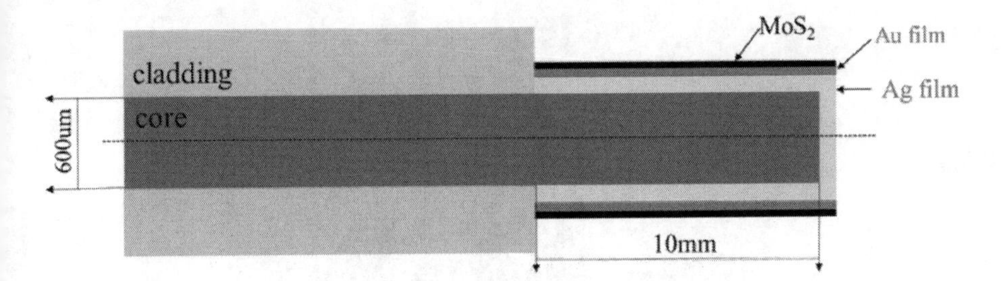

FIGURE 8.2 Gold/silver/MoS_2-functionalized unclad optical fiber sensor structure. Reprinted with permission from Optics and Lasers in Engineering. Copyright, 2020, Elsevier (Wang, Jiang, Niu, and Fan 2020).

and lung cancer. Recently, Wang, Jiang, Niu, and Fan (2020) fabricated a molybdenum disulfide (MoS_2) nanosheet Au/Ag-immobilized optical fiber SPR sensor as shown in Fig. 8.2. For this, first cladding from one end of 10 mm of a plastic optical fiber (600 μm, core diameter) is removed. Thereafter, coating of Ag film is followed by Au film for SPR sensing and then MoS_2 (as a 2D material) is immobilized to enhance the biocompatibility of a sensor probe.

SPR is a recent technique used for real-time monitoring and determines the concentration of biomolecules. According to the SPR principle, transposing in the refractive-index (RI) due to biological interlinkage around a sensing medium helps in monitoring and determining biomolecules. Various noble metal substrates, such as gold (Au)/silver (Ag), support the distribution of surface plasmon polariton in the visible range due to excellent optical properties. Au material is mostly preferred due to its superior oxidation resistance as well as excellent anti-corrosion ability in adverse environmental circumstances but analyte interactions are poor on Au and restrain its sensitivity (Wu, Chu, Koh, and Li 2010). Thus, 2D-material functionalization is carried out to enhance the sensitivity of SPR sensors. The unique structures and large surface of 2D materials enhance the adsorption of biological analytes and help in sensitivity improvement.

Wu, Chu, Koh, and Li (2010) fabricated the SPR biosensor consisting of 2D material, in which they used a graphene sheet coated as a biomolecular perception agent over the Au thin-film to increase biomolecules' adsorption over Au film. The proposed method is based on the attenuated total reflection (ATR) detected in the RI changes around the sensing medium that occur due to adsorption of biomolecules. They have also proved that graphene with an Au thin-film-based SPR biosensor is $(1 + 0.025\ L) \times \gamma$ times more sensitive than an Au-film-based biosensor, where L is graphene layers and γ is the biomolecules' adsorption factor over graphene ($\gamma > 1$). In Fig. 8.3, the first layer is a prism, whose RI is $n_1 = n_{prism} = 1.723$, the second layer is Au ($n_2 = 0.1726 + i3.4218$), the third layer is graphene ($n_3 = 3 + i1.149106$), and then $n_4 = 1.33$ (water). Here, $z_0 = 100$ nm is the layer of biomolecules dissolved in water. In this case, graphene absorbs the biomolecules stronger than the Au surface because graphene has carbon-based ring structures and π-stacking interactions within hexagonal cells of graphene. This process is also called enhanced adsorption efficiency and helps in the enhancement of overall sensing performance.

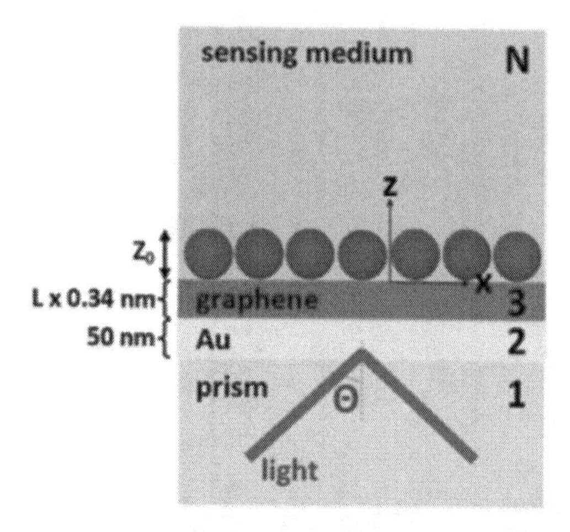

FIGURE 8.3 Gold/graphene-coated prism-based SPR biosensor. Reprinted with permission from Optics Express. Copyright, 2010, OSA (Wu, Chu, Koh, and Li 2010).

Liu et al. (2020) proposed a D-shaped photonic crystal fiber (PCF)–based SPR sensor for detection of refractive indices and investigated using the finite element method (FEM). Because PCF has many advantages, such as good tunability, vast mode area, outrageous birefringence, and tensile multiparameter sensing capability, that help in the development of prominent sensors. They have functionalized the PCF using indium tin oxide (ITO) for adding the plasmonic properties in a proposed sensor. ITO is cost-effective, has good conductivity, works better in the infrared range, and has variable photoelectric properties. The proposed design is shown in Fig. 8.4, in which one side of PCF is coated with ITO. ITO helps in shifting the sensing ability in the near-infrared range. For fabrication, die-cast and stack/draw methods are used to fabricate the PCF and then it has been polished to get the flat surface using a wheel polishing technique and the ITO coating. The PCF is composed of two kinds of air holes of diameter d_1 and diameter d_2 and the distance between the air holes is pitch distance (***). After fabrication of a sensor probe, it has been spliced with a source and a detector for final measurement using a single-mode fiber (SMF). When analyte is combined with the sensing surface, then varied the RI surrounding the sensor and wavelength shift occurs that help in determination of analyte concentration.

Similarly, pressure sensors have great potential in the utilization of wearable electronics. Thus, Yue et al. (2018) developed an MXene sponge and used it in a piezoresistive sensor with insulating polyvinyl alcohol (PVA) nanowires as a spacer for measurement of pressure distribution as shown in Fig. 8.5. It also helps in real-time detection of human physiological signals such as pulses, respiration, and joint movement.

As we know, Au/Ag are prominent for development of SPR sensors. But, material loss, oxidation tendency, and corrosion of the metals limit the plasmonic properties. Au is preferred as the most suitable material due to its better chemical stability, better optical performance, and good resistance to corrosion and oxidation. Au also consists of the wide resonance peak and reduces the accuracy of sensing but

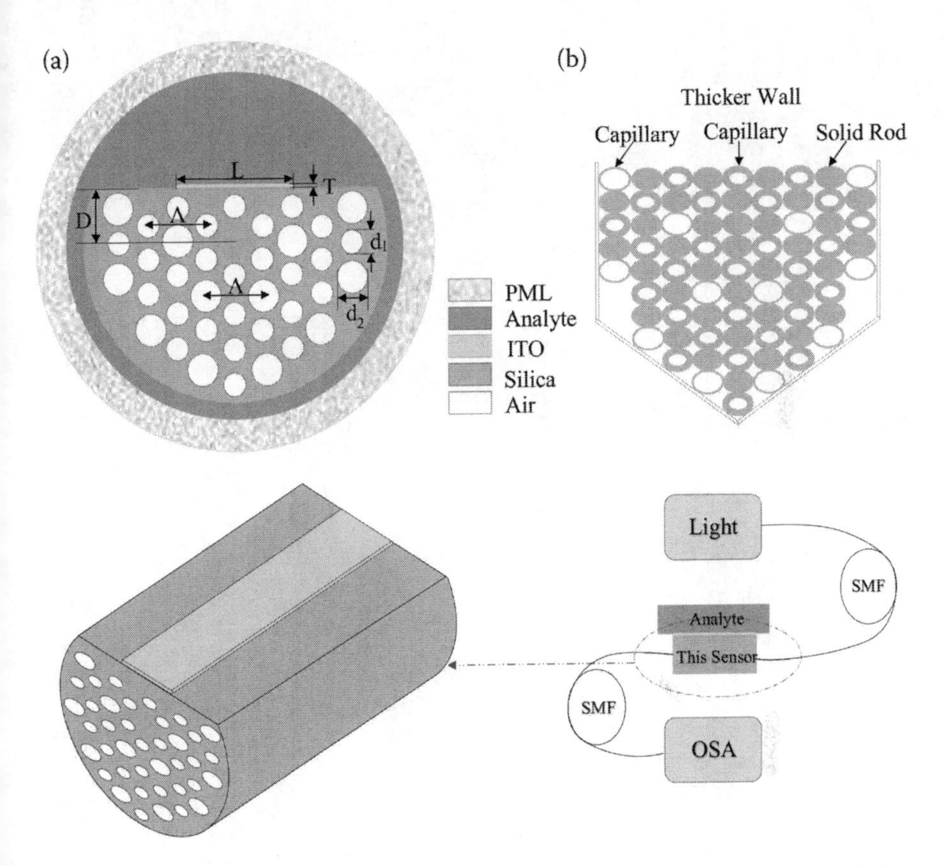

FIGURE 8.4 a) D-shaped photonic crystal fiber (PCF)–based SPR sensor. Reprinted with permission from Optics Communications. Copyright, 2020, Elsevier (Liu et al. 2020).

Au is lossy in nature and limits the binding capability of biomolecules. Besides, Ag film exhibits the highest resolution, but the sensitivity of Ag is lower compared to Au. Also, the stability of Ag is low and oxidizes easily; these limitations can be overcome by using the graphene over Ag or bimetallic combination (Au-Ag).

These are also not used in the long term due to their limited sensitivity; thus, AlaguVibisha et al. (2020) have used the 2D materials over a bimetallic layer of Cu–Ni for the investigation of the plasmonic property. Copper (Cu) is cheaper than Ag and Au and its conductivity is also good, but oxidation of Cu is high and that limits the plasmonic property. Thus, nickel (Ni) is used to prevent the oxidation of Cu, which works as a ferromagnetic metal and exhibits good magnetic and magneto-optical properties. A Kretschmann configuration-based sensor is shown in Fig. 8.6. It consists of five layers of Bk7 prism/Cu/Ni layers and 2D materials. Here, the p-polarized light of a wavelength of 633 nm is emitted on a prism interface at a higher critical angle; thereafter, the reflected signal is observed through the photodetector. They have shown that with this type of plasmonic structure, sensitivity is enhanced while maintaining the FWHM and this type of structure has huge applications in detecting the biomolecules. Table 8.1 shows the collision and plasma

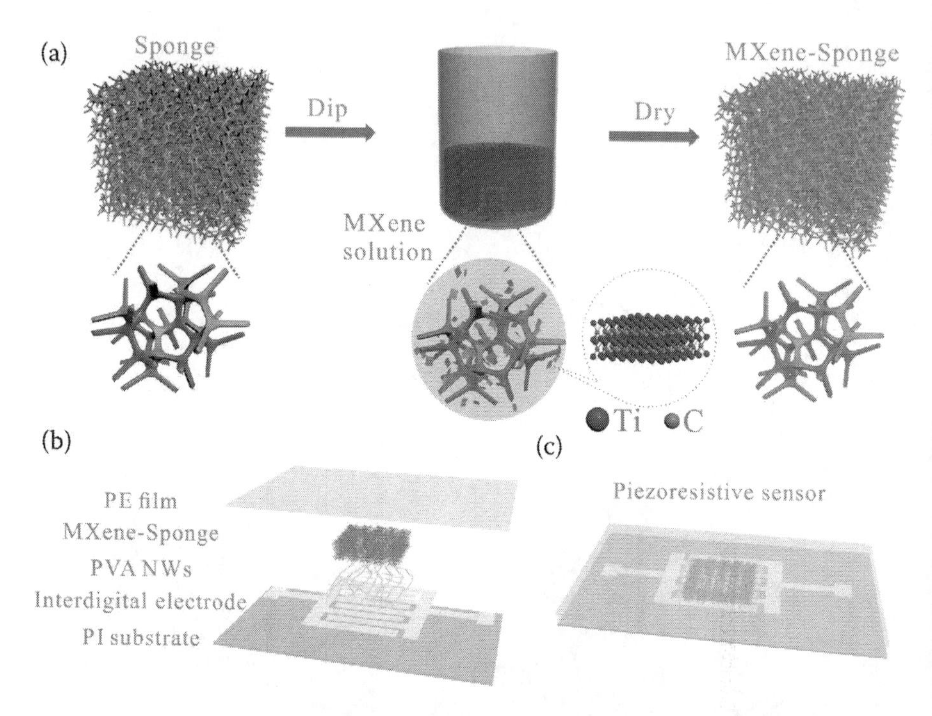

FIGURE 8.5 Fabrication steps of MXene-sponge-based biosensor. Reprinted with permission from Nano Energy Journal. Copyright, 2018, Elsevier (Yue et al. 2018).

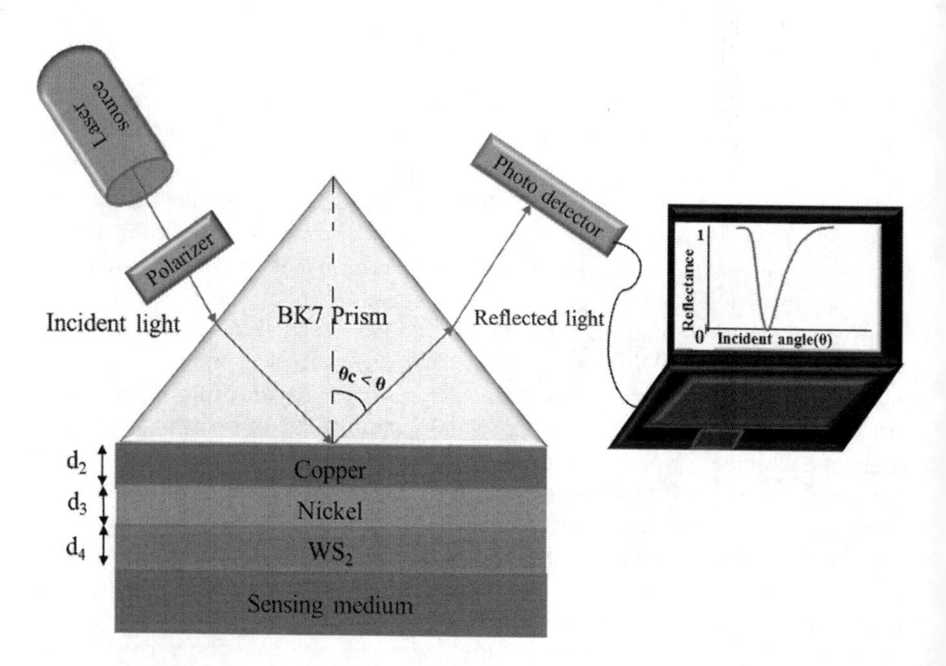

FIGURE 8.6 Cu–Ni–WS_2/prism-based SPR biosensor. Reprinted with permission from Optics Communications. Copyright, 2020, Elsevier (AlaguVibisha et al. 2020).

TABLE 8.1
Collision and Plasma Wavelength of Cu and Ni (AlaguVibisha et al. 2020)

Parameter	Cu	Ni
λ_c (m)	4.0852×10^{-5}	2.8409×10^{-5}
λp (m)	1.3617×10^{-7}	2.5381×10^{-7}

TABLE 8.2
The Thickness and Refractive Index of 2D Materials at λ = 633 nm (AlaguVibisha et al. 2020)

2D Materials	Thickness of Monolayer (nm)	Refractive Index
MoS_2	0.65	5.08 + 1.1723i
WS_2	0.80	4.9 + 0.3124i
$MoSe_2$	0.70	4.62 + 1.0063i
WSe_2	0.70	4.55 + 0.4332i
Graphene	0.34	3.0 + 1.1491i

wavelength of Cu and Ni and Table 8.2 consists of the refractive index of 2D materials for a different thickness of λ = 633 nm.

Similarly, Wang et al. (2019) proposed a titanium oxide (TiO_2)/Au film-coated multimode-fiber single-mode fiber-multimode fiber (MSM)–based SPR sensor for RI sensing, as shown in Fig. 8.7. To get the hetero-core structure, 10 mm of SMF (8.2/125 μm) is spliced between the two MMF (62.5/125 μm). Thereafter, they have introduced a D-type shape on a fabricated MSM structure and coated it with Au film for SPR phenomena and then further sensitivity has been improved by adding the TiO_2 film. They have also investigated the effect of different thicknesses of TiO_2 film on sensing performance and found that the resonant wavelength shifts towards an infrared direction, as shown in Fig. 8.8. It has also been observed that a redshift occurs in the case of thicker TiO_2 film and exhibits more RI sensitivity.

Singh, Mishra, and Gupta (2013) proposed the RI sensor using an SPR-based optical fiber structure. They have immobilized the Cu/oxides bilayers over an optical fiber to improve the sensing.

They have used the 20-cm length of polymer-coated silica (PCS) fiber with a 600-μm core diameter. Then, the cladding from 1 cm of optical fiber is removed using a sharp blade to enhance the excretion of evanescent waves (EWs) outside the sensing area. Thereafter, the unclad part of the fiber was cleaned using nitric acid, followed by a rinse using DI water and acetone. Then, the cleaned sensing area was coated using copper (Cu) followed by an oxide layer, as shown in Fig. 8.9. The experimental setup for RI sensing is shown in Fig. 8.10. The light is launched through the tungsten-halogen

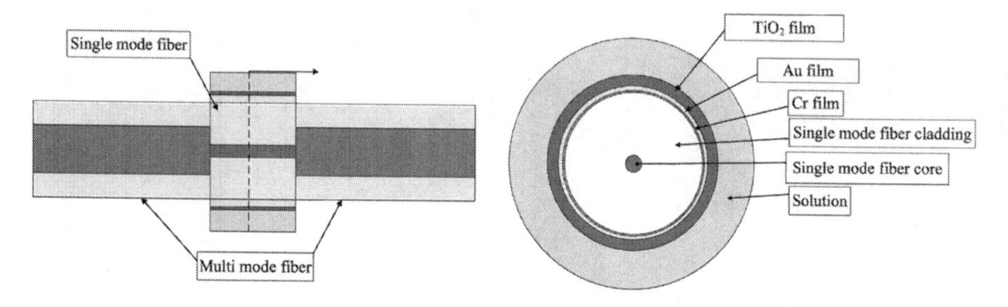

FIGURE 8.7 a) Core-mismatch-based optical fiber SPR sensor. Reprinted with permission from Optics Communications. Copyright, 2019, Elsevier (Wang et al. 2019).

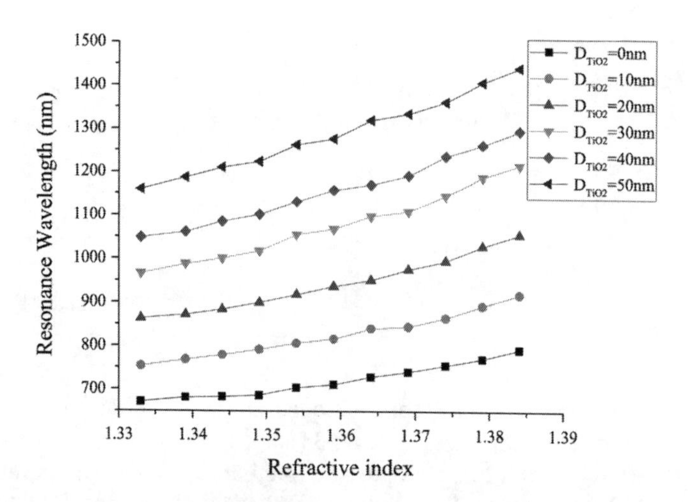

FIGURE 8.8 Change in resonance wavelength concerning refractive index at different TiO_2 film thickness. Reprinted with permission from Optics Communications. Copyright, 2019, Elsevier (Wang et al. 2019).

lamp (THL) and the sensor probe was kept in a flow cell, where there is an inlet and outlet to add and remove the analytes. Thereafter, the output port of the sensor probe is attached with a spectrometer to observe the signal variation in the computer.

In this chapter, 2D-material-based SPR sensors for the detection of various glucose, uric acid, ascorbic acid, nucleic acid, proteins, and microorganisms are discussed. Thereafter, the chapter has been summarized with the opportunities and challenges of 2D materials in clinical diagnosis.

8.2 DETECTION OF GLUCOSE

Glucose is found in the human body in the form of carbohydrates. Its concentration plays a very important role in human beings. Hypoglycemia arises due to the deficiency of glucose, and excessive glucose causes several severe diseases such as diabetes, cardiovascular, and hypertension (Duan, Liu, Wang, and Su 2019). There

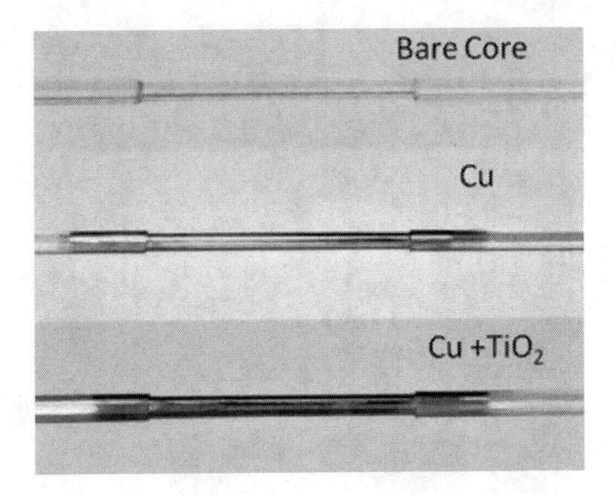

FIGURE 8.9 Optical fiber-based SPR sensor probe. Reprinted with permission from Sensors and Actuators A: Physical Journal. Copyright, 2013, Elsevier (Singh, Mishra, and Gupta 2013).

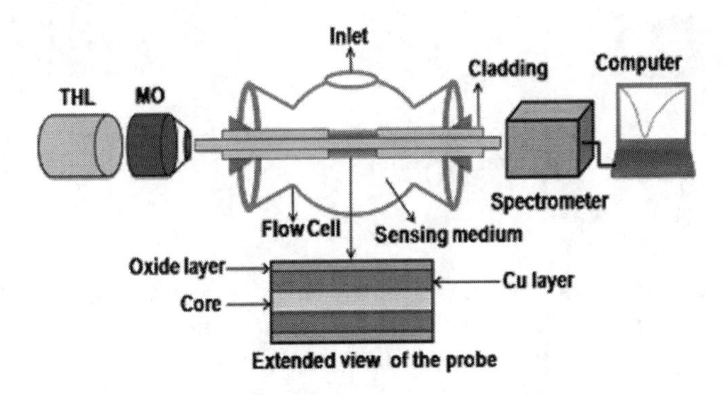

FIGURE 8.10 Experimental setup for SPR-based fiber optic sensor. Reprinted with permission from Sensors and Actuators A: Physical Journal. Copyright, 2013, Elsevier (Singh, Mishra, and Gupta 2013).

are several 2D materials such as WS_2 (Duan, Liu, Wang, and Su 2019; Lin et al. 2014), MoS_2 (Huang, Yao, Liang, and Qiu 2014; Yu et al. 2020), and GO (Yang, Zhu, et al. 2020) have been used for the development of glucose sensors. Duan, Liu, Wang, and Su (2019) proposed a glucose sensor based on WS_2 quantum dots (QDs) in which a redox reaction occurs between various valence states of iron. In glucose sensing, it mainly produces H_2O_2 due to GOx, where Fe^{2+} is oxidized by H_2O_2 into Fe^{3+}. This Fe^{3+} quenches the fluorescence of WS_2-QDs, as shown in Fig. 8.11.

Thereafter, 2D materials have been used for development of SPR sensors. Yu et al. (2020) developed Au-film-coated D-shaped optical fiber–based SPR sensor for glucose detection in the human body. For the fiber sensor structure, SMF-28 (8.2/125 μm) has been used. The D-shape structure is formed with the help of polished paper and then chromium and Au film are used for the SPR effect. Thereafter, using the composites of 2D materials, MoS_2-graphene over an optical fiber using a

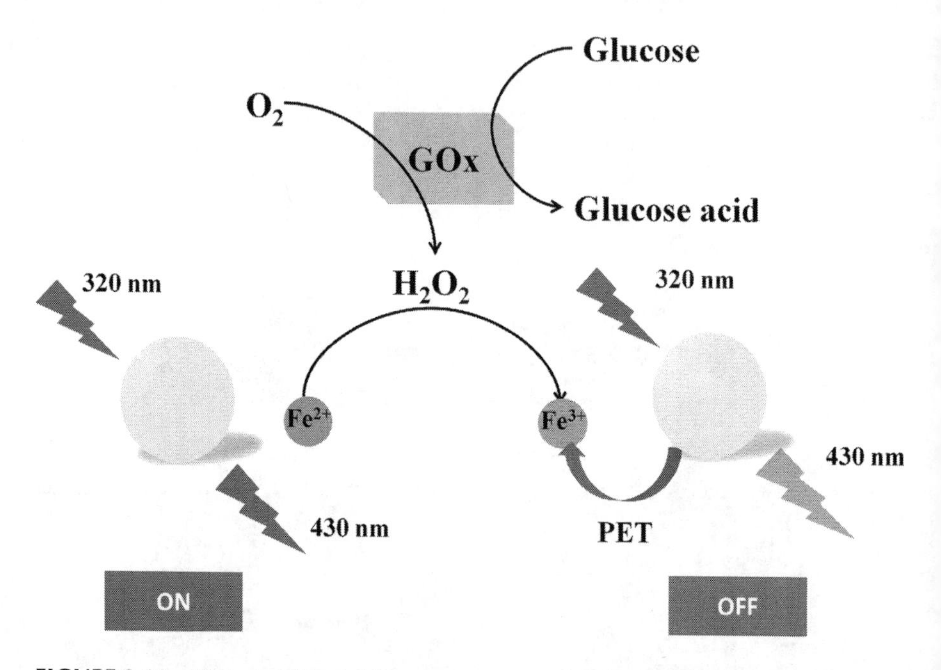

FIGURE 8.11 2D-material-based fluorescence sensor for glucose detection. Reprinted with permission from Journal of Luminescence. Copyright, 2019, Elsevier (Duan, Liu, Wang, and Su 2019).

chemical vapor deposition (CVD) for further improvement of sensing performance is shown in Fig. 8.12. To add the specificity, pyrene 1 boronic acid (PBA) has been used over a nanomaterial-coated fiber structure that mainly reacts with glucose and exhibits a high selectivity with detection of glucose in the serum.

Recently, Yang, Zhu, et al. (2020) proposed a glucose sensor using a GOx/GO/AuNPs-functionalized tapered optical fiber. As shown in Fig. 8.13, they have fabricated the tapered optical fiber waist diameter of 25 μm with a waist length of 6 mm that mainly works as a sensing zone.

Thereafter, AuNPs have been immobilized using the protocol mentioned in Fig. 8.14 (Yang, Zhang, et al. 2020). Before coating, a fiber structure needs to be

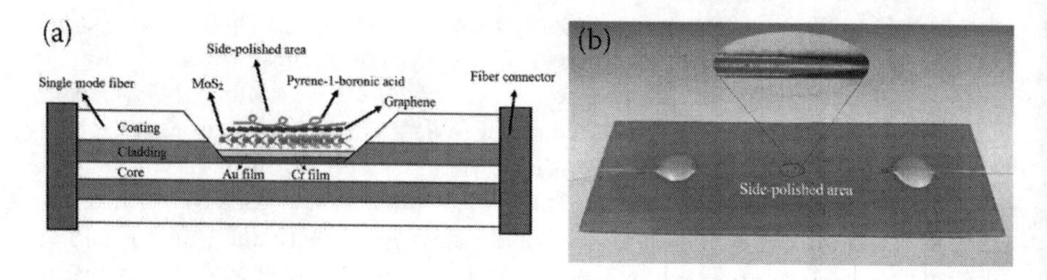

FIGURE 8.12 SPR sensor using D-shaped optical fiber. Reprinted with permission from Talanta. Copyright, 2019, Elsevier (Yu et al. 2020).

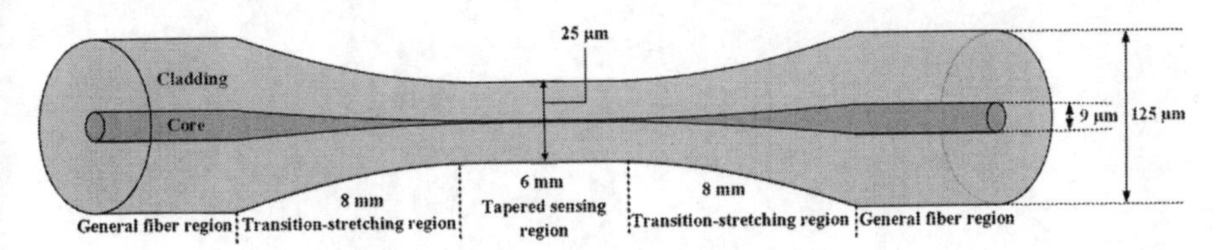

FIGURE 8.13 Representation of optical-tapered fiber. Reprinted with permission from Optik. Copyright, 2020, Elsevier (Yang, Zhu, et al. 2020).

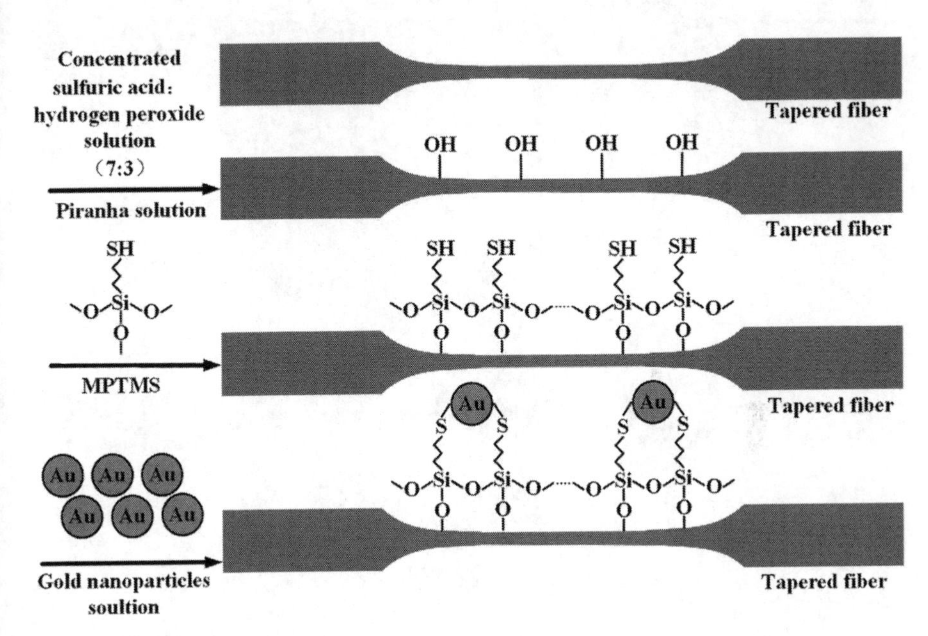

FIGURE 8.14 Gold nanoparticles' immobilization over a tapered fiber. Reprinted with permission from Plasmonics Journal. Copyright, 2019, Springer (Yang, Zhang, et al. 2020).

clean, using a three-step process: (i) in acetone for 20 min, (ii) in Piranha solution for 30 min, and (iii) DI water. Thereafter, GO has been coated over a fiber structure using the annealing method. Then, it needs to be kept in an ethanolic MPTMS solution for 12 h that works as an adhesive for immobilization of AuNPs on a fiber structure. Thereafter, unbound MPTMS can be removed using a rinse with ethanol and dry the sensor structure using nitrogen gas. Further, fiber was dipped in AuNPs aqueous for its immobilization. Thereafter, unbound particles of Au can be removed through a rinse of ethanol and dried with nitrogen gas.

After immobilization of nanomaterials over a tapered optical fiber structure, glucose oxidase (GOx) has been functionalized using a protocol mentioned in the schematic shown in Fig. 8.15 (Yang, Zhang, et al. 2020). For this, it is first kept in an MUA solution to produce the carboxyl groups over a fiber structure. Then, these carboxyl groups can be re-activated using EDC/NHS and kept in a GOx solution for enzyme coating. The schematic for the final measurement of a glucose solution is

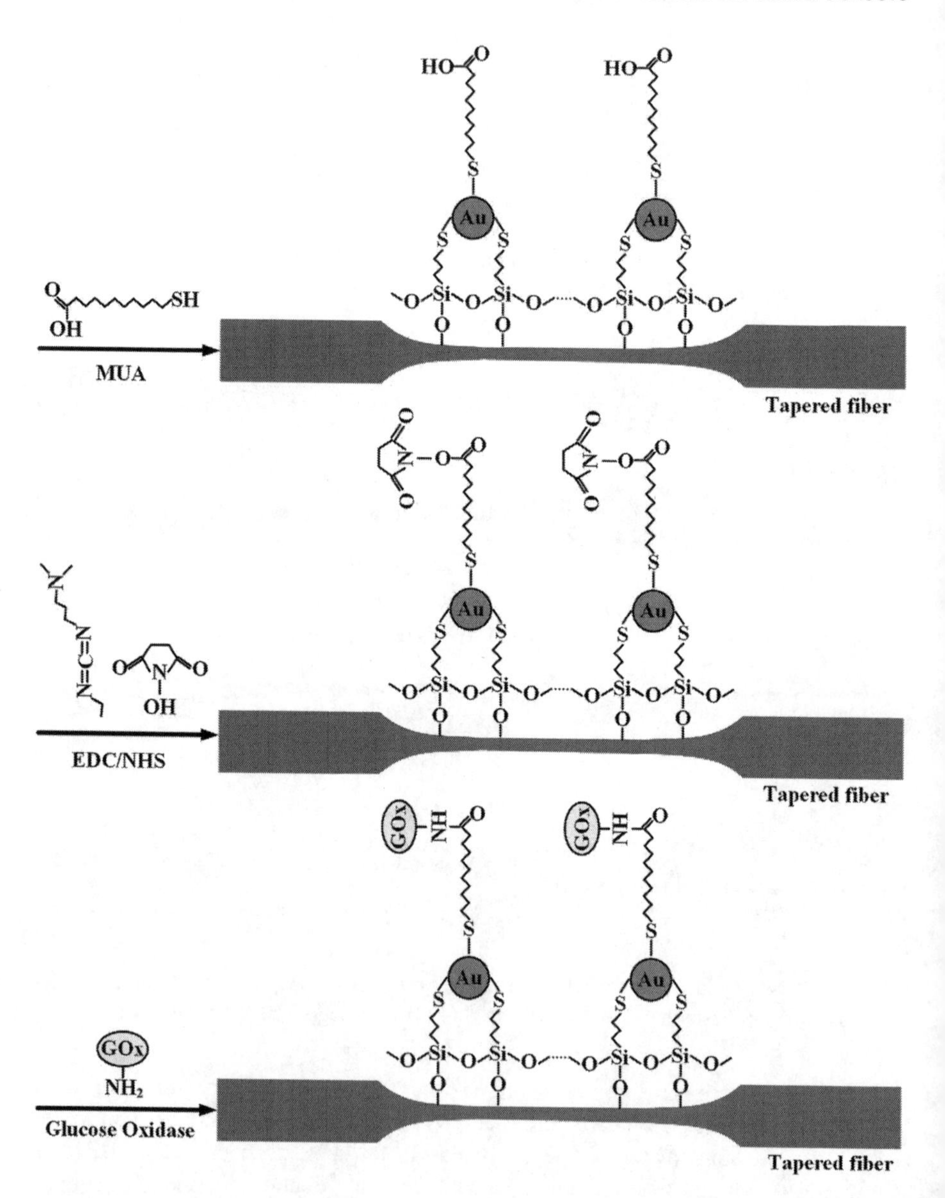

FIGURE 8.15 Steps of enzyme functionalization process. Reprinted with permission from Plasmonics Journal. Copyright, 2019, Springer (Yang, Zhang, et al. 2020).

shown in Fig. 8.16. Here, light is launched through the tungsten-halogen light source and the spectrum is measured through the spectrometer connected through the computer. The sensor probe is spliced through the source and the spectrometer through the fusion splicer. Here, it can be clearly observed that the tapered structure is first coated with GO and then AuNPs followed by GOx. Here, GOx plays a role in enhancing the specificity by producing H_2O_2 in the presence of glucose that

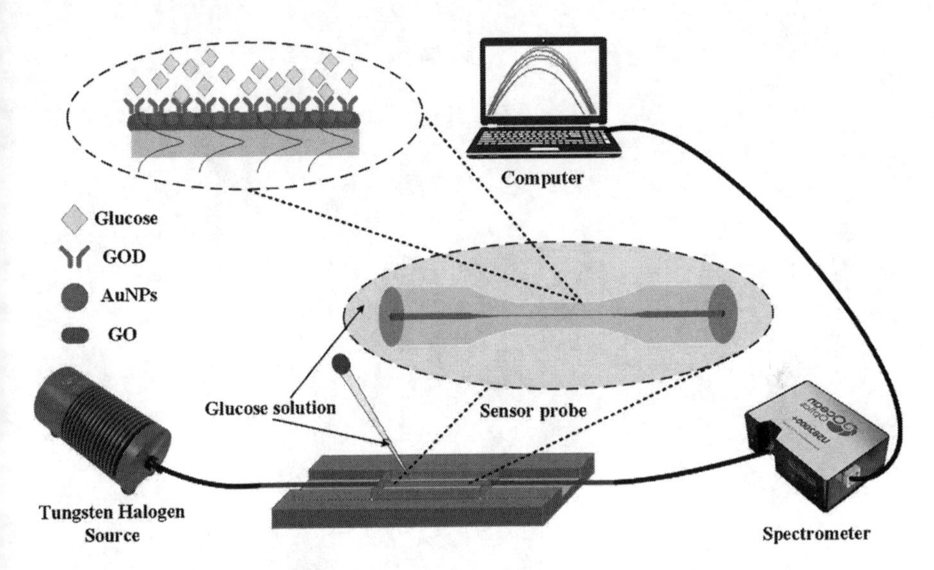

FIGURE 8.16 Schematic of glucose measurement process. Reprinted with permission from Optik Journal. Copyright, 2020, Elsevier (Yang, Zhu, et al. 2020).

changes the RI of a surrounding sensing area and further changes the resonance wavelength based on the glucose concentration.

8.3 DETECTION OF URIC ACID

Uric acid (UA) is also an important biomolecule found in human bodies. It also needs to be monitored regularly because a deficiency of UA leads to diabetes mellitus and sclerosis; excessive UA causes more severe diseases such as arthritis, gout, kidney, cardiovascular, and neurological diseases. Thus, Durai, Kong, and Badhulika (2020) developed a 2D material, a WS_2-based electrochemical non-enzymatic sensor for UA detection. But this sensor is limited due to its low sensitivity/specificity. Later on, Singh et al. (2019) proposed a tapered optical fiber–based LSPR sensor using uricase/GO/AuNPs-immobilization for UA detection. They have fabricated the tapered optical fiber structure of a 40 μm waist diameter. Then, it is coated with around 10 nm AuNPs for the LSPR phenomena, followed by a coating of GO for further improvement in sensing performance. Uricase enzyme is used to increase the selectivity of the sensor probe. The proposed sensor is working in a wide range of 10–800 μM that lies in the UA found in human bodies. Chemical reactions and a complete schematic for fabrication of the sensor probe are shown in Fig. 8.17.

Thereafter, Kumar, Singh, et al. (2020) proposed a micro-ball fiber structure for UA detection. The fusion splicer has been used to fabricate a fiber ball structure, as shown in Fig. 8.18. To increase the sensitivity, the fiber ball is coated with AuNPs followed by GO; thereafter, uricase has been used to increase its specificity. Its linearity range lies in the range of 10 μM–1 mM UA concentrations.

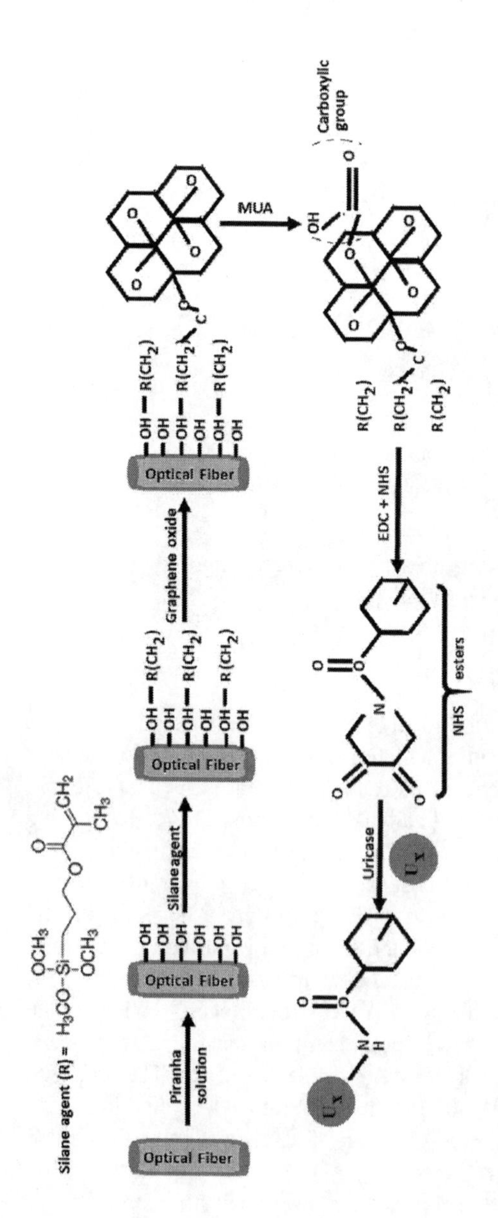

FIGURE 8.17 Fabrication process of graphene oxide-based sensor for uric acid detection. Reprinted with permission from Optical Fiber Technology Journal. Copyright, 2020, Elsevier (Singh et al. 2019).

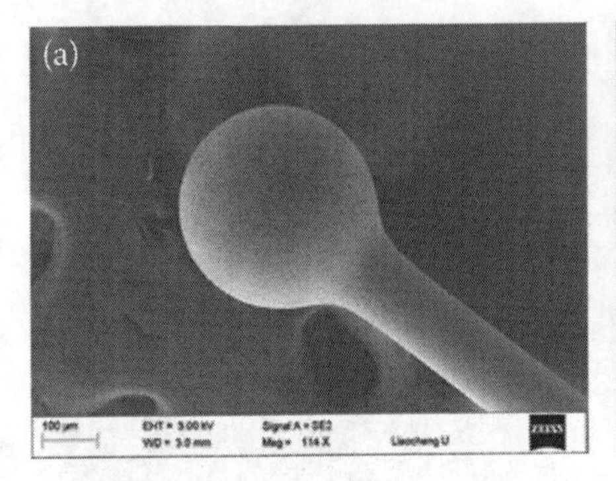
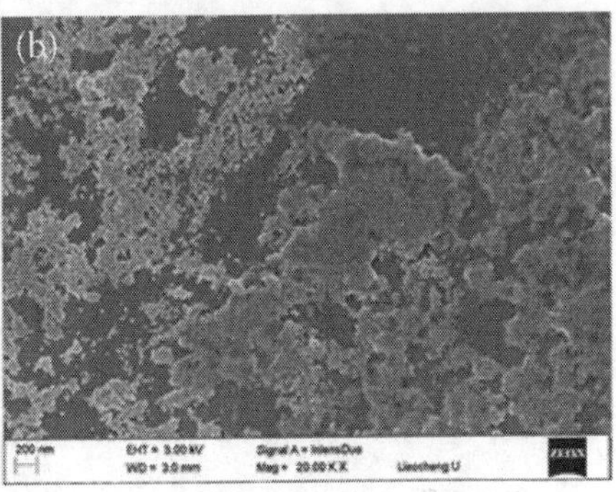

FIGURE 8.18 Micro-ball fiber-based sensor for uric acid detection. Reprinted with permission from IEEE Sensors Journal. Copyright, 2020, IEEE (Kumar, Singh, et al. 2020).

8.4 DETECTION OF ASCORBIC ACID

Ascorbic acid (AA), also known as vitamin C, is found as an antioxidant in many soft drinks as well as in fruits, foods, etc. It is also found in the serum (blood) of human bodies in the range of 40–120 μM. It helps in the development of bones, cartilage, and cell growth and also helps with recovery from burns and injuries. A deficiency of AA leads to stomach throes and scurvy. Murugan et al. (2021) have used 2D MXene for the detection of AA as well as UA and dopamine. For UA detection, Zhu et al. (2020) proposed an LSPR sensor using a multi-tapered fiber structure for AA detection. They have used the AuNPs for LSPR and then used the GO to increase the biocompatibility of the sensor probe. To add the specificity property of a sensor probe, it has been functionalized with ascorbate oxidase. They also fabricated three different types including four-tapered, five-tapered, and eight-tapered structures that have been fabricated to analyze the performance of tapered structures, as shown in Fig. 8.19. In this, the length of a tapered part is 1 mm and each tapered area is separated with a 1 mm distance. The waist diameter of each tapered fiber is uniform, i.e., 40 μm.

The SEM images of GO/AuNPs-immobilized tapered structures are shown in Fig. 8.20(a–c). Fig. 8.20d shows the GO and AuNPs on the surface of the nanomaterials' immobilized structure and its material composition has been analyzed through the SEM-EDS and shown in Fig. 8.20e. In this, carbon (C) is due to graphene and Au is due to AuNPs and silicon (Si), and oxygen (O_2) is due to optical fiber composition (silica, SiO_2).

8.5 PROTEIN DETECTION

In conventional SPR sensors, expensive equipment is needed, such as a sputtering system or rotation vacuum evaporation, for metal coating over an optical fiber structure. Nowadays, these coatings can be performed using a chemical method for the fabrication of SPR sensors. In this method, there is no need for expensive equipment as film thickness can be also easily controlled using the proper chemical reactions. As we know, silver (Ag) film-based SPR sensors exhibit a narrow

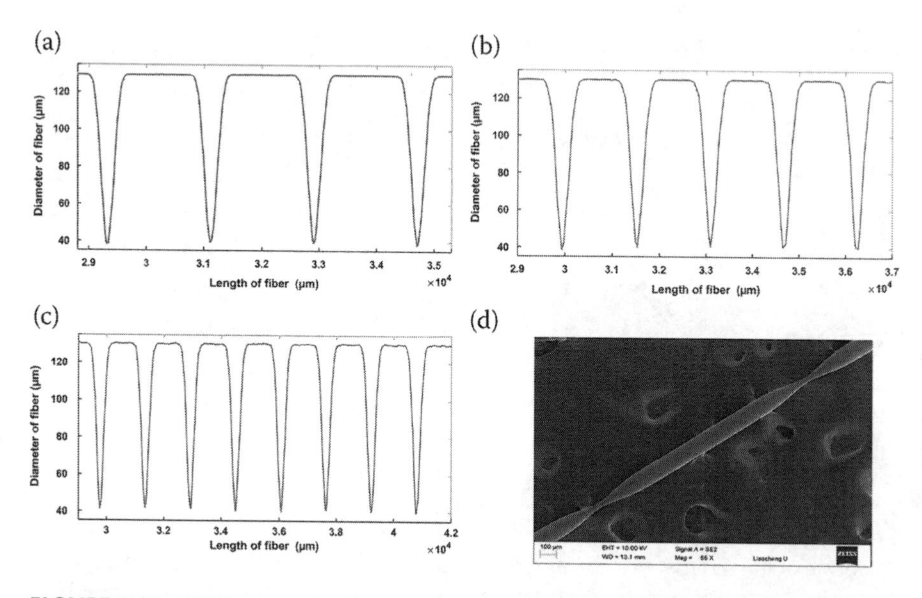

FIGURE 8.19 Different types of tapered optical fiber and its SEM image. Reprinted with permission from Optics & Laser Technology Journal. Copyright, 2020, Elsevier (Zhu et al. 2020).

resonance peak and excellent detection accuracy but Ag is oxidized easily. Thus, graphene and its derivative GO, are the options to prevent Ag from oxidation. These carbon derivatives do not allow oxygen molecules to enter into the Ag surface and prevent oxidation. GO added some excellent properties such as biocompatibility, high selectivity, and solubility. Wang and Wang (2018) developed a GO/Ag-immobilized PCS fiber for human IgG detection. For this, first a fiber was kept in an acetone volatile chemical for partial cladding removal and then used the chemical method for Ag functionalization over a fiber. The complete schematic for GO/Ag-coating is shown in Fig. 8.21. For GO immobilization, the Ag-PCS fiber is dipped into mercapto ethylamine to form the amino group on the Ag film, and then a covalent bond is created between the epoxy group of GO (0.5 mg mL^{-1}) and amino group at the surface of the Ag film. The unbounded GO was removed by gently washing the fiber surface with DI water. Thus, it helps to get the GO/Ag-immobilized PCS fiber. Thereafter, antibody functionalization is required to add the specificity over the sensor probe. For this, first 2D-material-coated fiber was dipped into a mixture of EDC/NHS for 20 min; thereafter, the sensor was rinsed through PBS (phosphate buffer saline). Then, the fiber can be functionalized with Protein A (200 μg mL^{-1}) and followed by goat anti-human IgG for 2–2 h, respectively. Thereafter, the fiber was immersed in a 1% bovine serum albumin (BSA) solution for 1 h to restrict the remaining carboxyl groups. Thus, nanomaterials' immobilization and antibody functionalization are completed and the sensor probes are ready for human IgG detection.

Finally, measurement of analytes can be performed through an experimental setup, in which light is launched using a tungsten-halogen lamp (400–1,000 nm)

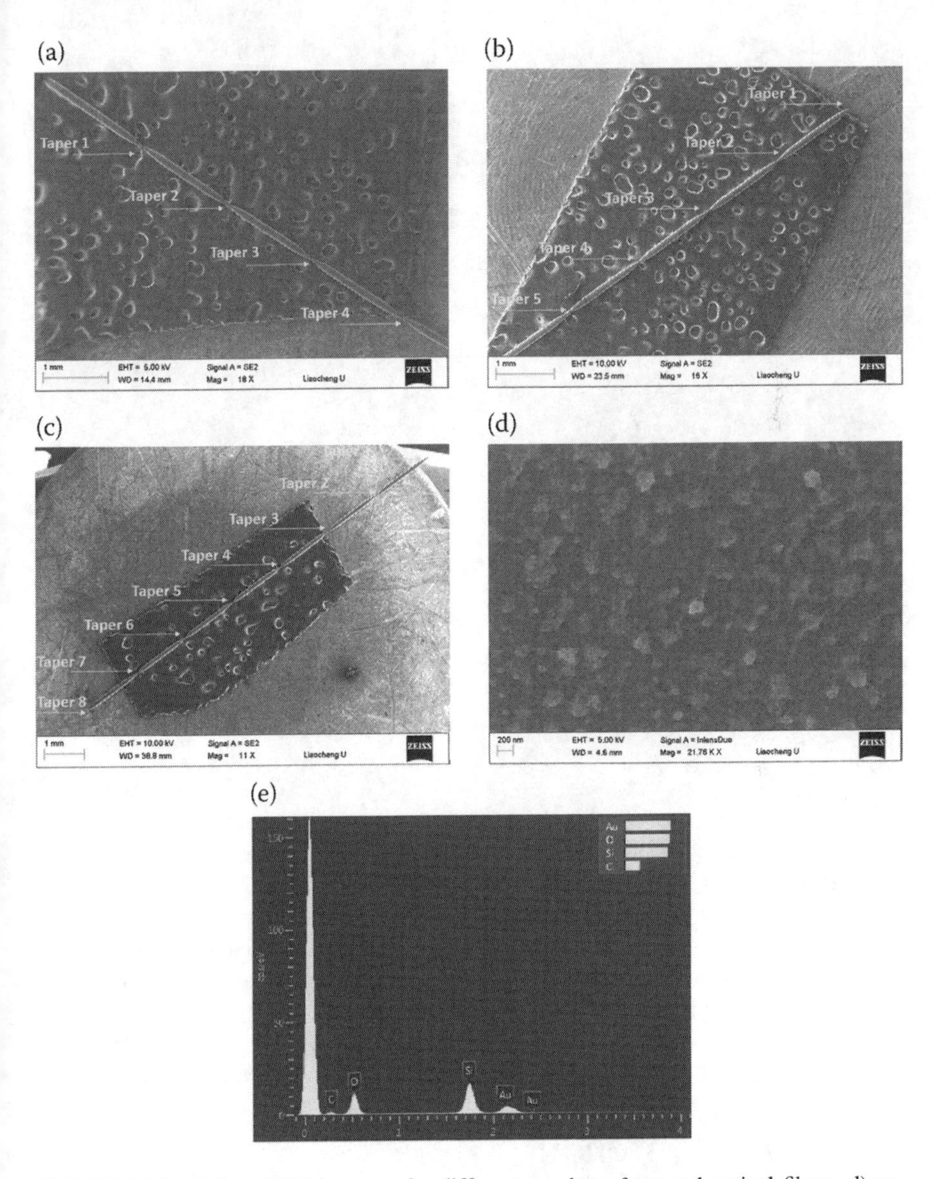

FIGURE 8.20 (a,b,c) SEM images of a different number of tapered optical fibers, d) nanoparticles' immobilized sensor, and e) EDX image to confirm the functionalized nanomaterials. Reprinted with permission from Optics & Laser Technology Journal. Copyright, 2020, Elsevier (Zhu et al. 2020).

light source. The output is collected through the spectrometer, which is connected through the computer to observe the sensing signal. Here, a 2 × 1 fiber coupler is used to connect the input source, spectrometer, and sensor probe, as shown in Fig. 8.22. Thereafter, different concentrations of human IgG solutions were measured and shown in Fig. 8.23. The resonance wavelength was measured at the

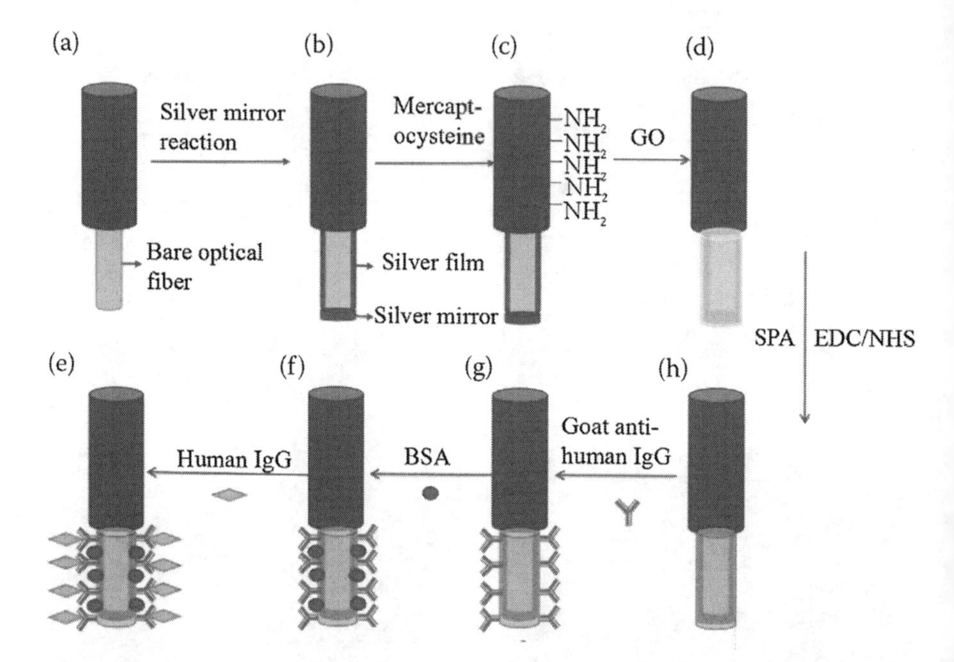

FIGURE 8.21 Fabrication steps of optical fiber sensor for protein detection. Reprinted with permission from Sensors and Actuators B: Chemical. Copyright, 2018, Elsevier (Wang and Wang 2018).

interval of 10 min for 70 min. It has been found that the wavelength shift was saturated with respect to time for all concentrations, i.e., our target solution (human IgG) was completed and combined with the goat anti-human IgG.

Huang, Yao, Liang, and Qiu (2013) show the SPR sensor for concanavalin A (ConA) protein detection. ConA is a plant lectin that can bind with membrane receptors and lead to a proliferation of cells containing glucose and mannose residues. For this work, dextran (Dex) encapsulated AuNPs (Dex-AuNPs) synthesized as an amplification agent and then deposited the GO over Au film, followed by the functionalization of phenoxy-derivatized dextran (DexP) as shown in Fig. 8.24.

8.6 DNA/RNA DETECTION

As we know, Au film is not very sensitive for biomolecule detection due to its poor absorption and limits in the performance of the biosensor. Thus, immobilization of AuNPs on a substrate increases the reaction and generated LSPR that improves the sensitivity. Due to this reason, AuNPs are widely used in the development of LSPR sensors for the detection of antibodies, proteins, and nucleic acids. But, somehow, this sensitivity is also limited, and need to use the 2D nanolayers of MoS_2 and graphene over AuNPs to increase the growth of the affinity layer and for enhancement of adsorption of biomolecules. MoS_2 as a TMD material exhibits a strong light-matter interaction and shows similar properties to graphene, but MoS_2 has more absorption per atomic layer (5%) than graphene (2.3%). Thus, El Barghouti, Akjouj, and Mir

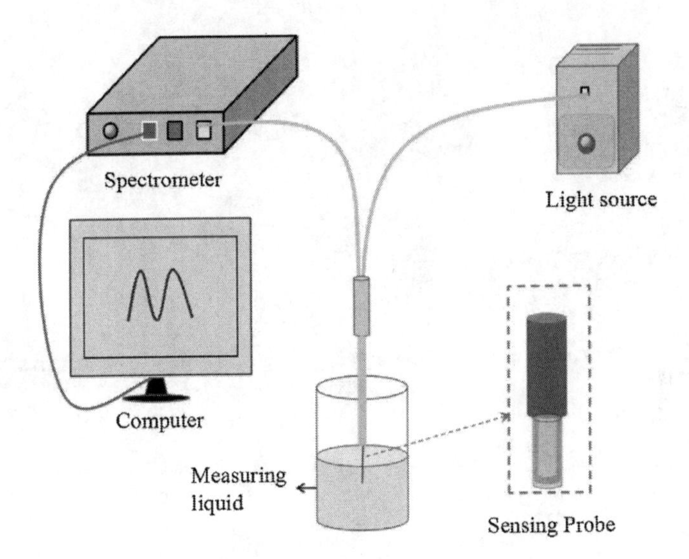

FIGURE 8.22 Experimental setup for protein detection. Reprinted with permission from Sensors and Actuators B: Chemical. Copyright, 2018, Elsevier (Wang and Wang 2018).

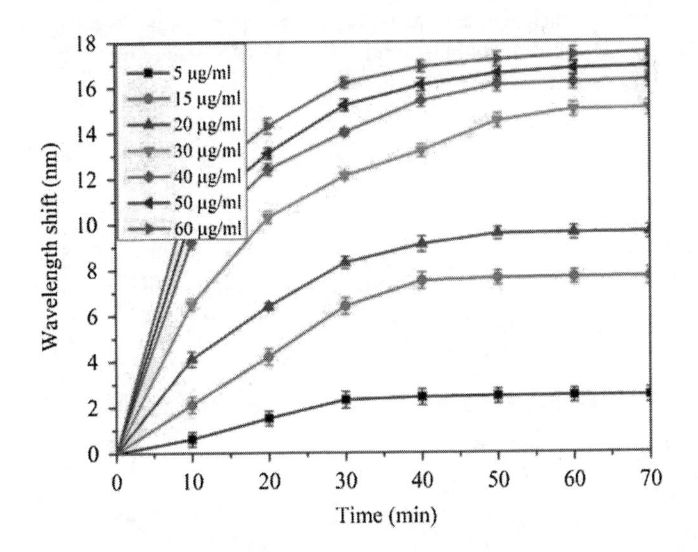

FIGURE 8.23 Wavelength variation with respect to different concentrations of human IgG. Reprinted with permission from Sensors and Actuators B: Chemical. Copyright, 2018, Elsevier (Wang and Wang 2018).

(2020) investigated the MoS_2–graphene hybrid structure-based LSPR biosensor for molecule detection and measured the absorption and sensitivity of the proposed sensor. They found that due to the addition of MoS_2–graphene over gold nanoparticles (AuNPs), or only the addition of MoS_2, or graphene over AuNPs enhances the sensitivity. They have also shown that a lower diameter of AuNPs is more sensitive than a

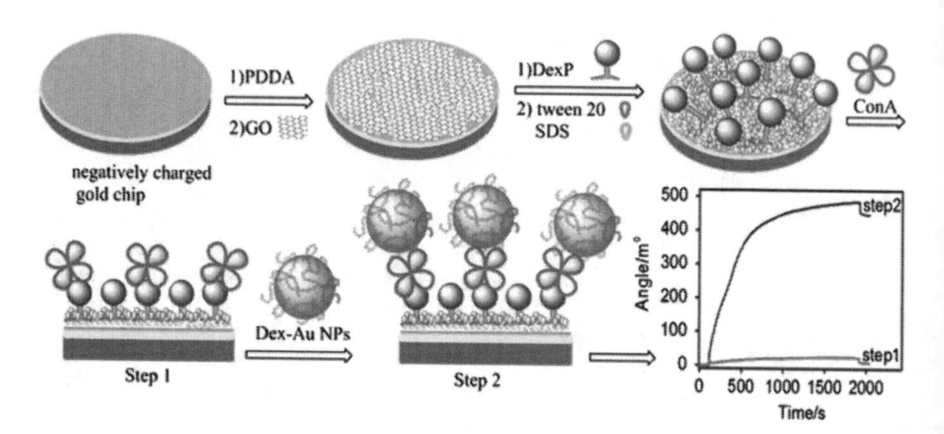

FIGURE 8.24 Fabrication steps of GO-based SPR sensor for ConA detection. Reprinted with permission from Biosensors and Bioelectronics. Copyright, 2013, Elsevier (Huang, Yao, Liang, and Qiu 2013).

higher diameter of AuNPs. Further, an LSPR-proposed sensor has been used for DNA (deoxyribonucleic acid) hybridization, as shown in Fig. 8.25a. The proposed model consists of a substrate/AuNPs/MoS_2/graphene/DNA probe and PBS solution. The RI varied at the sensing surface due to a change in absorption of molecules from a single-strand DNA (ssDNA) to a double-strand DNA (DNA); mathematically it can be represented as (El Barghouti, Akjouj, and Mir 2020):

$$n_a(DNA) = n_a(PBS) + c_a \frac{d_n}{d_c} \tag{8.1}$$

where $n_a(PBS)$ and $n_a(DNA)$ are RI sensing mediums with PBS and DNA molecules, respectively. c_a is DNA concentration and d_n/d_c is a change in RI due to DNA capture. Thus, the RI of a sensing medium after a DNA addition is:

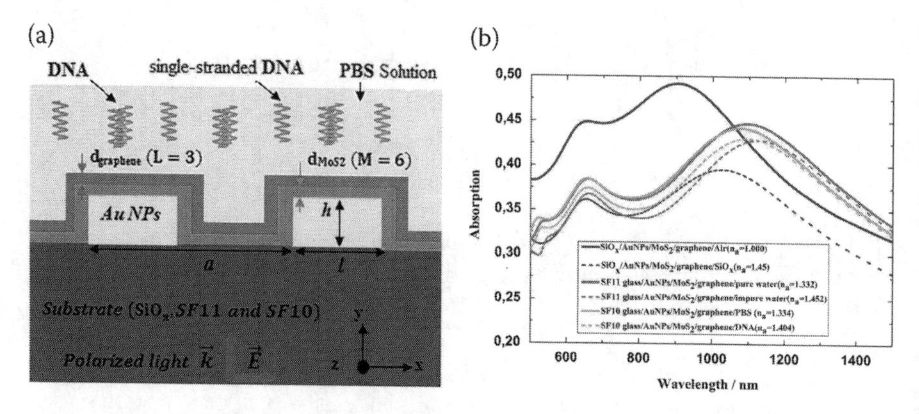

FIGURE 8.25 MoS_2-based SPR biosensor for detection of DNA hybridization. Reprinted with permission from Optics & Laser Technology Journal. Copyright, 2020, Elsevier (El Barghouti, Akjouj, and Mir 2020).

$$n_a(DNA) = 1.334 + n_a \tag{8.2}$$

where n_a is the RI change that occurs due to a biological interaction (DNA) in the sensing surface. Fig. 8.25b shows the comparative plot with the addition of DNA biomolecules.

Similarly, Nie et al. (2017) developed a 2D-material-based SPR sensor for the detection of microRNA (miRNA). For this, AuNPs decorated with MoS_2 have been used to enhance the sensitivity. To add the specificity, thiol-modified DNA oligonucleotide probes were functionalized over nanomaterial-immobilized sensor structures. In this work, MoS_2 nanosheets are used as reducing agents for the synthesis of AuNPs-MoS_2 nanocomposites using a hydrothermal method, as illustrated in Fig. 8.26. In this process, sodium carboxymethyl cellulose (CMC, 50 mM, 400 mL) and chloroauric acid (24.28 mM, 197 mL) are added with an aqueous solution of MoS_2 nanosheets (0.08 mg mL^{-1}, 5 mL^{-1}) and then heated under vigorous stirring. Thereafter, the centrifuged nanocomposite solution is stored in a 2–8°C refrigerator. Here, the complementary sequences of the target miRNA and thiol-modified probe are immobilized on an Au film to capture the DNA. In this case, AuNPs help in LSPR and then the SPR phenomena are generated due to the Au film. Here, MoS_2-loaded AuNPs help in the enhancement of SPR signals compared to AuNPs. Due to the addition of 2D material, the performance of sensors such as sensitivity, reproducibility, and selectivity increases, which is very useful in medical applications.

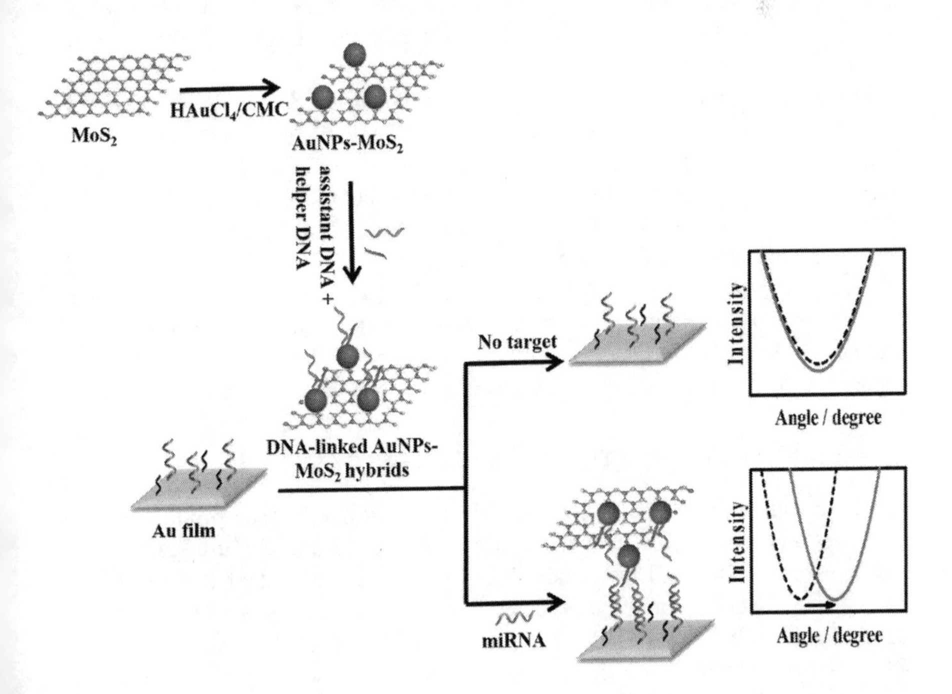

FIGURE 8.26 AuNPs-MoS_2 nanocomposite-based SPR biosensor. Reprinted with permission from Analytica Chimica Acta Journal. Copyright, 2017, Elsevier (Nie et al. 2017).

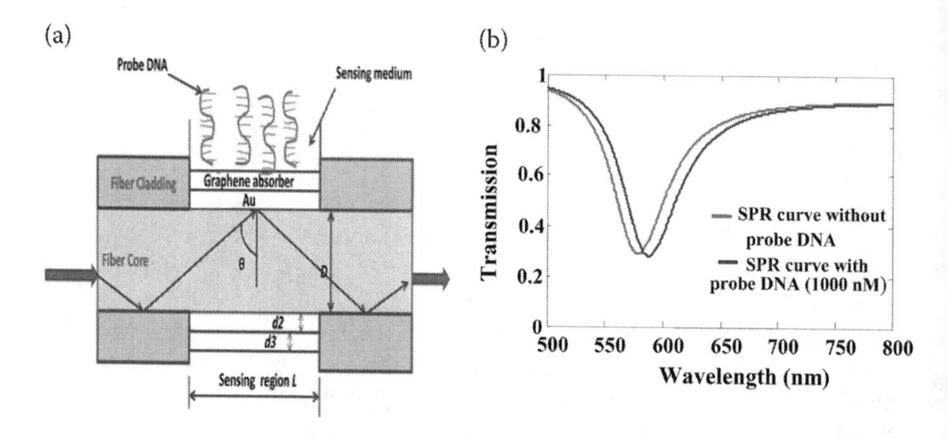

FIGURE 8.27 Optical fiber–based SPR sensor for DNA hybridization detection. Reprinted with permission from Optics Communications. Copyright, 2017, Elsevier (Shushama, Rana, Inum, and Hossain 2017).

Shushama, Rana, Inum, and Hossain (2017) numerically investigated the DNA hybridization process using a graphene-based Au film immobilized optical fiber structure, as shown in Fig. 8.27. The change in wavelength can be easily observed from a bare sensor in which there are no DNA molecules as well as a sensor with probe DNA. The significant wavelength change can be easily observed that is due to the change in RI surrounding the sensing surface. It can also be observed that complementary DNA strands show more wavelength variation in comparison to mismatched DNA strands.

Similarly, Rahman et al. (2017) proposed an Ag-MoS_2-graphene-immobilized optical fiber structure for DNA hybridization. To fabricate this sensor structure, cladding is removed from the middle of the sensor. Thereafter, the Ag film is coated over an unclad fiber structure, and then a MoS_2 monolayer is followed by a graphene absorber, as shown in Fig. 8.28a. It also found that the resonance angle can be varied due to the adsorption of a probe DNA (in Fig. 8.28b).

Again, Rahman, Anower, and Abdulrazak (2019) proposed using an Ag film and phosphorene-based optical fiber sensor and also used the 2D materials such as graphene, MoS_2, $MoSe_2$, WS_2, and WSe_2 to increase the figure of merit and the sensitivity of sensing the DNA hybridization, as shown in Fig. 8.29.

8.7 DETECTION OF OTHER BIOMOLECULES

Zhang et al. (2019) developed an MXene-modified screen-printed electrode (MXene/SPE)-based sensor for detection of two commonly used drugs: acetaminophen (ACOP) and isoniazid (INZ). An excess amount of these drugs may damage the liver in some cases. Thus, this electrochemical sensor detected and differentiated the ACOP and INZ, simultaneously. Ti_3C_2Tx was drop-cast over an SPE surface that works as a signal enhancing matrix for sensing. This sensor is used for the determination of individual concentrations of INZ and ACOP in human fluids, as shown in Fig. 8.30.

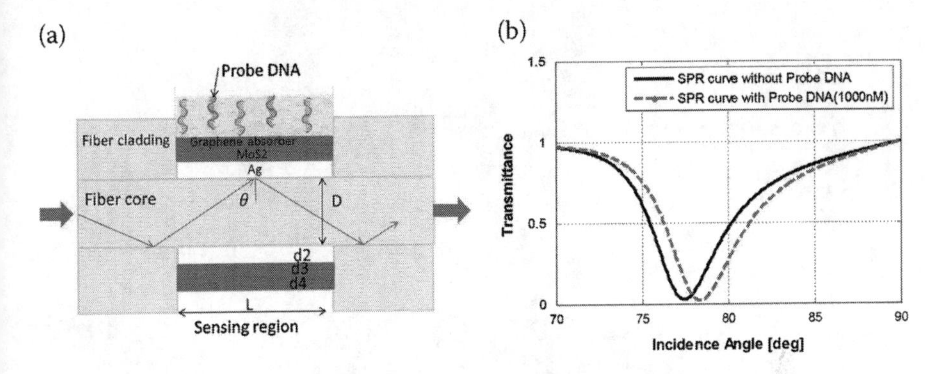

FIGURE 8.28 Ag-MoS_2-graphene-based optical fiber sensor for detection of DNA hybridization. Reprinted with permission from Optik. Copyright, 2017, Elsevier (Rahman et al. 2017).

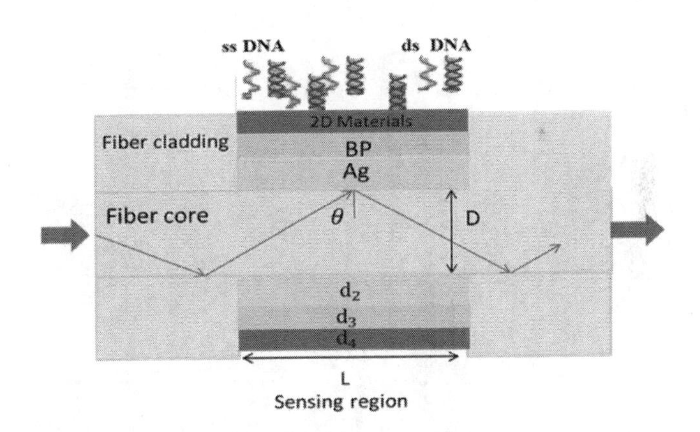

FIGURE 8.29 Phosphorene-based optical fiber SPR sensor for DNA hybridization. Reprinted with permission from Photonics and Nanostructures—Fundamentals and Applications Journal. Copyright, 2017, Elsevier (Rahman, Anower, and Abdulrazak 2019).

Thereafter, Song et al. (2019) developed an electrochemical sensor using MXene nanocomposites for the detection of organophosphorus pesticides (OPs). The development process of the proposed biosensor for monitoring the methamidophos is shown in Fig. 8.31. For this, MXene/Au NPs and MnO_2/Mn_3O_4 are functionalized over a glassy-carbon-electrode (GCE); thereafter, an AChE-Chit solution was deposited over it.

8.8 DETECTION OF MICROORGANISMS

The detection of microorganisms such as cells and bacteria is also important for clinical applications. Singh et al. (2020) developed the SMF-spliced multi-core

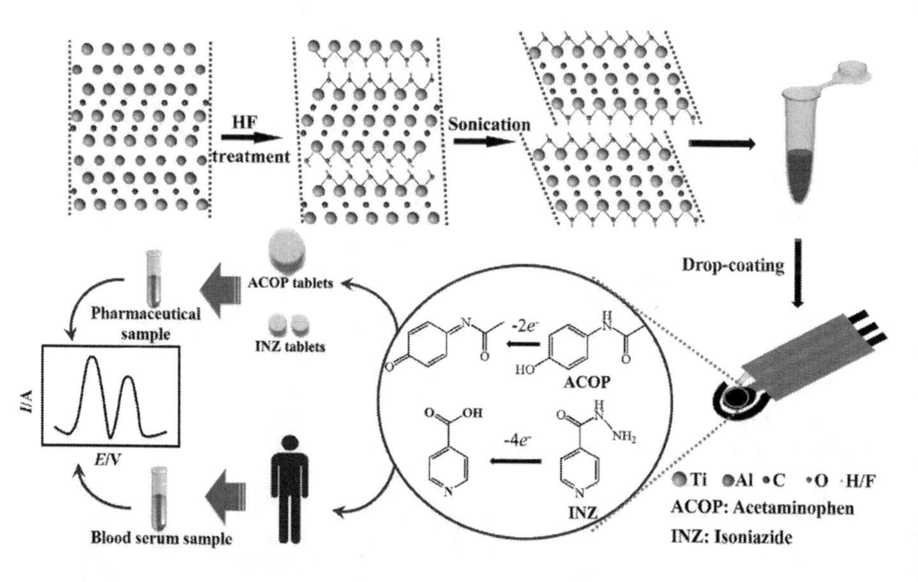

FIGURE 8.30 Synthesis process of MXene and its application. Reprinted with permission from Biosensors and Bioelectronics Journal. Copyright, 2019, Elsevier (Zhang et al. 2019).

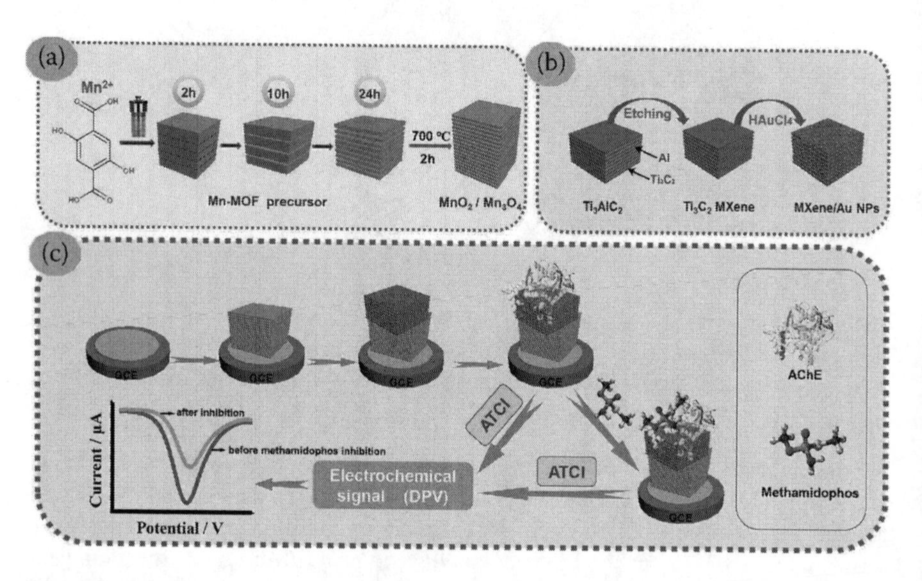

FIGURE 8.31 Fabrication steps of MXene-based SPR biosensor. Reprinted with permission from Journal of Hazardous Materials. Copyright, 2019, Elsevier (Song et al. 2019).

fiber (MCF) sensor probe for cancer cell detection. They have immobilized the AuNPs over an SMF-MCF sensor structure to execute the LSPR phenomena. Thereafter, GO and copper-oxide nanoflowers (CuO-NFs) are used to enhance further sensitivity. Further, the structure is coated with 2-deoxy-D-glucose

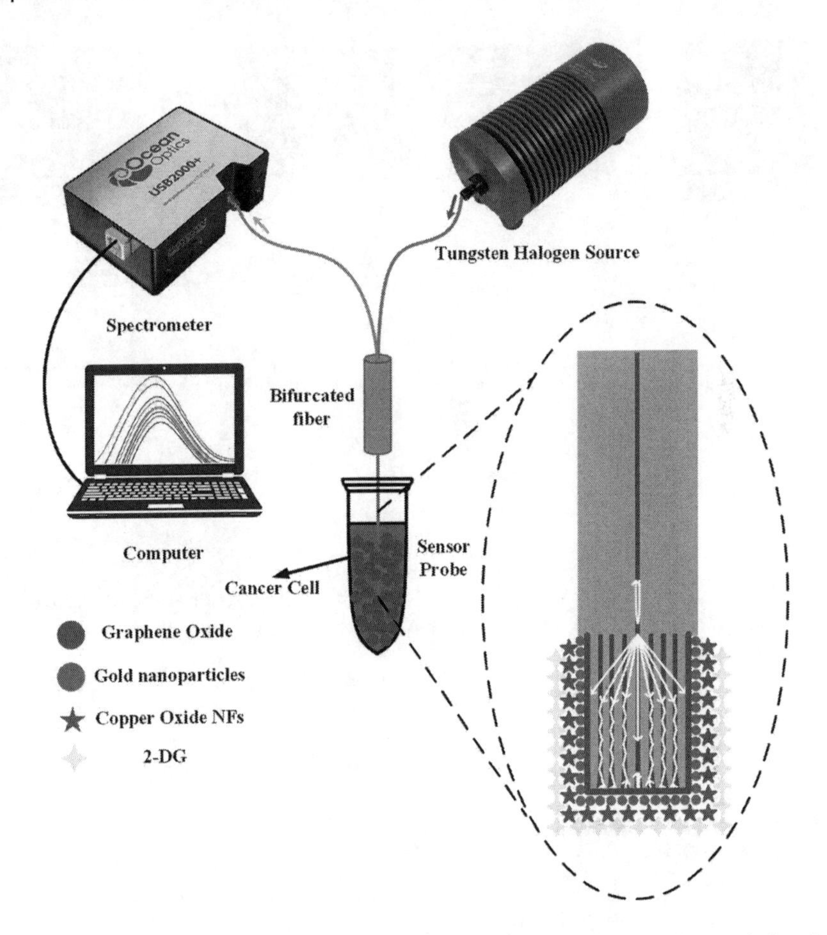

FIGURE 8.32 Cancer cell detection using multicore fiber. Reprinted with permission from Biosensors & Bioelectronics Journal. Copyright, 2020, Elsevier (Singh et al. 2020).

(2-DG) for specific detection of cancer cells, as shown in Fig. 8.32. For detection, six different cell lines such as MCF-7, Hepa 1–6, A549, HepG2 (cancerous cells), and NCF, LO2 (cell lines from normal tissues) are cultured in specific environmental conditions.

Similarly, Kumar, Guo, et al. (2020) proposed an LSPR sensor for *Shigella* bacteria detection. They have used the 2D material MoS_2 for further improvement of sensitivity. To make it specific, *Shigella*-specific oligonucleotides are used. *Shigella* is a bacterial species that causes several diseases such as fever, diarrhea, vomiting, cramping, etc., in human beings; 10 CFU mL^{-1} to 100 CFU mL^{-1} (CFU, colony-forming unit) of *Shigella* is enough to cause an intestinal infection. Thus, a working proposed sensor lies in the range of $1–10^9$ CFU mL^{-1} and the detection limit is found to be very low (around 2 CFU mL^{-1}). To fabricate an SMF-MCF structure-based sensor probe, a fusion splicer has been used. The core diameters of MCF and SMF are 9 μm and 6.1 μm, respectively, but the cladding diameter of both

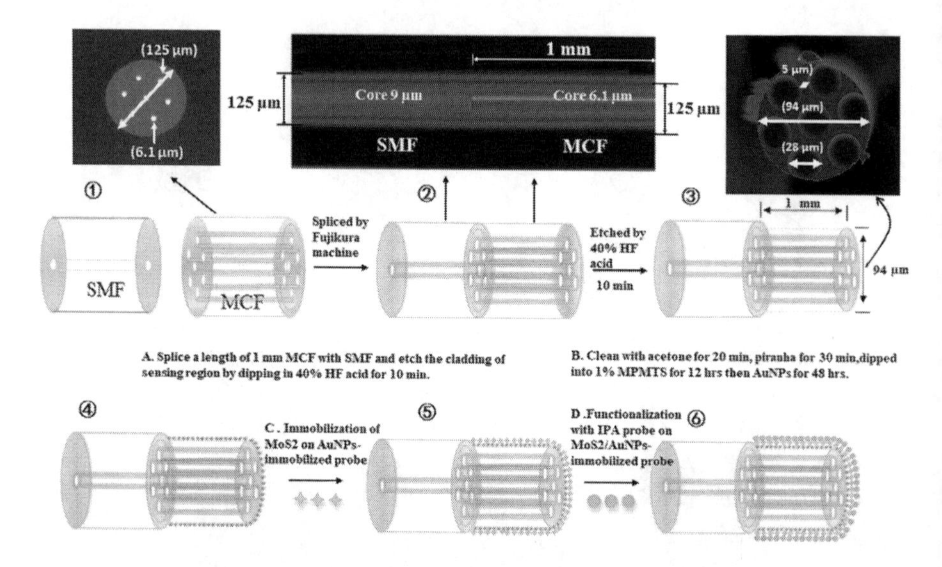

FIGURE 8.33 Procedure of fabrication of MoS_2-based LSPR sensor probe. Reprinted with permission from Journal of Lightwave Technology. Copyright, 2020, IEEE (Kumar, Guo, et al. 2020).

fibers is the same; thus, the splicer works on the cladding alignment technique, as shown in Fig. 8.33.

MCF's core is highly doped and etched with 40% hydrofluoric acid (HF) in this work. Etching helps in the enhancement of coupling of the modes and excretion of the evanescent waves. Thereafter, an acetone and piranha solution were used for preliminary cleaning of the fiber structure, before the coating nanoparticles. Thereafter, AuNPs and MoS_2 were immobilized using a standard protocol. Finally, a nanoparticle-coated sensor structure was functionalized using the specific bacterial probe for *Shigella* detection.

To culture the *Shigella sonnei*, a proper media is required that contains 1.25 g peptone, 0.75 g beef powder, and 1.25 g NaCl in 250 mL of DI water as well as 3.75 g of agar (for agar media). Then, bacteria were added to the media and kept in a shaker incubator at 200 rpm and 37°C temperature overnight. The complete schematic of the bacterial culture is shown in Fig. 8.34. Thereafter, 50 μL of cultured bacteria were added into an agar plate and again kept in an incubator for 48 h. Then it was counted using a conventional surface plate count method. Thereafter, a solution was serially diluted into lower concentrations using bacterial media and used for sensing purposes.

Kaushik, Tiwari, Pal, and Sinha (2019) developed MoS_2/Au-immobilized optical fiber sensor for *E. coli* detection, with a size is around 1–2 μm in length and 0.5–1 μm in diameter. The complete fabrication schematic is shown in Fig. 8.35. For this, first the multimode fiber (MMF) has been taken and etched with 48% HF acid to remove the cladding.

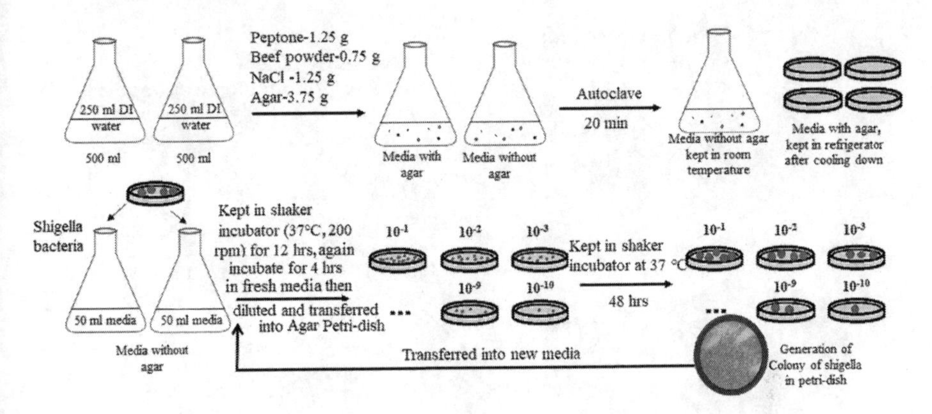

FIGURE 8.34 Culture process of *Shigella sonnei* bacteria. Reprinted with permission from Journal of Lightwave Technology. Copyright, 2020, IEEE (Kumar, Guo, et al. 2020).

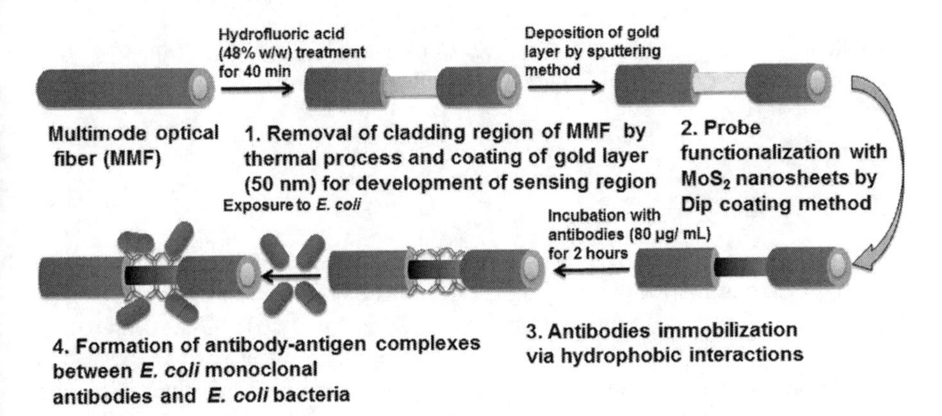

FIGURE 8.35 Fabrication steps of MoS_2 nanosheets-based SPR sensor. Reprinted with permission from Biosensors & Bioelectronics Journal. Copyright, 2019, Elsevier (Kaushik, Tiwari, Pal, and Sinha 2019).

Thereafter, the Au layer is coated using the sputtering method and then the MoS_2 nanosheets are immobilized using the dip-coating method. To add the specificity in the sensor structure, it has been functionalized with antibodies (80 µg mL^{-1}, size ~10 nm) via hydrophobic interactions. For measurement, samples of *E. coli* are prepared in the range of 1000–8000 CFU mL^{-1} that formed the antibody-antigen complexes of antibodies and *E. coli* bacteria. An experimental setup to detect the *E. coli* consists of the tungsten-halogen light source and spectrometer, as shown in Fig. 8.36. For this, first noise signals are removed by measuring the dark and blank reference and then readings are taken in the presence of different concentrations of *E. coli* solutions and observed in the spectrum in a laptop connected through the spectrometer.

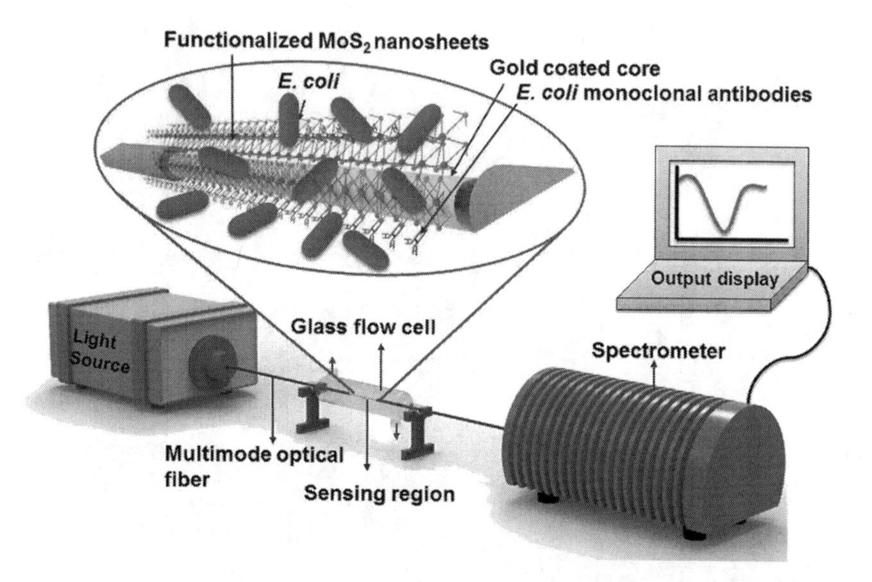

FIGURE 8.36 Experimental setup for *E. coli* bacteria detection. Reprinted with permission from Biosensors & Bioelectronics Journal. Copyright, 2019, Elsevier (Kaushik, Tiwari, Pal, and Sinha 2019).

8.9 SUMMARY

This chapter contains an overview of 2D-material-based SPR sensors for biomedical applications. Various biosensing applications of 2D materials, such as GO, g-C_3N_4, BN, B NSs, transition metal oxides (such as MoO_3, WO_3, and MnO_2), silicate clays, TMD (e.g. MoS_2, WS_2), BP, LDHs, antimonene, SnTe NSs, and MXenes, are discussed in this chapter and were not explored until now. These are emerging biosensing materials and there is a huge scope in the near future due to their peculiar electronic and optical properties. Due to these reasons, these are widely used in bioimaging, fluorescence, SERS, ECL, colorimetric, as well SPR for development of biosensors. Further, 2D material-based SPR sensors for the detection of uric acid, glucose, and ascorbic acid have been explored. These 2D materials are also used in *in vitro* applications such as cell detection and DNA/RNA detection.

Nowadays, SPR biosensing is very popular because it is real time, cost effective, and does not use any labels. Recently, it has been explored that, with the help of 2D materials, SPR sensing can be highly sensitive and effective for bio/chemomolecules detection. It magnifies the signals and is useful for the detection of macromolecules at a low detection limit in the range of femtomolar. These materials easily couple with Au and silver materials and along with the advantages of biocompatibility and prevention of metal surface oxidation. These sensors are useful for point-of-care applications and can be used for online monitoring with the help of an optical fiber.

REFERENCES

AlaguVibisha, G., Jeeban Kumar Nayak, P. Maheswari, N. Priyadharsini, A. Nisha, Z. Jaroszewicz, K. B. Rajesh, and Rajan Jha. 2020. "Sensitivity Enhancement of Surface Plasmon Resonance Sensor Using Hybrid Configuration of 2D Materials over Bimetallic Layer of Cu–Ni." *Optics Communications* 463: 125337. 10.1016/j.optcom.2020.125337.

Deshmukh, Kalim, Tomáš Kovářík, and S. K. Khadheer Pasha. 2020. "State of the Art Recent Progress in Two Dimensional MXenes Based Gas Sensors and Biosensors: A Comprehensive Review." *Coordination Chemistry Reviews* 424: 213514. 10.1016/j.ccr.2020.213514.

Duan, Xinhe, Qing Liu, Guannan Wang, and Xingguang Su. 2019. "WS2 Quantum Dots as a Sensitive Fluorescence Probe for the Detection of Glucose." *Journal of Luminescence* 207: 491–6. 10.1016/j.jlumin.2018.11.034.

Durai, Lignesh, Chang Yi Kong, and Sushmee Badhulika. 2020. "One-Step Solvothermal Synthesis of Nanoflake-Nanorod WS2 Hybrid for Non-Enzymatic Detection of Uric Acid and Quercetin in Blood Serum." *Materials Science and Engineering: C* 107: 110217. 10.1016/j.msec.2019.110217.

El Barghouti, Mohamed, Abdellatif Akjouj, and Abdellah Mir. 2020. "MoS_2–Graphene Hybrid Nanostructures Enhanced Localized Surface Plasmon Resonance Biosensors." *Optics & Laser Technology* 130: 106306. 10.1016/j.optlastec.2020.106306.

Hu, Huawen, Ali Zavabeti, Haiyan Quan, Wuqing Zhu, Hongyang Wei, Dongchu Chen, and Jian Zhen Ou. 2019. "Recent Advances in Two-Dimensional Transition Metal Dichalcogenides for Biological Sensing." *Biosensors and Bioelectronics* 142: 111573. 10.1016/j.bios.2019.111573.

Huang, Jingwei, Yuqing He, Jun Jin, Yanrong Li, Zhengping Dong, and Rong Li. 2014. "A Novel Glucose Sensor Based on MoS_2 Nanosheet Functionalized with Ni Nanoparticles." *Electrochimica Acta* 136: 41–6. 10.1016/j.electacta.2014.05.070.

Huang, Chun-Fang, Gui-Hong Yao, Ru-Ping Liang, and Jian-Ding Qiu. 2013. "Graphene Oxide and Dextran Capped Gold Nanoparticles Based Surface Plasmon Resonance Sensor for Sensitive Detection of Concanavalin A." *Biosensors and Bioelectronics* 50: 305–10. 10.1016/j.bios.2013.07.002.

Kaushik, Siddharth, Umesh K. Tiwari, Sudipta S. Pal, and Ravindra K. Sinha. 2019. "Rapid Detection of Escherichia coli Using Fiber Optic Surface Plasmon Resonance Immunosensor Based on Biofunctionalized Molybdenum Disulfide (MoS_2) Nanosheets." *Biosensors and Bioelectronics* 126: 501–9. 10.1016/j.bios.2018.11.006.

Kumar, S., Z. Guo, R. Singh, Q. Wang, B. Zhang, S. Cheng, F. Z. Liu, C. Marques, B. K. Kaushik, and R. Jha. 2020. "MoS_2 Functionalized Multicore Fiber Probes for Selective Detection of Shigella Bacteria Based on Localized Plasmon." *Journal of Lightwave Technology* 168: 112557. 10.1109/JLT.2020.3036610.

Kumar, S., R. Singh, G. Zhu, Q. Yang, X. Zhang, S. Cheng, B. Zhang, B. K. Kaushik, and F. Z. Liu. 2020. "Development of Uric Acid Biosensor Using Gold Nanoparticles and Graphene Oxide Functionalized Micro-Ball Fiber Sensor Probe." *IEEE Transactions on NanoBioscience* 19 (2): 173–82. 10.1109/TNB.2019.2958891.

Lin, Tianran, Liangshuang Zhong, Zhiping Song, Liangqia Guo, Hanyin Wu, Qingquan Guo, Ying Chen, FengFu Fu, and Guonan Chen. 2014. "Visual Detection of Blood Glucose Based on Peroxidase-Like Activity of WS_2 Nanosheets." *Biosensors and Bioelectronics* 62: 302–7. 10.1016/j.bios.2014.07.001.

Liu, Chao, Jianwei Wang, Famei Wang, Weiquan Su, Lin Yang, Jingwei Lv, Guanglai Fu, Xianli Li, Qiang Liu, Tao Sun, and Paul K. Chu. 2020. "Surface Plasmon Resonance (SPR) Infrared Sensor Based on D-Shape Photonic Crystal Fibers with ITO Coatings." *Optics Communications* 464: 125496. 10.1016/j.optcom.2020.125496.

Murugan, Nagaraj, Rajendran Jerome, Murugan Preethika, Anandhakumar Sundaramurthy, and Ashok K. Sundramoorthy. 2021. "2D-Titanium Carbide (MXene) Based Selective Electrochemical Sensor for Simultaneous Detection of Ascorbic Acid, Dopamine and Uric Acid." *Journal of Materials Science & Technology* 72: 122–31. 10.1016/j.jmst.2020.07.037.

Nie, Wenyan, Qing Wang, Xiaohai Yang, Hua Zhang, Zhiping Li, Lei Gao, Yan Zheng, Xiaofeng Liu, and Kemin Wang. 2017. "High Sensitivity Surface Plasmon Resonance Biosensor for Detection of MicroRNA Based on Gold Nanoparticles-Decorated Molybdenum Sulfide." *Analytica Chimica Acta* 993: 55–62. 10.1016/j.aca.2017.09.015.

Özkan, Abdullah, Necip Atar, and Mehmet Lütfi Yola. 2019. "Enhanced Surface Plasmon Resonance (SPR) Signals Based on Immobilization of Core-Shell Nanoparticles Incorporated Boron Nitride Nanosheets: Development of Molecularly Imprinted SPR Nanosensor for Anticancer Drug, Etoposide." *Biosensors and Bioelectronics* 130: 293–8. 10.1016/j.bios.2019.01.053.

Rahman, M. Saifur, Md Shamim Anower, and Lway Faisal Abdulrazak. 2019. "Utilization of a Phosphorene-Graphene/TMDC Heterostructure in a Surface Plasmon Resonance-Based Fiber Optic Biosensor." *Photonics and Nanostructures—Fundamentals and Applications* 35: 100711. 10.1016/j.photonics.2019.100711.

Rahman, M. Saifur, M. S. Anower, Md Khalilur Rahman, Md Rabiul Hasan, Md Biplob Hossain, and Md Ismail Haque. 2017. "Modeling of a Highly Sensitive MoS_2-Graphene Hybrid Based Fiber Optic SPR Biosensor for Sensing DNA Hybridization." *Optik* 140: 989–97. 10.1016/j.ijleo.2017.05.001.

Shushama, Kamrun Nahar, Md Masud Rana, Reefat Inum, and Md Biplob Hossain. 2017. "Graphene Coated Fiber Optic Surface Plasmon Resonance Biosensor for the DNA Hybridization Detection: Simulation Analysis." *Optics Communications* 383: 186–90. 10.1016/j.optcom.2016.09.015.

Singh, Ragini, Santosh Kumar, Feng-Zhen Liu, Cheng Shuang, Bingyuan Zhang, Rajan Jha, and Brajesh Kumar Kaushik. 2020. "Etched Multicore Fiber Sensor using Copper Oxide and Gold Nanoparticles Decorated Graphene Oxide Structure for Cancer Cells Detection." *Biosensors and Bioelectronics* 168: 112557. 10.1016/j.bios.2020.112557.

Singh, Sarika, Satyendra K. Mishra, and Banshi D. Gupta. 2013. "Sensitivity Enhancement of a Surface Plasmon Resonance Based Fibre Optic Refractive Index Sensor Utilizing an Additional Layer of Oxides." *Sensors and Actuators A: Physical* 193: 136–40. 10.1016/j.sna.2013.01.012.

Singh, Lokendra, Ragini Singh, Bingyuan Zhang, Shuang Cheng, Brajesh Kumar Kaushik, and Santosh Kumar. 2019. "LSPR Based Uric Acid Sensor Using Graphene Oxide and Gold Nanoparticles Functionalized Tapered Fiber." *Optical Fiber Technology* 53: 102043. 10.1016/j.yofte.2019.102043.

Song, Dandan, Xinyu Jiang, Yanshan Li, Xiong Lu, Sunrui Luan, Yuanzhe Wang, Yan Li, and Faming Gao. 2019. "Metal–Organic Frameworks-Derived MnO_2/Mn_3O_4 Microcuboids with Hierarchically Ordered Nanosheets and $Ti3C_2$ MXene/Au NPs Composites for Electrochemical Pesticide Detection." *Journal of Hazardous Materials* 373: 367–76. 10.1016/j.jhazmat.2019.03.083.

Su, Shao, Qian Sun, Xiaodan Gu, Yongqiang Xu, Jianlei Shen, Dan Zhu, Jie Chao, Chunhai Fan, and Lianhui Wang. 2019. "Two-Dimensional Nanomaterials for Biosensing Applications." *TrAC Trends in Analytical Chemistry* 119: 115610. 10.1016/j.trac.2019.07.021.

Wang, Qi, and Bo-Tao Wang. 2018. "Surface Plasmon Resonance Biosensor Based on Graphene Oxide/Silver Coated Polymer Cladding Silica Fiber." *Sensors and Actuators B: Chemical* 275: 332–8. 10.1016/j.snb.2018.08.065.

Wang, Qi, Xi Jiang, Li-Ye Niu, and Xiao-Chen Fan. 2020. "Enhanced Sensitivity of Bimetallic Optical Fiber SPR Sensor Based on MoS_2 Nanosheets." *Optics and Lasers in Engineering* 128: 105997. 10.1016/j.optlaseng.2019.105997.

Wang, Tao, Mingjiang Zhang, Kun Liu, Junfeng Jiang, Yuanhao Zhao, Jinying Ma, and Tiegen Liu. 2019. "The Effect of the TiO2 Film on the Performance of the Optical Fiber SPR Sensor." *Optics Communications* 448: 93–7. 10.1016/j.optcom.2019.05.023.

Wu, L., H. S. Chu, W. S. Koh, and E. P. Li. 2010. "Highly Sensitive Graphene Biosensors Based on Surface Plasmon Resonance." *Optics Express* 18 (14): 14395–400. 10.1364/OE.18.014395.

Yang, Qingshan, Xia Zhang, Santosh Kumar, Ragini Singh, Bingyuan Zhang, Chenglin Bai, and Xipeng Pu. 2020. "Development of Glucose Sensor using Gold Nanoparticles and Glucose-Oxidase Functionalized Tapered Fiber Structure." *Plasmonics* 15 (3): 841–8. 10.1007/s11468-019-01104-7.

Yang, Qingshan, Guo Zhu, Lokendra Singh, Yu Wang, Ragini Singh, Bingyuan Zhang, Xia Zhang, and Santosh Kumar. 2020. "Highly Sensitive and Selective Sensor Probe using Glucose Oxidase/Gold Nanoparticles/Graphene Oxide Functionalized Tapered Optical Fiber Structure for Detection of Glucose." *Optik* 208: 164536. 10.1016/j.ijleo.2020.164536.

Yu, Haixia, Yang Chong, Penghao Zhang, Jiaming Ma, and Dachao Li. 2020. "A D-Shaped Fiber SPR Sensor with a Composite Nanostructure of MoS_2-Graphene for Glucose Detection." *Talanta* 219: 121324. 10.1016/j.talanta.2020.121324.

Yue, Yang, Nishuang Liu, Weijie Liu, Mian Li, Yanan Ma, Cheng Luo, Siliang Wang, Jiangyu Rao, Xiaokang Hu, Jun Su, Zhi Zhang, Qing Huang, and Yihua Gao. 2018. "3D Hybrid Porous Mxene-Sponge Network and Its Application in Piezoresistive Sensor." *Nano Energy* 50: 79–87. 10.1016/j.nanoen.2018.05.020.

Zhang, Yu, Xiantao Jiang, Junjie Zhang, Han Zhang, and Yingchun Li. 2019. "Simultaneous Voltammetric Determination of Acetaminophen and Isoniazid using MXene Modified Screen-Printed Electrode." *Biosensors and Bioelectronics* 130: 315–21. 10.1016/j.bios.2019.01.043.

Zhu, Guo, Niteshkumar Agrawal, Ragini Singh, Santosh Kumar, Bingyuan Zhang, Chinmoy Saha, and Chandrakanta Kumar. 2020. "A Novel Periodically Tapered Structure-Based Gold Nanoparticles and Graphene Oxide—Immobilized Optical Fiber Sensor to Detect Ascorbic Acid." *Optics & Laser Technology* 127: 106156. 10.1016/j.optlastec.2020.106156.

9 Future Trends of Emerging Materials in SPR Sensing

9.1 PRINCIPLE OF SENSORS

Surface plasmon resonance (SPR) is a coherent delocalized electron oscillation that exists at the interface between any two materials with a real part of a dielectric constant or emissivity sign shift on the interface like a metal-dielectric interface. There are evanescent waves (EWs) at both sides of the interface (Fig. 9.1).

So, decay on both sides of the interface wave only propagates in the longitudinal direction by solving Maxwell equations. Where, E is an electric field and B is a magnetic field.

$$\nabla . \mathrm{E} = 0 \tag{9.1}$$

$$\nabla . \mathrm{B} = 0 \tag{9.2}$$

$$\nabla \times \mathrm{E} = \frac{\partial B}{\partial t} \tag{9.3}$$

$$\nabla \times \mathrm{B} = \frac{1}{c^2} \frac{\partial E}{\partial t} \tag{9.4}$$

We find longitudinal oscillations are quantized, which depend on the permittivity of metals and dielectrics for the angular frequency of lights and their speed. The relations are shown as

$$k_x = \frac{\omega}{c} \sqrt{\frac{\varepsilon_d . \varepsilon_m}{\varepsilon_d + \varepsilon_m}} \tag{9.5}$$

where ω is the angular frequency, c represent the speed of light, and ε_d and ε_m are the permittivity of dielectric and metals, respectively.

Also, the range of frequency is assigned as

$$\omega_{SP} < \frac{\omega_P}{\sqrt{1 + \varepsilon_d}} \tag{9.6}$$

DOI: 10.1201/9781003190738-9

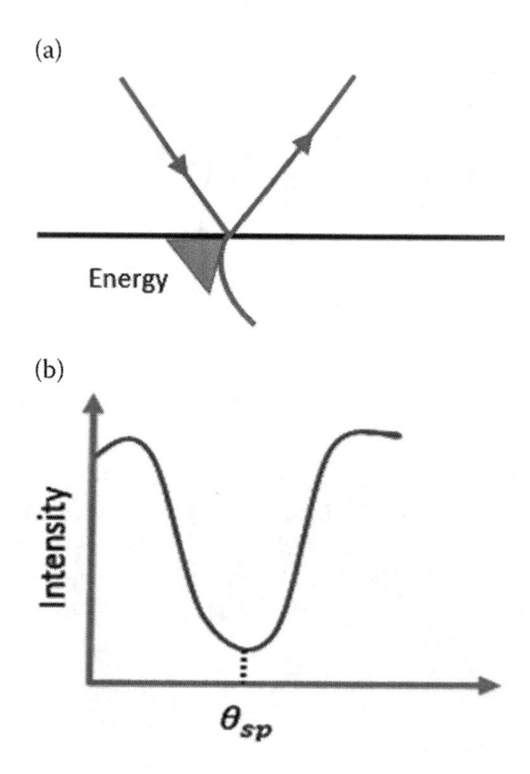

FIGURE 9.1 (a–b) The energy portion and surface plasmon angle for intensity.

where ω_P is the plasmon frequency.

It is not possible to absorb surface plasmon directly from light coming from a source, because the k_x matching is not possible. Fig. 9.2 shows the depth in intensity at a specific angle, where the surface plasmon gets excited, which is known as the angle of resonance. This resonance depends on the dielectric constant's permittivity. The modified structure with metals and dielectric makes it sensible and can be used as a sensor for different applications. It is also possible to make SPR-based sensors for the field of physical sensing, biomedical sensing, mining industries, structural health monitoring system, and many more. Due to the use of physical-chemical properties of SPR sensors, 2D materials such as graphene and transition metal dichalcogenide (TMD) are commonly used in plasmonic sensing because of their attractive physical-chemical characteristics. SPs come in two varieties: localized surface plasmons (LSPs) and propagating surface plasmon polaritons (SPPs), both of which are fundamental components of plasmonic sensors (Zhou et al. 2020). Structure-based $MoSe_2$ is proposed (Jiang, Zhao, and Zhang 2020), one of the most studied TMDs, which has drawn a lot of interest in various fields due to its 2D-layered structures and electrochemical properties. $MoSe_2$ nanosheets were found to have a high surface activity and good absorption, improving the efficiency of the electro-catalysis process. Dral and ten Elshof (2018) developed metal oxide nanoflakes in two dimensions (2D) that are used for active sensing components to detect analytes, radiation, and gases. Since they have broad surface areas for crystal

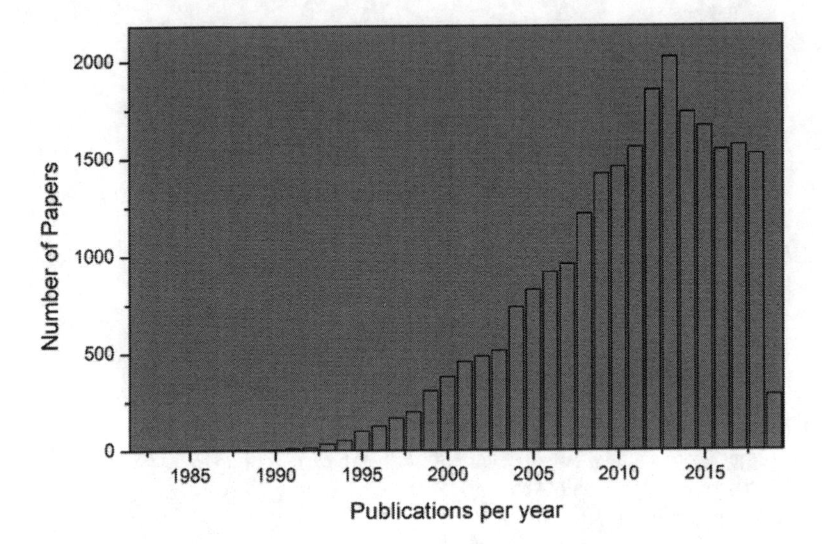

FIGURE 9.2 The diagram shows the number of papers that were published annually in PubMed on surface plasm resonance (SPR) technology. Reprinted with permission from Biosensors and Bioelectronics journal. Copyright, 2019, Elsevier (Patil et al. 2019).

facet technology, 2D material geometries are especially appealing for sensing applications. Patil et al. (2019) developed the concept for the signal produced by plasmon material that could be amplified by graphene-based nanocomposites, increasing the sensitivity of molecular detection from Femto to attomolar levels. A typical SPR sensor operates in one of two modes: angular interrogation or wavelength interrogation. Kumar, Yadav, Kushwaha, and Srivastava (2020) proposed a new structure of SPR sensors based on a BK7 prism/ZnO/Ag/$BaTiO_3$/TMDs (TMDCs, mostly MoS_2 and WS_2)/graphene. The sensitivity of this configuration is 180° RIU^{-1}, compared to 174° RIU^{-1} and 157° RIU^{-1} for mono-layer graphene and MoS_2, respectively. They have also shown that the SPR sensor with a graphene layer, on the other hand, has a smaller full-width half maximum (FWHM), a higher efficiency parameter, and better detection accuracy. Exponential growth was found in the field of SPR sensors by the database of PubMed until 2019; the methodology developed and published around 24,148 papers.

Most 2D metal-oxide nanoflake papers that have been produced are proof-of-concept studies, but still have yet to produce competitive results. The bar graph of publications based on 2D metal-oxide nanoflakes thickness up to 50 nm was used (Fig. 9.3).

Between the metal surface of a sensor with a particular biorecognition element (BRE) and a sensing medium, either liquid or air/vacuum, the SPR phenomenon occurs. Whenever a molecule unique to this element's site/scaffold/receptor is recognized, the metal's surface changes, shifting an angle as shown in Fig. 9.4(a); the shifting occurs due to changes in the refractive index (RI) at the surface of the metal. The incident light's dispersion relation matches that of the surface plasmon at the resonant wavelength, causing the reflectance to dip as shown in Fig. 9.4(b).

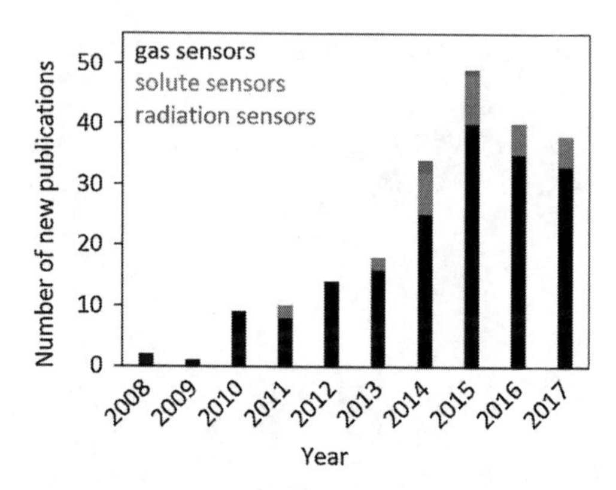

FIGURE 9.3 Indicator for the number of new sensors with 2D metal-oxide nanoflakes published per year as active sensing elements with a thickness of up to 50 nm. Reprinted with permission from Sensors and Actuators B: Chemical Journal. Copyright, 2018, Elsevier (Dral and ten Elshof 2018).

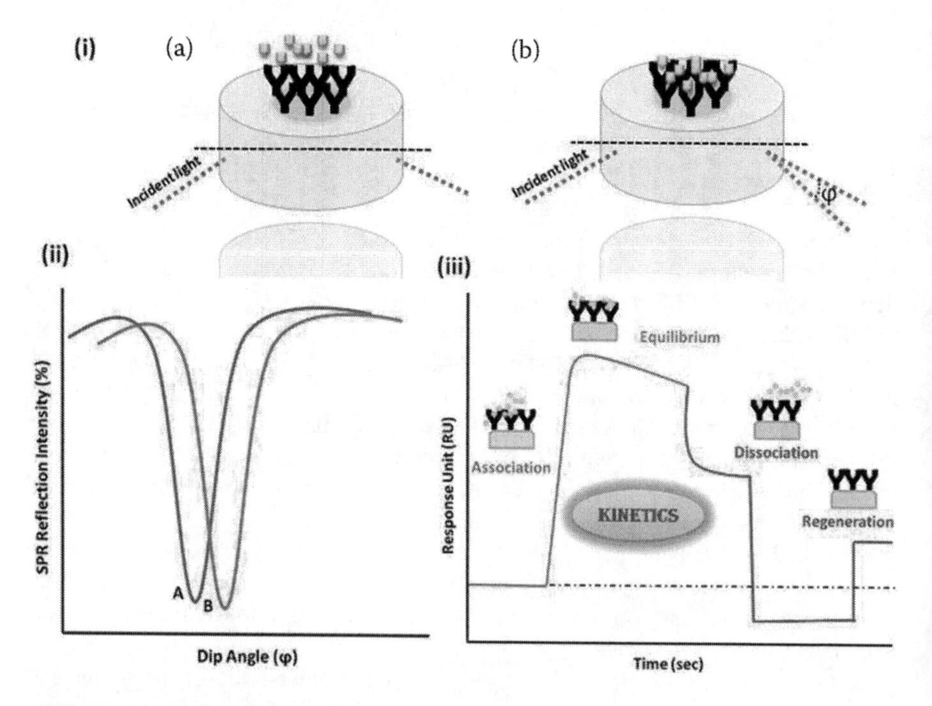

FIGURE 9.4 Surface plasmon resonance is depicted in this diagram. A metal surface sensor with a site for molecular recognition links to the molecule of interest: a) sensor shift is observed at a certain angle of an incident b) when the impulse of the sensed surface shifts from the detector to c) association-dissociation kinetics. Reprinted with permission from Biosensors and Bioelectronics journal. Copyright, 2019, Elsevier (Patil et al. 2019).

TABLE 9.1
Summary of Methods for Preparing 2D Nanomaterials (Zhou et al. 2020)

Preparation Method		2D Nanomaterials	Advantages	Disadvantages
Top-down method	Mechanical exfoliation	Graphene, TMDs, Phosphorene	High quality, large area	Low production yield
	Sonication exfoliation	TMDs, Phosphorene, Antimonene	High production yield and dispersity	Low quality and poor uniformity
	Intercalation exfoliation	Graphene, TMDs	High quality, large area, high production yield	Poor uniformity
	Chemical etching	MXene	High hydrophilicity and dispersity	Limited to MXene
Bottom-up method	CVD	Graphene, h-BN, TMDs	High quality, large area, high production yield	Stringent experimental condition
	Hydrothermal or solve thermal reaction	TMDs	High production yield	Small area and poor uniformity
	Double decomposition reaction	MOFs	Simplicity and high production yield	Limited to specific material
	Hydrolysis	2D metallic oxides	Ability to prepare amorphous 2D metallic oxides	Limited to specific material
	Thermal decomposition	TMDs	In-situ deposition	Limited to specific material

The real-time-based sensing can be possible by converting dip sensing into adsorption and desorption kinetics w.r.t. The sensor will show the shift in dip angle as a function of time in Fig. 9.4(c). The SPR sensitivity is mainly based on the surface of the sensor, which changes RI when associated events take place on the surface, by altering phase, wavelength, and other features of the applied light (Table 9.1).

9.1.1 2D Nanomaterials: Structures and Their Properties

Graphene: Graphene is a single layer or a few layers of sp^2-bonded carbon atoms with exceptional physicochemical properties including ultrahigh electrical and thermal conductivity, mechanical strength, unique surface area and stability, and controllable Fermi energy. Graphene oxide (GO) is a derivative of graphene with a variety of surface radicals, including epoxy, hydroxyl, and carboxyl, that give it a high chemical affinity for target molecules. Furthermore, the content and form of surface radicals can be changed to modify the electron structure and chemical properties of GO.

H-BN: 2D h-BN has a hexagonal and atomically thin structure similar to graphene, and it exhibits many of the same properties as graphene, including high thermal conductivity and mechanical power. The electron distribution in a B–N bond is less homogeneous than in a C–C bond, and the electron is more centered on the N atom's BN and outperforms graphene in terms of thermal and chemical stability. As a result, h-BN can be an ideal shielding material for active metal nanostructures in plasmonic sensors.

TMDs: The semi-conductive TMDs, including zero-gap graphene and wide-gap h-BN, have narrow and tunable bandgap energy, which is useful in optoelectronic and electronic applications. TMDs have the general chemical formula MX_2, where M is a transition metal atom and X is a chalcogen atom. One intermediate sublayer of transition metals is sandwiched between two sublayers of chalcogen atoms to form single-layered TMDs.

Group* Ⅴ*A 2D Nanomaterials: Five valent electrons are hybridized into four sp^3 orbitals in the group of VA atoms. A lone pair of electrons occupies one of the orbitals. The remaining semi-filled orbitals form bonds with neighboring atoms. The noncoplanar sp^3 orbitals cause the structure of group VA 2D nanomaterials to pucker (-phase) or buckle (-phase). Phosphorene is a monolayer or several-layer black phosphorene (BP) exfoliated from bulk black phosphorus in the process. It is the first and most studied category of VA 2D nanomaterial. Due to its low in-plane symmetry, BP exhibits high anisotropy in electrical, optical, and thermal properties in both directions, resulting in an anisotropic response in plasmonic sensing.

MXene: MXene is a new class of 2D nanomaterials that includes transition metal carbides, nitrides, and carbo nitrides and has the universe formula $M_{n+1}X_nT_x$, where M denotes the early d-transition metal atom, X is the N and/or C atom, and T is the surface terminated functional group, such as –OH, O, and –F. MXenes exhibited remarkable optical properties, such as tunable broadband absorption and intense surface plasmon, suggesting that they have great potential for optoelectronic and plasmonic applications.

2D Metallic Oxides: The layered structure of certain metallic oxides, such as α-MoO_3, is expressed by α-MoO_3. Via van der Waals (VdW) interaction, the multilayer α-MoO_3 is created by stacking the double layers. Because of the weak interaction, different ions can easily intercalate in the interlayer gap, causing an oxygen atom to be lost and the formation of sub-stoichiometric molybdenum trioxide (MoO_{3-x}).

The plasmon resonance of 2D a-MoO_3 is tunable in the entire visible region as a result of this, which could be useful for its use in low-dimensional sensing and optical systems (Table 9.2)

9.1.2 Architectures and Synthesis Process of 2D Nanoflakes

Architecture of 2D Nanoflakes: Both individually and in 3D structures, 2D nanoflakes can be used as active sensing components. Individual nanoflakes are difficult to treat, so they're often used for solute sensing in suspended form (Fig. 9.5). Fig. 9.6 shows a few-nm thin, flexible nanoflake and a stack of rigid nanoflakes with a thickness of 10–15 nm. The majority of studies that use 2D nanoflakes as active

TABLE 9.2
Summary of 2D Nanomaterials in Atom Structure and Properties (Zhou et al. 2020)

2D Nanomaterials	Atom Structures	Bandgap	Electron/Hole Mobility	Stability	Refractive Index
Graphene	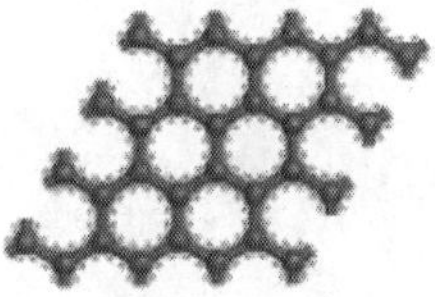	0 eV	~10^4 cm^2 V^{-1} s^{-1} (E)	Good	3 + 1.149106i (λ = 633nm)
h-BN monolayer	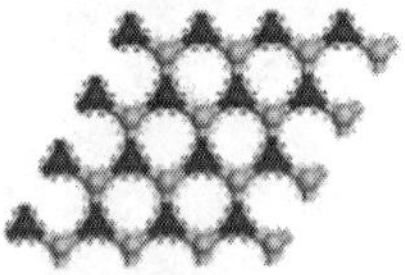	5.97 eV	–	Excellent	NR
2H-MoS_2 monolayer	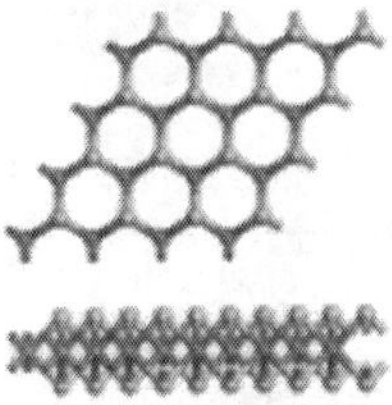	1.8 eV (D)	30 cm^2 V^{-1} s^{-1} (E)	Good	5.0805 + 1.1723i (λ = 633 nm)
Antimonene	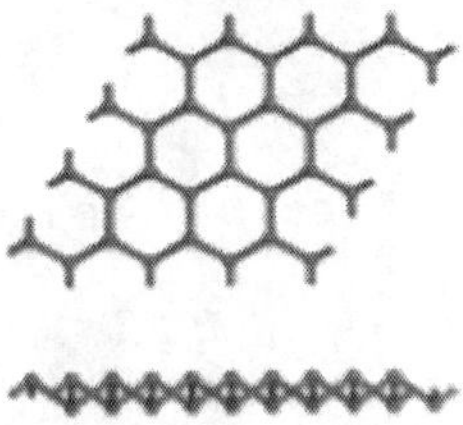	2.28 eV (ID)	630 cm^2 V^{-1} s^{-1} (E)1737 cm^2 V^{-1} s^{-1} (H)	Not good	NR
Phosphorene	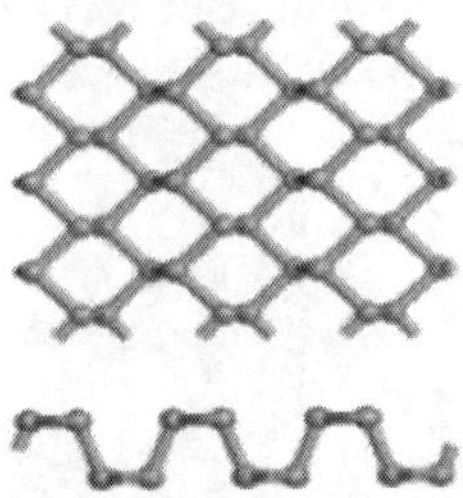	2.0 eV (D)	286 cm^2 V^{-1} s^{-1} (H)	Poor	3.531–0.04087i (λ = 633 nm)
α-MoO_3	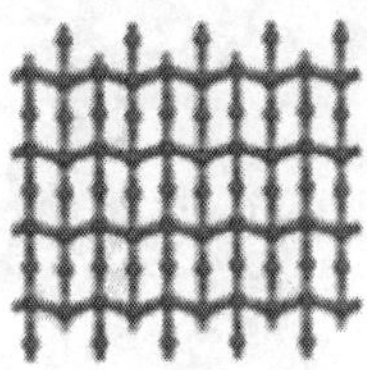	3.15 eV*	1100 cm^2 V^{-1} s^{-1} (E)	Good	NR

(Continued)

TABLE 9.2 (Continued)
Summary of 2D Nanomaterials in Atom Structure and Properties (Zhou et al. 2020)

2D Nanomaterials	Atom Structures	Bandgap	Electron/Hole Mobility	Stability	Refractive Index
$Ti_3C_2T_x$		NR	2.6 ± 0.7 cm^2 V^{-1} s^{-1} (E)	Good	2.38 + 1.33i (λ = 633 nm) 2.64 + 1.00i (λ = 532 nm)

Here, D represents direct bandgap, ID is indirect bandgap, and NR means "not reported."

FIGURE 9.5 TEM images: (a) and SEM images of (b to g) of different types (single and assembled) of nanoflakes. a) Nanoflake flexible; b) stack of rigid nanoflakes; c, d) flower-like assemblies; e) tree-like assembly; f) walls; and g) forest. Reprinted with permission from Sensors and Actuators B: Chemical Journal. Copyright, 2018, Elsevier (Dral and ten Elshof 2018).

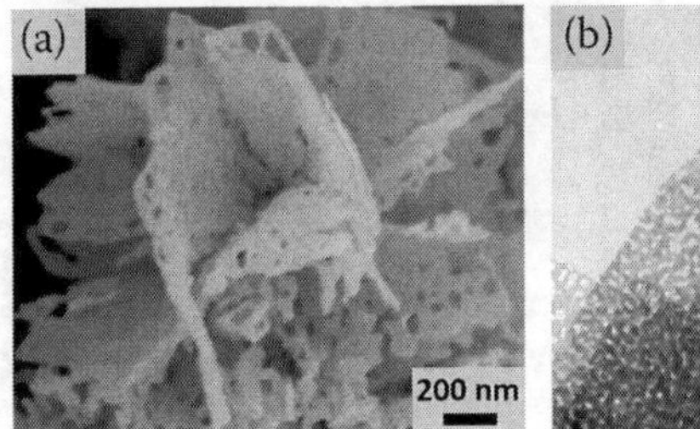

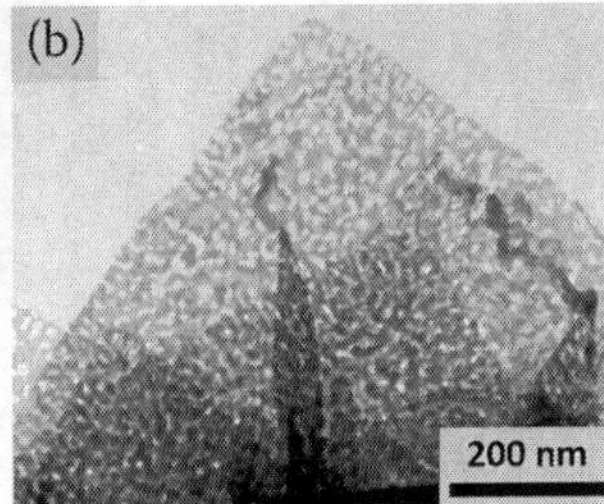

FIGURE 9.6 a) SEM representation of meso and macropores for SnO_2 nanoflakes in two dimensions. Reprinted with permission from Sensors and Actuators B: Chemical Journal. Copyright, 2018, Elsevier (Dral and ten Elshof 2018).

sensing elements do so by embedding them in porous 3D structures (Dral and ten Elshof 2018).

Synthesis of 2D Nanoflakes: The synthesis method decides the shaped 2D nanoflakes' crystalline structures (single-crystalline or polycrystalline), which can have a major impact on sensor performance. The majority of 2D nanoflakes are made using hydrothermal or solvothermal methods. Metal oxide can develop spontaneously in the desired structure for 2D-in-3D hierarchical architectures.

9.1.3 Different Composition of Sensing Layer

Jiang, Zhao, and Zhang (2020) proposed a composite base $MoSe_2$ structure for enhancement of a sensing parameter of an SPR sensor and it is very effective for gas sensing. This structure is a combination of Au-$MoSe_2$ and their synthesis done for air and ammonia. The following reactions were used to create a novel gas sensor based on a combination of Au/$MoSe_2$ and a single layer of MoS_2 piezoelectric nano-generator for NH_3 detection.

$$O_2(gas) \rightarrow O_2(\mathrm{ads})$$

$$O_2(\mathrm{ads}) + e^- \rightarrow O_2^-(\mathrm{ads})$$

$$4NH_3(\mathrm{gas}) + 5O_2(\mathrm{ads}) \rightarrow 4NO + 6H_2O + 5e^-$$

Nanoparticles (NPs) and the Schottky interface between Au and $MoSe_2$ working mechanism for Au (5.4 eV) have been stuck in the air at the $MoSe_2$ surface with electrons passing from $MoSe_2$ to Au, adding electrons to O_2 to form O_2^-. The produced O_2^- is going to respond with the NH_3 for electron release that responds to NH_3 detection. The author proposed a new SPR biosensor with a BK7 prism/ZnO structure with 10 nm thick of ZnO base of the prism. On top of the ZnO layer, there is a 54-nm silver (Ag) layer. Above the silver layer is a $BaTiO_3$ layer with a thickness of 5 nm. The design is based on three different configurations that each use WS_2, MoS_2, and graphene to optimize ZnO, Ag, and $BaTiO_3$ thickness. In the

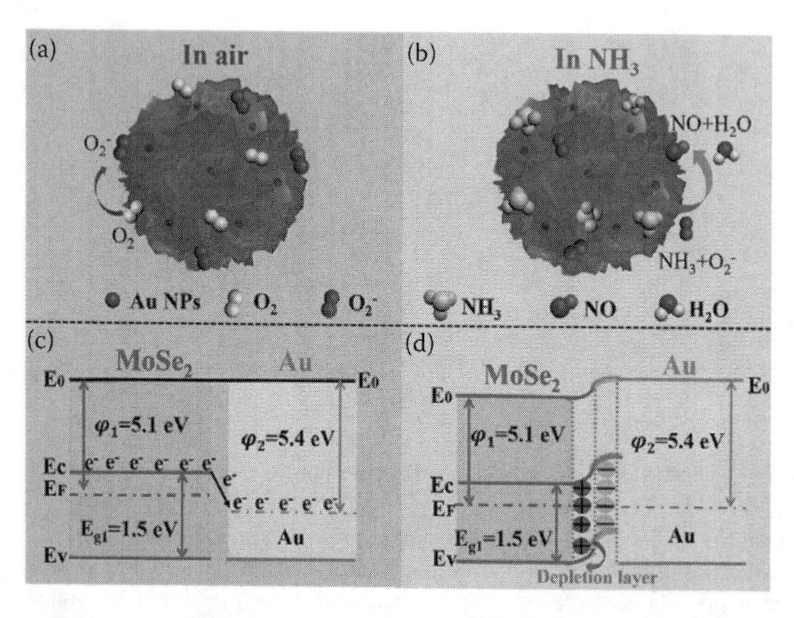

FIGURE 9.7 Au-$MoSe_2$ composite structure mechanism for the gas-sensing and energy band scheme (a, c) in the air and (b, d) for NH_3 gas. Reprinted with permission from Micro-Electronic Engineering Journal. Copyright, 2020, Elsevier (Jiang, Zhao, and Zhang 2020).

prism-ZnO-Ag-$BaTiO_3$-WS_2-sensing medium, the proposed configuration shows the combination of oxides, metals, and graphene: BK7-ZnO-Ag-$BaTiO_3$-WS_2 and water as sensing mediums with a thickness of ZnO with 10 nm, Ag with 54 nm, $BaTiO_3$ with 5 nm, and WS_2 with L (no. of WS_2 layers) x 0.80 nm layers. Also, L is showing the number of layers for WS_2 as a sensing medium if water is used (Fig. 9.7 and 9.8).

This new proposed structure, ZnO-Ag-$BaTiO_3$-WS_2-graphene-MoS_2, shows a great biosensor ability and other industrial applications. With the number of graphene layers, the FWHM continues to grow (Fig. 9.9).

9.2 APPLICATION IN PHYSICAL SENSING

Physical sensing is also one of the vital fields in the application of SPR-based sensors with 2D materials. Much of the research consists of different ideas about the sensing parameter of the different fields like strain sensor, alcohol sensor, humidity sensor, piezoresistive sensor, and RI sensor with different 2D materials (Kanmani et al. 2019). For the optimization process of SPR properties, the consequences of TiO_2 on the performance of the different layer thicknesses of Ag, namely 20 nm, 30 nm, and 40 nm, were proposed. For alcohol-sensing applications, a silver coating on a side-polished optical fiber is used. Metals like silver (Ag) and gold (Au) are extremely useful in a number of applications, including SPR sensors. The efficiency of their sensitivity or precision of detection in different applications is better in the case of Au, which indicates the quality response. That's why silver has the highest

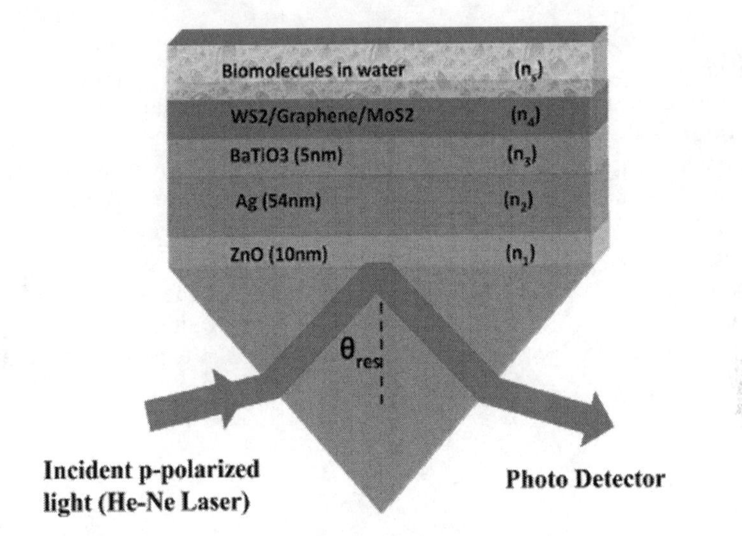

FIGURE 9.8 Proposed configuration: BK7 glass prism-ZnO-Ag-BatiO$_3$-WS$_2$ and sensing medium (water) with an optimized thickness for ZnO with 10 nm, Ag with 54 nm, BatiO$_3$ with 5 nm, and WS$_2$ with L × 0.80 nm. Reprinted with permission from Sensors and Actuators Reports Journal. Copyright, 2020, Elsevier (Kumar, Yadav, Kushwaha, and Srivastava 2020).

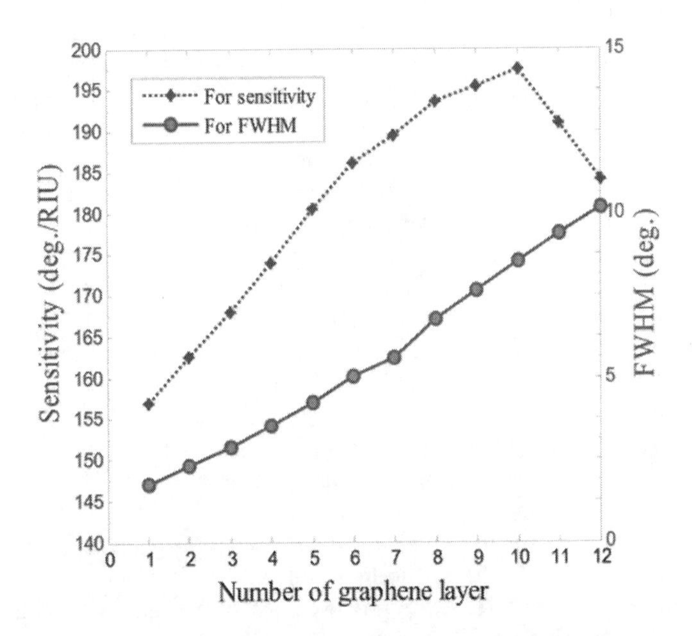

FIGURE 9.9 Variation in SPR biosensor sensitivity and FWHM with different numbers of graphene layers is optimal 10 nm of ZnO, 54 nm of Ag, and 5 nm of BaTiO$_3$ thicknesses. Reprinted with permission from Sensors and Actuators Reports Journal. Copyright, 2020, Elsevier (Kumar, Yadav, Kushwaha, and Srivastava 2020).

ε_r / ε_i ratio, in which ε_r and ε_i describe the real and imaginary parts of permittivity. On the cladless optical fiber, with a beam electron evaporating machine (EB43-T) at a high vacuum pressure of 10^{-5}–10^{-7}, Torr, a thin silver layer, is added after installation of the laterally polished optical fibers. Then, titanium dioxide was dried at the ambient temperature for 3 h by a drop-casting process on the Ag thin layer. The films used in this study had a standard thickness of about 20 nm. Isopropyl is combined with varying quantities of distilled water to make 20%, 40%, and 60% isopropyl for sensor preparation. A refractometer can be used to determine the isopropyl solution's RI in the range of 1.333–1.3597 on the polished surface of the sensor optical fiber. The surface plasmon wave propagation constant (κ_{SP}) for the metal-dielectric interface that was calculated using the experimental results is as follows:

$$k_{sp} = \frac{\omega}{c}\left(\frac{\varepsilon_m \varepsilon_c}{\varepsilon_m + \varepsilon_c}\right)^{1/2} \tag{9.7}$$

The dielectric constants for the metal and dielectric mediums are defined by ε_m and ε_c, respectively. ω is defined as incident light's frequency, and c is the light velocity. Different analytes for different RI are shown in Fig. 9.10.

Sharma, Kaur, and Popescu (2020) proposed different 2D materials/heterostructures in the visible spectral area of fiber-optic SPR humidity sensor. When various 2D graphene materials WS_2, BlueP-MoS_2, BlueP-WS_2, MoS_2, and WSe_2 are considered for different examinations. RH is approached in two ways: angular interrogation and intensity interrogation. Different 2D materials are present in the power loss spectrum characteristics according to simulation performance (i.e., SPR, PLMax, and FWHM) and thus in the whole combined performance factor (CPF) output parameter of the sensor probe RH. Under angular interrogation and strength interrogation techniques, it can be as low as 0.00068% RH. 2D monolayers, such as WS_2 and graphene, have excellent optical properties that enable them to detect humidity with high sensitivity and stability (even at very high percent RH magnitudes) and fast recovery. The suggested structure is as follows: A p-polarized light ("on axis") is recorded to occur in the middle of a multi-mode fiber at a corner of "α" represents the angle θ of the appropriate guided modes propagating through the fiber (inside the fiber). When the consistent propagation of the evanescent wave or the surface plasmon match, the resonance occurs. At the particular angle of SPR, also known as the α SPR angle, their condition of resonance is satisfied. At resonance, the power loss (PL) is greatest. Furthermore, as the RH value changes, the PL curve moves to a different angle (as seen in Fig. 9.11).

Consider (one at a time) the thickness of various 2D materials described in Table 9.3, as well as their changes in RI (Table 9.4).

In presence of 2D materials, the proposed sensor can detect humidity in any environment. Leonardi et al. (2018) made a suggestion based on nanosheets of 2D-WS_2, where the humidity sensor has been designed for a highly sensitive room temperature. The sensor reacts with an RH limit of almost three magnitude orders. The sensor showed a high moisture sensitivity at room temperature with a current

(a)

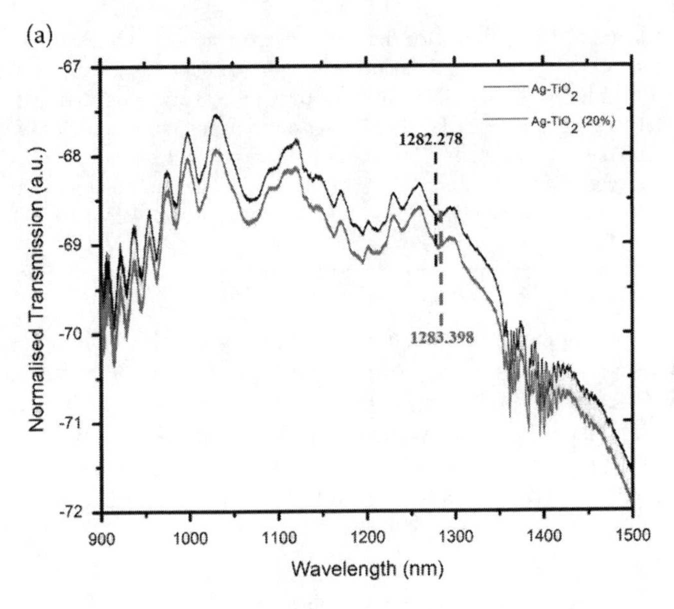
Ag-TiO2
Ag-TiO2 (20%)
1282.278
1283.398
Normalised Transmission (a.u.)
Wavelength (nm)
-67
-68
-69
-70
-71
-72
900
1000
1100
1200
1300
1400
1500

(b)

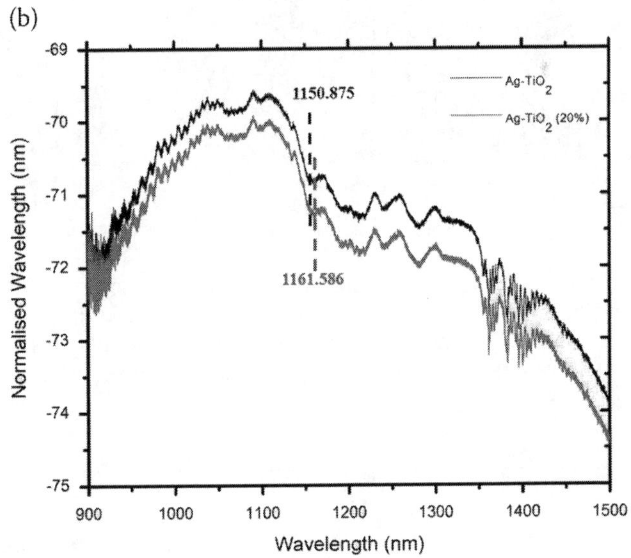
Ag-TiO2
Ag-TiO2 (20%)
1150.875
1161.586
Normalised Wavelength (nm)
Wavelength (nm)
-69
-70
-71
-72
-73
-74
-75
900
1000
1100
1200
1300
1400
1500

(c)

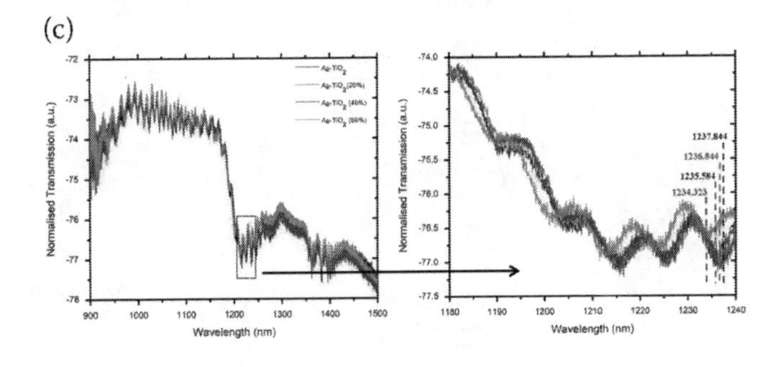
Normalised Transmission (a.u.)
Wavelength (nm)
1237.844
1236.844
1235.584
1234.323

FIGURE 9.10 (a, b) This presents the transmission spectrum of the SPR sensor in the range 900–1,500 nm, as an Ag wavelength feature with a thickness of 20 nm and 40 nm, with the TiO_2 layer. Reprinted with permission from Optical Fiber Technology Journal. Copyright, 2019, Elsevier (Kanmani et al. 2019). c) SPR sensor transmission spectrum as a function of the Ag (30 nm)-TiO_2 wavelength. This arrow indicates an extension of Ag, Ag-TiO_2, DI resonant wavelength, 20%, 40%, and 60% of alcohol within the range of 900–1,500 nm. Reprinted with permission from Optical Fiber Technology Journal. Copyright, 2019, Elsevier (Kanmani et al. 2019).

increase of approximately three magnitude orders, as the RH value ranged from 8% to 85% with rapid reaction and recovery times of 140 s and 30 s, respectively. To improve SPR sensitivity, Rahman et al. (2020) have suggested a heterostructure for $PtSe_2$ and 2D materials. Two exclusive SPR biosensors, combining the newly formed TMDs (e.g., $PtSe_2$), with some 2D materials (e.g., MoS_2, graphene, and WS_2) on standard biosensors based on SPR, were proposed. Complete reflection,

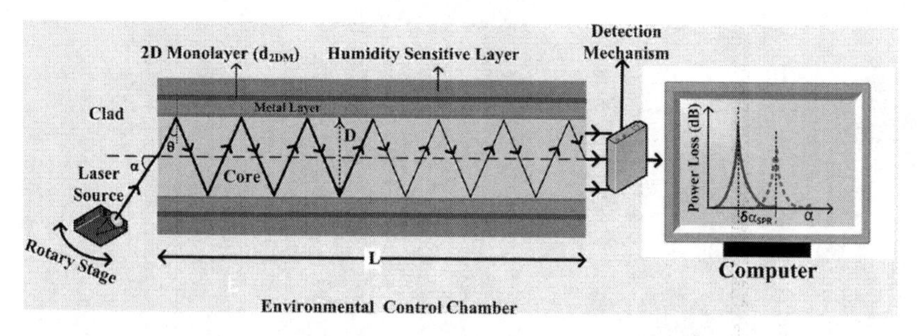

FIGURE 9.11 Proposed diagram of a humidity sensor based on SPR. The SPR-based sensor uses multimodal L/D fiber as 25, L, and D, representing both fiber length and core diameter. In the ambient chamber used to monitor RH measurements. Reprinted with permission from Optical Materials Journal. Copyright, 2020, Elsevier (Sharma, Kaur, and Popescu 2020).

TABLE 9.3
2D Materials Used in Simulation in λ = 632.8 nm are of Complex RI and a Monolayer Thickness (d_{2DM}) (Sharma, Kaur, and Popescu 2020)

SL. No.	2D Monolayer	d_{2DM} (nm)	RI (λ = 632.8 nm)
1	Graphene	0.34	2.7302 + 1.3555i
2	WS_2	0.80	4.8854 + 0.3142i
3	Blue P-WS_2	0.75	2.5185 + 0.1734i
4	Blue P-MoS_2	0.75	2.8070 + 0.2120i
5	WSe_2	0.70	4.4183 + 0.6052i
6	MoS_2	0.71	5.2244 + 1.0831i

TABLE 9.4
LOD Comparison to the SPR Humidity Sensor Previously Reported (Sharma, Kaur, and Popescu 2020)

S.No.	Sensor's Structure	RH Range (%)	LOD
1.	Side-polished fiber sensor	10–85	3%
2.	D-shaped fiber sensor with Au grating (λ = 1,525–1,610 nm)	0–80	3.49%
3.	Chalcogenide prism based SPR sensor (λ = 632.8 nm)	0–100	0.01%
4.	Side-polished fiber sensor with WS_2 film overlay (λ = 1,550 nm)	35–85	0.475%
5.	Absorption-based fiber sensor	20–90	0.06%
6.	Ag/TiO_2 plasmonic structures with side-polished fiber sensor	50–90	0 .028%
7.	Proposed work	0–100	0.0041% 0.00068%

sensitivity (factor) (QF)-, and quality factor (Q) attenuation is used to monitor the sensors. The sensor output is studied with a sensitivity, a value (FOM), an attenuated total reflection (ATR), and quality factor (QF). Of this, 1.67 times as many sensors as traditional Au/Ag sensors dependent on a FOM or QF comparable, respectively, were found with the structures Au-$PtSe_2$-WS_2 and Ag-$PTSe_2$-WS_2. Ag-Au is 50 nm in thickness and the WS_2 layers are calibrated for four Au-based sensors and two Ag-based sensors. In both cases, $PtSe_2$ can be used with an optimized diameter of 2 nm.

As described in Fig. 9.12, Kretschmann's configuration is dependent on the proposed sensor. The sensor has an SPR active layer (50 nm) on a prism coupler, e.g., Ag and Au. The dispersion profile Au/Ag is measured with the Drude-Lorentz wavelength (λ) model:

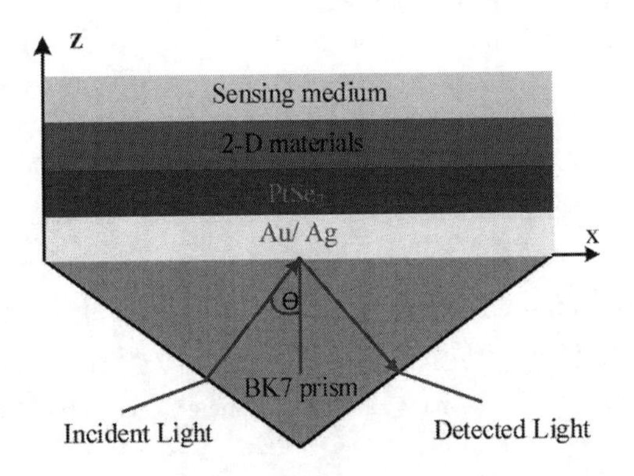

FIGURE 9.12 Scheme illustration of the SPR biosensor proposed. Reprinted with permission from Optical Materials Journal. Copyright, 2020, Elsevier (Rahman et al. 2020).

$$n_{metal}(\lambda) = \left(1 - \frac{\lambda^2{}_*\lambda^c}{\lambda_p^2(\lambda^c + \lambda_* i)}\right)^{1/2} \tag{9.8}$$

The wavelengths of the collision and plasma metal wavelength defined in Table 9.5 are indicated by λ_c and λ_p in Eq. 9.8. The BK7 prism is the architecture for usage combined with the plane monochromatic of polarized light, due to its low RI, offers high sensitivity. A 633 nm wavelength laser He–Ne is the cause of the light source incident. The RI value of BK7 for this wavelength of light is measured as 1.5151.

The proposed systems, both Ag-$PtSe_2$-WS_2 and Au-$PtSe_2$-WS_2, show high sensitivity of 187° RIU^{-1} and 194° RIU^{-1} and give 1.36 and 1.67 times greater sensitivity than traditional Au and Ag sensors. Zheng et al. (2020) has offered graphene-based wearable physical sensors for piezoresistive use and has attracted great attention because of their large range for individual health monitoring, interfaces between humans and machines, robots, sports, and therapies. Nanowire, nanoparticle, nanocarbon, carbon black, carbon nano tubing, and graphene materials have been explored to build elastomer substrates with delicate piezoresistive sensors. Graphene and its derivatives, in particular, are considered the most suitable candidates as wearable sensors; the flexible sensors are made of 1D fibrous, 2D planar, and 3D interconnected cellular graphene architectures. The tactile pressures, temperatures, and physiological strands are discussed as are their unique geometry, lightweight, flexibility, and transportation. Four main mechanisms, including the geometry effect, changes in band structure, change of the area of contact, and tunneling effect are included in piezoresistive sensings. Graphene-based strain, temperature, and pressure sensors are categorized and compared in terms of their structures and performance. Artificial electronic skin, human health monitoring, activity detection, HMIs, and wearable entertainment are all examples of potential practical applications. In graphics and modular electronics, there is advancement in a broad variety of new applications in biomedicine, health care, robotics, artificial intelligence, and entertainment technologies (Fig. 9.13).

The exciting mechanical, physical, visual, and transport characteristics of the processing material methods and modern sensing methods and the flexibility of graphic piezoresistive physical sensors such as tension sensors, pressure sensors, and temperature sensors were quickly created by graphene. Zainuddin et al. (2019) proposed the SPR sensor, which is made of side-polished optical fiber with the

TABLE 9.5
Dispersion Coefficients Used in Drude–Lorentz to Evaluate the Gold and Silver RI (Rahman et al. 2020)

SL. No.	Name of the Metal	Plasma Wavelength (in meter)	Collision Wavelength (in meter)
1.	Gold (Au)	1.6826×10^{-7}	8.9342×10^{-6}
2.	Silver (Ag)	1.4541×10^{-7}	17.614×10^{-6}

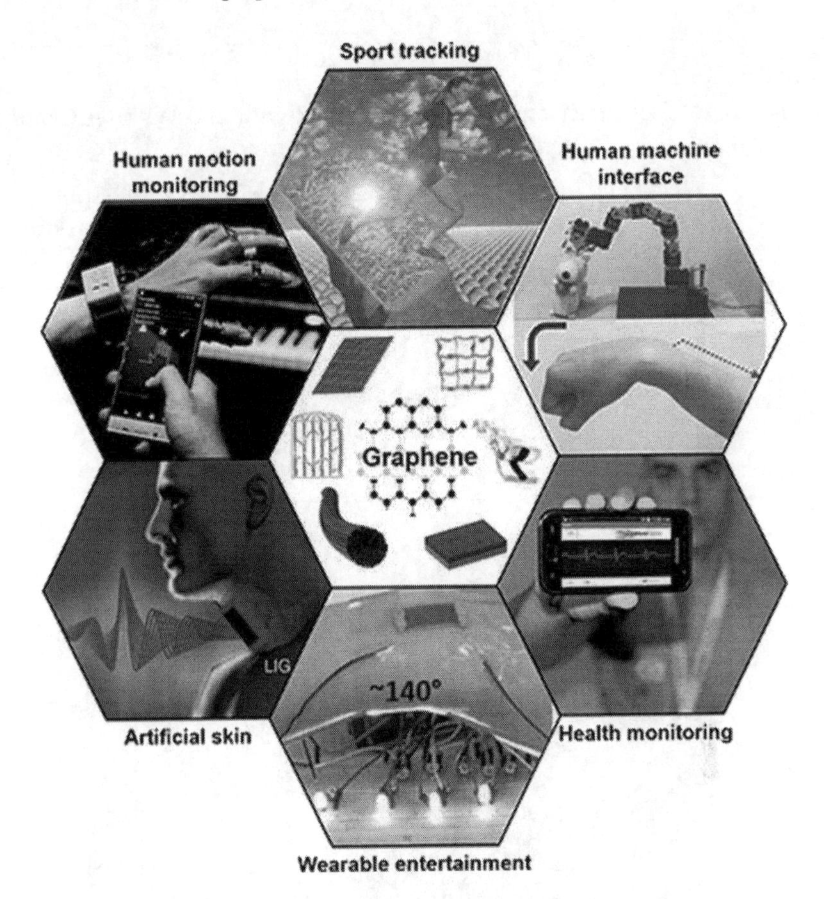

FIGURE 9.13 Graphene-based sensor integration in monitoring human motion, sports tracking, human-machine interface, health monitoring, wearable entertainment, and artificial skins. Reprinted with permission from Materials Today Journal. Copyright, 2020, Elsevier (Zheng et al. 2020).

cladding removed symmetrically and coated with various thicknesses of embedded silver film. The 40-nm-thick Ag application has also been covered in fiber without cladding, showing a higher sensitivity to ~2166 nm RIU^{-1} with distilled water and with alcohol 208.333 nm RIU^{-1}. The transmitting wavelength was measured as 460 nm and 530 nm in the two fiber conditions in the active sensing region as a length of 3 mm worked at a wavelength from 300 to 1,100 nm. Table 9.6 shows a different sensitivity to their different silver coating thickness.

The WS_2-metal-WS_2-graphene, heterostructure-design-based SPR RI sensor was proposed by Dey, Islam, and Park (2021). The technique for modeling and developing the forward sensor was carried out by finite-differential time-domain (FDTD). The proposed optimized sensor structure has a high sensitivity of 208° RIU^{-1} with a precision of 1.12 for detection and a consistency factor of 223.66 RIU^{-1}, superior to current 2D SPR materials (Fig. 9.14).

TABLE 9.6
Comparison of Sensitivity and SNR in Silver SPR with and Without Cladding Layer (Zainuddin et al. 2019)

Ag Thickness (nm)	Loss (dB)	RI	Δ n_s	λ_{res} (nm)	$\Delta\lambda_{res}$ (nm)	$\Delta\lambda_{1/2}$ (nm)	SNR	Sensitivity (nm RIU^{-1})
40	0.65	1.330	−0.012	474.5	−2.5	−5	−0.5	−208.333
		1.3450		477				
50	0.65	1.3330	−0.012	474	−0.5	−6.5	−0.077	−41.667
		1.3450		474.5				
40	1.80	1.3330	−0.012	524	−26	−25	−1.04	−2166.667
		1.3450		550				

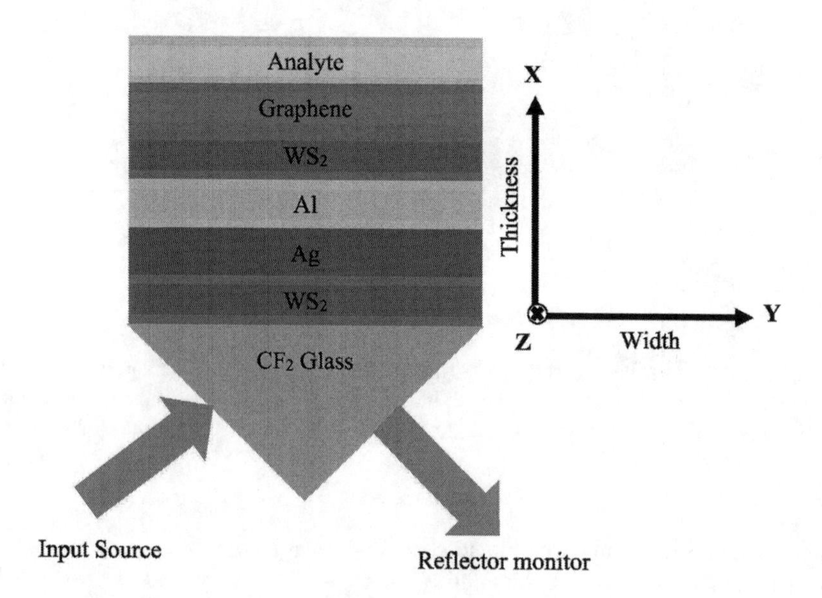

FIGURE 9.14 Schematic diagram for SPR biosensor based on WS_2-metal-WS_2-graphene (Dey, Islam, and Park 2021).

9.3 APPLICATION IN MINING INDUSTRIES

Standard gas sensors are designed to gather information in physical and chemical terms from gas analytics and provide signals. Highly sensitive materials with gaseous analysis produced by gaseous stimuli and viable methods to incorporate them into sensor devices for the transformation of the next-generation gas sensors with electric transductions. Yao, Li, and Xu (2021) proposed the electrically transmitted gas sensor of metal-organic frameworks and their derivatives. MOFs and their derivatives from metal-organic frameworks (also called MOFs) have vast applications as porous materials due to their diverse nature. In contrast to electrically transduced sensors, conductive composites, or MOFs (dynamic or porous MOFs), have partially overcome the sensor production barriers; conducting MOFs are providing new avenues for sensor development. The gas sensors, MOF basic and MOF derived, can be classified into five different groups. Gas sensors are translated electrically. The composite, porous, and electrical properties may be modulated in a monitored form by MOF materials derived from carbon materials (for example carbon-nanotubes (CNTs), amorphous carbon, graphene (G), metal compounds (for example, MMOxs, metal sulfides (MSs)) (Fig. 9.15).

Chemical sensors may be the pellets, the type of film, or the single type of crystal. Film chemical resistors are the most popular shape for basic research and industrial devices. Fig. 9.16 shows two other types: the Figaro-Taguchi (TGS) indirect-heated sensor and the (MEMS) microelectromechanical system. The indirectly heated chemical resistor is usually composed of sensing film, a metal electrode insulation tube, and a Nickel-Chromium (Ni-Cr) wire. In comparison to the indirectly heated chemical insulation, a metal electrode pair tube, generate nano-micro technology a relatively small microelectromechanical system type and a Ni-Cr inserted in the tube heating wire. Compared to the indirectly heated chemical

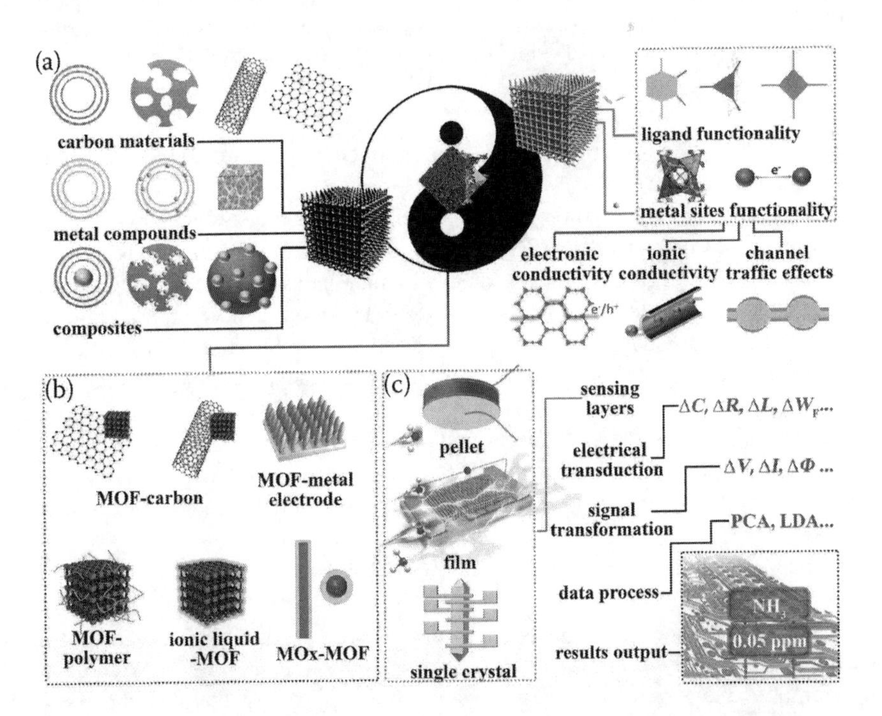

FIGURE 9.15 a) The materials derived are MOF and MOF; b) the MOF composite; and c) the process of sensing gases' signal transformation (I—current, V—voltage, E—power potential) and capacitance changes are presented. Reprinted with permission from Coordination Chemistry Reviews Journal. Copyright, 2021, Elsevier (Yao, Li, and Xu 2021).

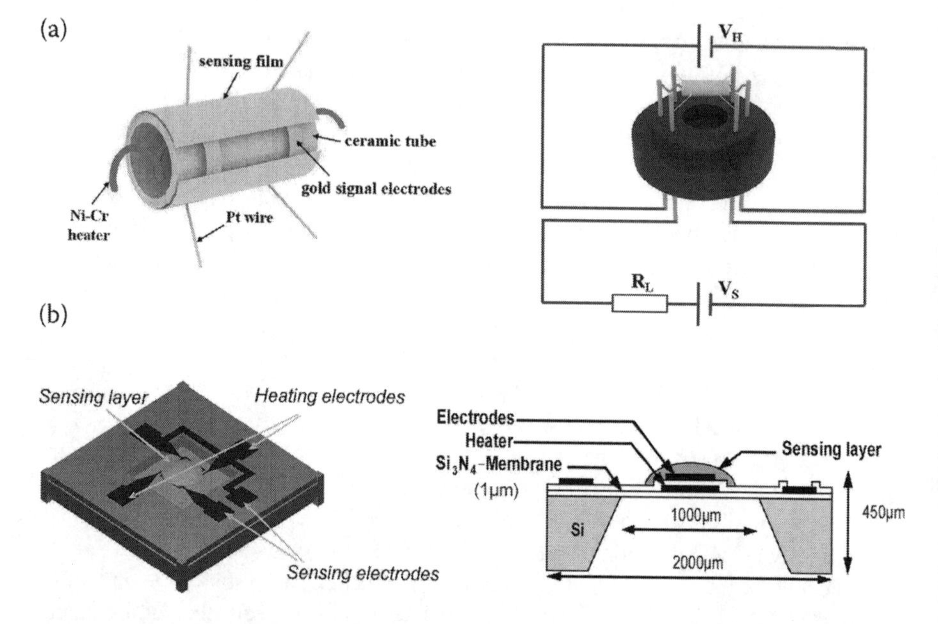

FIGURE 9.16 Typical sensing elements of chemiresistive: a) the indirectly heated form and testing circuits of the well-known Type Figaro-Taguchi sensor (TGS). b) (MEMS) 3D view and side view. Reprinted with permission from Coordination Chemistry Reviews Journal. Copyright, 2021, Elsevier (Yao, Li, and Xu 2021).

storage device, nano-micro technologies produce a relatively small microelectromechanical system type.

Zrelli and Ezzedine (2018) proposed a real-time monitoring system based on optical and wireless sensor network (WSN) sensors for underground mining. Uses of these sensors are in underground tunnels in the mining industry (UMMS—underground mining monitoring system). Differences in stress vibration, temperature, and moisture may be identified and located within the introduced architecture. For UMMS, the proposed optical sensor(s) are employed in this work (Raman, Brillouin, Fabery-Perot, and Bragg) (Fig. 9.17).

In coal mines, FBG stress sensors and acceleration sensors are used to monitor stress changes and seismic signals. FBG sensors are linked with fluctuations of temperature, strain, and also sense pressure changes summarized by the equation

$$\Delta\lambda_B = c_e \varepsilon + c_T T + c_P P \tag{9.9}$$

Here, an FBG sensor shows different variations where $\Delta\lambda_B$ is the wavelength; C_T, Ce, and C_p are temperature, strain, and coefficients of the pressure of FBG. Change in vibration and strain in underground mining may oddly threaten the lives of miners. Consequently, this work focused on strain change measurements with a fixed range of temperature (25°C and 60°C). Results show that, when the heat is less than 30°C and the pressure in underground mining varies within 2–15 MPa, we can confirm that in this case, the pressure varying reaches 1090 $\mu\varepsilon$. Thus, only

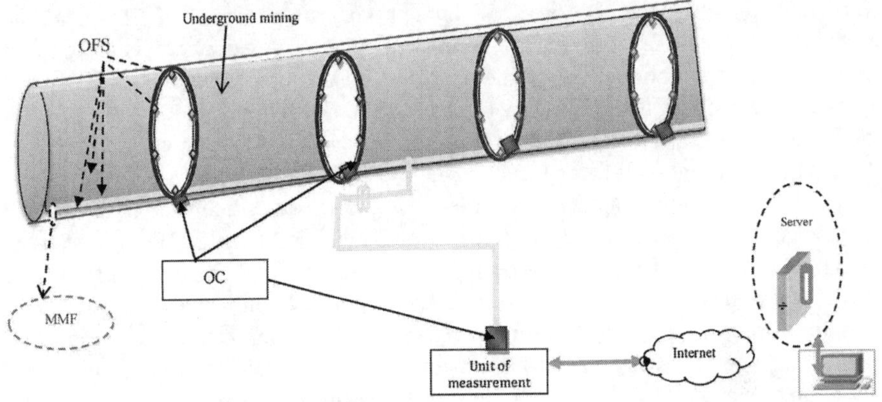

FIGURE 9.17 Proposed underground mining monitoring system; UMSM system. Reprinted with permission from Optik Journal. Copyright, 2018, Elsevier (Zrelli and Ezzedine 2018).

temperature and pressure are monitored, and based on that monitoring it is easy to conclude the strain changes in underground mining (Fig. 9.18).

Ashraf et al. (2020) provides a 2D energy storage and sensor interface property assessment. This paper deals with their structural morphologies of 2D-based heterostructures and also gives problems and solutions for grid malfunctions or interfacial synthesis methods. Finding graphene was the motivation for the theoretical and experimental researcher to synthesize another 2D material, similar to hexagonal boron nitride, silicon, and molybdenum disulfide (MoS_2). The construction of a

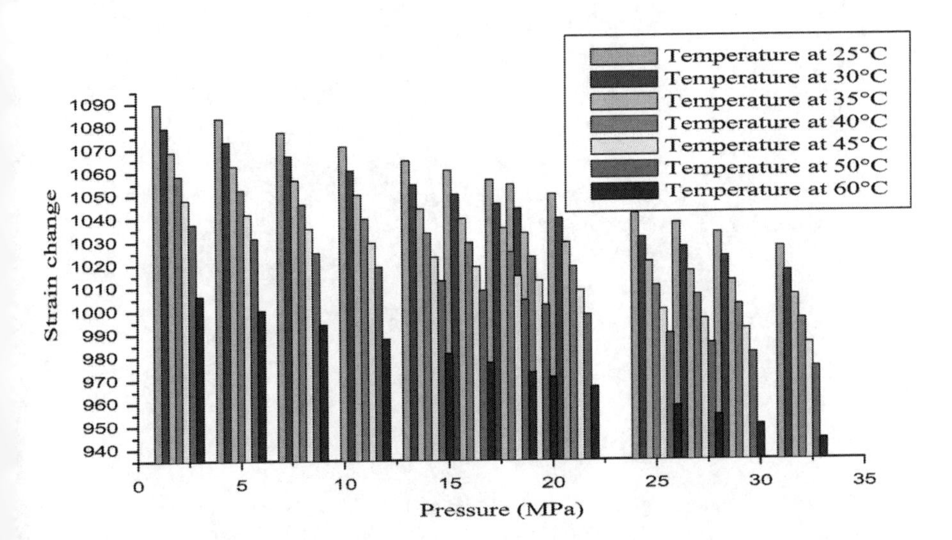

FIGURE 9.18 Change of strain at the set temperature and pressure. Reprinted with permission from Optik Journal. Copyright, 2018, Elsevier (Zrelli and Ezzedine 2018).

vertical heterostructure can be done by applying many types of 2D materials without causing the problem of grid misplacement, as layered materials are weakly connected to the phenomena of the van der Waals force with the various grid constant, covalently linked, unscrewing-free grids. This feature enables the creation of isolated, mixed, and highly different atomic levels without creating a gateway that explores a high-end heterostructure in van der Waals (Fig. 9.19).

In the hollow core of a microstructured fiber gas sensor, which is divided into two forms according to a sensory concept, Li, Yan, Dang, and Meng (2021) suggested the design and the use of interferometric type and absorbent type. Innovation structures have increased significantly the sensitivity and detection limit. However, it is difficult to eliminate cross-sensitivity and it is not possible to recognize the gas component. The absorption-based gas sensors have matured in recent years and have resulted in a more precise and effective detection. Improved sensing efficiency by introducing and optimizing various parameters of special fibers. It is more suitable for explosive and inflammable gas detection. Its low loss of transmission, high sensitivity, and other efficiency is superior to conventional fiber optics. This dissertation involves a thorough study of the various systems and technical principles (Table 9.7).

Ultrasensitive, resistive humidity sensor in the ambient temperature is based on the 2D-layered titanium carbide proposed by Zhang et al. (2021) for the investigation of water vapor adsorption on $Ti_3C_2T_x$ surfaces by using Fourier's transform infrared spectroscopy. Ti_3C_2Tx's surface function analysis suggests that the high moisture and sensitivity of the hydroxyl group are due to the high coverage rate on the Ti_3C_2Tx surface. MXenes unique structures and their surface properties

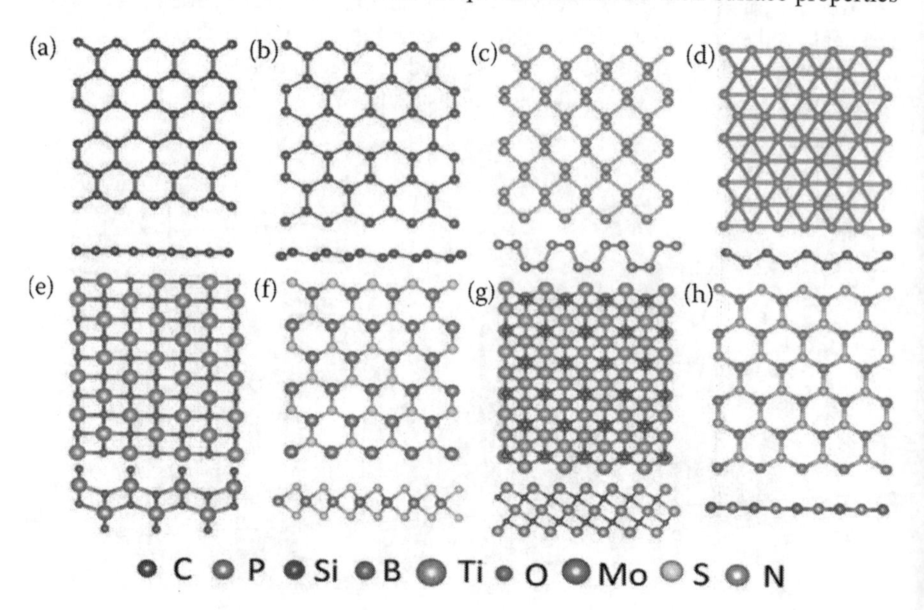

FIGURE 9.19 Single a) graphene with different layer, b) silicene, c) phosphorene, d) borophene, e) TiO_2, f) MoS_2, g)Ti_3C_2, and h) h-BN top and side view of atomic structure. Reprinted with permission from the Chinese Journal of Physics. Copyright, 2020, Elsevier (Ashraf et al. 2020).

TABLE 9.7
Main Technical Theory and Traditional Diagram of Structure. Reprinted with Permission from Optics and Laser Technology Journal. Copyright, 2021, Elsevier (Li et al. 2021)

Technology	Principle	Typical Structure Diagram
Interferometer Type		
Single FPI	The beam is mirrored in the welding surface and the open surface.	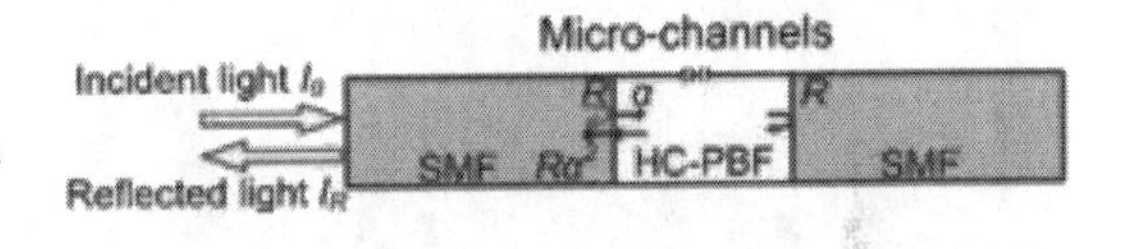
Cascaded FPIs	The beams are separately mirrored on both soldering surfaces and form three-beam stores that improve their sensitivity by forming and demodulating a coating.	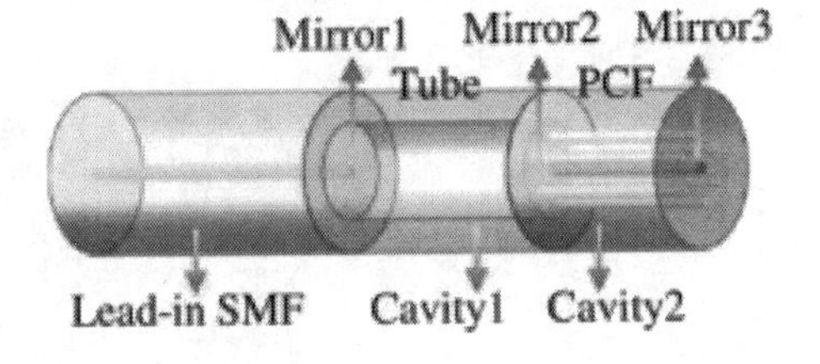
Coating Sensitive Materials	Change the cavity length using the sensitive material and the gas response (expansion or heating).	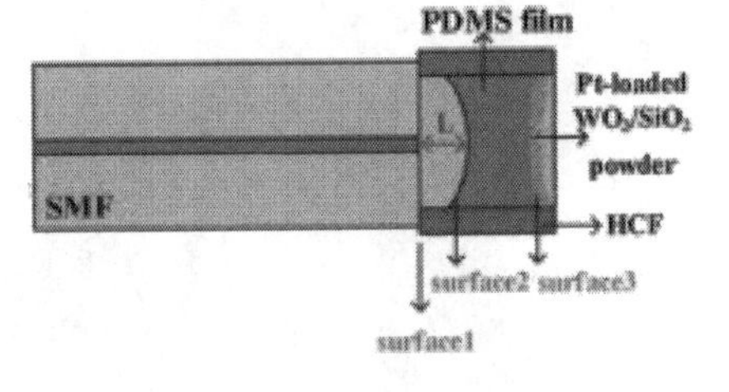
Mach–Zehnder Interferometer	In high-order modes, the welding point induces inter-mode disruption with the HCF and a second soldering point is coupled.	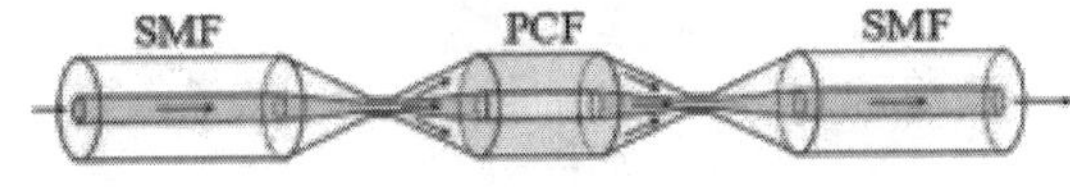
Sagnac Interferometer	The light is split into two beams and travels to	

(*Continued*)

TABLE 9.7 (Continued)
Main Technical Theory and Traditional Diagram of Structure. Reprinted with Permission from Optics and Laser Technology Journal. Copyright, 2021, Elsevier (Li et al. 2021)

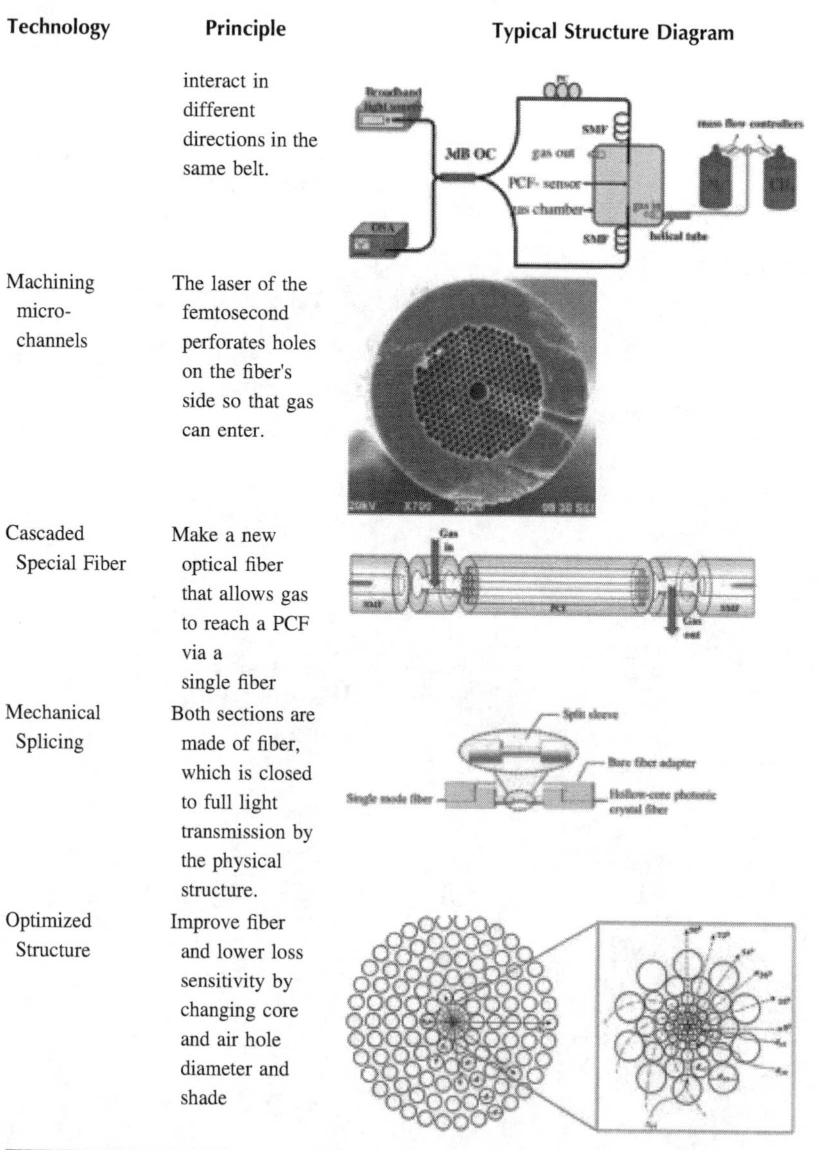

Technology	Principle	Typical Structure Diagram
	interact in different directions in the same belt.	
Machining micro-channels	The laser of the femtosecond perforates holes on the fiber's side so that gas can enter.	
Cascaded Special Fiber	Make a new optical fiber that allows gas to reach a PCF via a single fiber	
Mechanical Splicing	Both sections are made of fiber, which is closed to full light transmission by the physical structure.	
Optimized Structure	Improve fiber and lower loss sensitivity by changing core and air hole diameter and shade	

enable them to develop improved resistive moisture sensor performance. Umar et al. (2020) used a 2D SnO_2 gas sensor in the manufacture of gas sensors like H_2, CO, and C_3H_8 for the use of synthesized SnO_2 disks. A synthesized SnO_2 electrode at 400°C showed gas reactions of 14.7, 9.3, and 8.1 to H_2, CO, and C_3E_8, respectively. In contrast, rational reaction times of 4 s, 3 s, and 8 s and rehabilitation times were observed with H_2, CO, and C_3H_8 gases of 331 s, 201 s, and 252 s, respectively. The oxygen atom binding to the site of O_{2c} better controls the sensor conductance than the sites of O_{3c}.

Low conductivity results in the exposure of the gases' reduction as H_2, CO, and C_3H_8 for the semi-conductive n-type. These gases are oxidized when they are in communication contact with those electrons releasing oxygenated species that are returned to the SnO_2 conducting band. A SnO_2 disc-based gas-sensing mechanism is shown in Fig. 9.20. Kumar et al. (2020) have developed the Di-chalcogenide 2D transition metal and NO_2 gas sensors for metal oxide and were modified or worked on by different metals to improve different parameters for sensing, including selectiveness, sensitiveness, effectiveness, stability, or device life span. In TMD, many methods include chemical vapor deposition (CVD), exfoliation of the liquid stage (fluid phase), and exfoliation of electrochemical by Li-intercalation in the preparation of one or more layered TMD slides. Many TMDs (such as $MoTe_2$, MoS_2, and WS_2, for example) reflect semiconductor competencies, have a wide variety of bandwidths, and are more appropriate for electronic tools. MoS_2 focuses on applications for gas sensing in all TMD products (Fig. 9.21).

Compared to other rGO-based sensors, with a range of hetero joint functions, a larger surface area, and more sites of adsorption, the rGO-MoS_2-Cds nano-compound sensors displayed an excellent response by 25.7%.

9.4 APPLICATIONS IN BIOMEDICAL INDUSTRIES

KR setup based on a prism cannot be used because of high sample and volume needs. However, the FOSPR sensor (fiber-optic SPR) has the advantage of

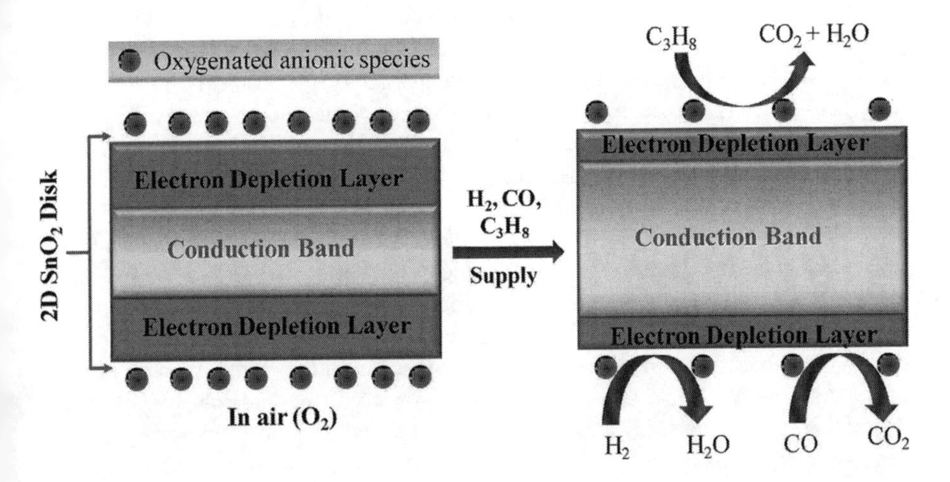

FIGURE 9.20 SnO_2 disc-dependent gas sensing mechanism. Reprinted with permission from International Journal of Hydrogen Energy Journal. Copyright, 2020, Elsevier (Umar et al. 2020).

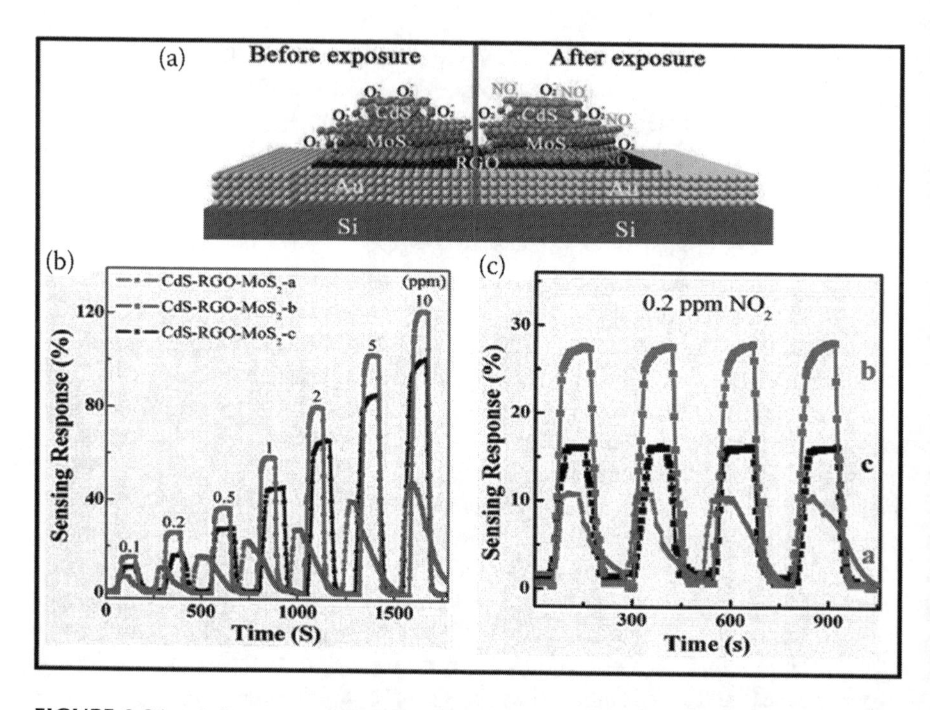

FIGURE 9.21 a) Sensor nanocomposite exposure (before and after) to NO_2, b) different concentrations (0.1–11 ppm) of NO_2 for different nanocomposites, and c) the same 0.2 ppm NO_2 gas-sensor response for the three nanocomposite sensors. Reprinted with permission from Materials Science in Semiconductor Processing Journal. Copyright, 2020, Elsevier (Kumar et al. 2020).

portability, size, ability to multiplex, remote sensing, and an important value (FOM). It is a good candidate for cortisol (steroid hormone), a biomarker for psychological stress and cardiovascular disorders, and is essential for controlling metabolism, immune response, obesity, weakness, and bone fragility of the human body because of the biosensitivity of SPR. Sharma, Kaur, and Marques (2020) have developed an SPR ultrasensitive cortisol detection 2D material/metal carbide base fiber sample. The design of the proposed sensors comprise the core, polymer-cladded glass, a layer of Ag in the form of 2D materials (WS_2, graphene, and MoS_2), and layered materials based on metal transitions (MXenes). The performance analysis for the sensing is performed about the merit figure and CPF. A thorough review of each sample version shows overall superior sensor efficiency, including a detection limit for all the sensor design variants of 15.7 fg mL^{-1} with the $Ti_3C_2O_2$-based sample. The testing of cortisol in tanks of fish with a range of concentrations between 0.01 and 7 ng mL^{-1} is a promising and robust task for the water world. Jia, Liao, Li, and Cai (2021) have updated the proposed resonance sensors for ultrasensitive detection of niobium disulfide mercury nanosheets. The proposed sensor is based on the high performance of niobium disulfide (NbS_2). The SPR sensitivity optimum value reached 193° RIU^{-1} range achieved through the simulation calculation of the NbS_2 nanosheets, a 38.1% higher over NbS_2 overlay than the case. More importantly, a very low concentration of Hg^{2+} has been detected and described as the NbS_2-based SPR sensor for sensing mechanism. Surprisingly, the sensors used to detect mercury are 13.55° μM^{-1} and the limit of detection is 1 pM. This SPR sensor sensitivity was revealed using HRTEM, TEM, and XPS with a value of 200 pM max for the Hg^{2+} sensor (Fig. 9.22).

Drinking water with strong ions of metal (Ag^+, Cr^{3+}, Hg^{2+}, etc.) had been injected into the sample unit of the NbS_2 sample system in the concentration of 10^{-6} M. Furthermore, Hg^{2+} potable water was pumped in the same manner from 10^{-12} M to 10^{-6} M. After 30 min of dipping, ultrapure water was used to wash chips and the SPR spectrum was noticed. Different Hg^{2+} concentrations were used to obtain SPR curves. Sluggish and halcyon with around 0.1 mL min^{-1} were controlled also by the speed of the metal ion aqueous solution. Yu et al. (2020) have developed an SPR sensor having D-shaped with a MoS_2 graphene composite glucose sensor. Modifications of graphene and CVD MoS_2 for MoS_2 graphene-based composite nanostructure on the sensor surface. Pyrene-1-boronic acid (PBA) on graphene was changed by α–α stacking interactions. The excellent image properties and PBA's ability to bind glucose molecules specifically of the MoS_2 graphene-based composite nanostructure have increased the glucose sensor performance. Sadeghi and Shirkani (2020) proposed a highly susceptible SPR mid-infrared biosensor based on graphene-gold for-biomolecules with a wide variety of biological cells. This sensor was made up of a graphene-gold grating structure with a gold substratum and a detection medium attached upwards. Graphene also located the SPs more and prevented the oxidation of metals. An SPR mode was observed at a wavelength of 2001 nm and, during the optimization process, a sensitivity of 1,782 nm RIU^{-1} and a quality factor of 21,214 for the refractive index 1.333 were obtained. The biomolecules and organic cells with an RI of 1.100 to 2.100 showed striking achievement for nearly all biomolecules. In particular, glucose was

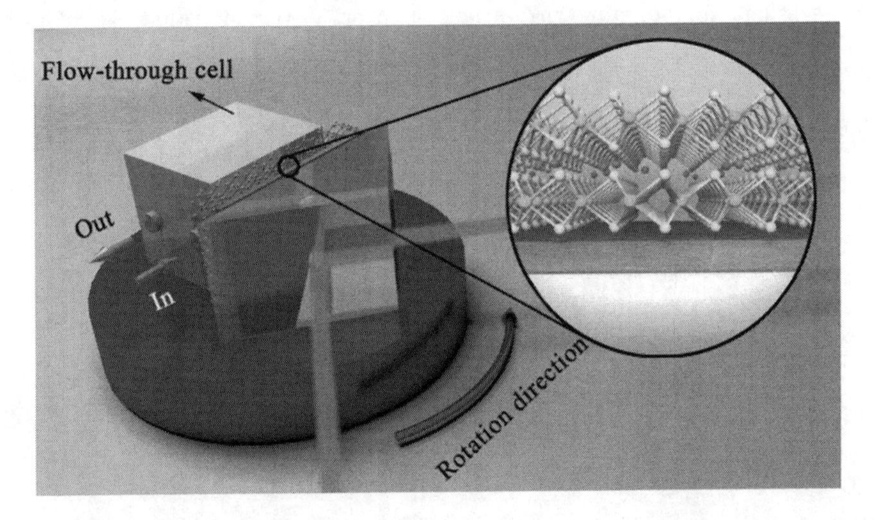

FIGURE 9.22 Schematic diagram based on NbS_2 nanosheets for heavy-metal ion detection sensors. Reprinted with permission from Journal of Alloys and Compounds. Copyright, 2021, Elsevier (Jia et al. 2021).

considered a significant biomolecule, which was determined to have 1,770 nm RIU^{-1} and 18,438 sensitivity and quality factors. Studies had also been carried out to measure Madin-Darby Canine Kidney (MDCK) and blood with quality factors of 21,214 and 22,250, respectively, up to Δn = 0:00005. Finally, the best value was obtained for the 1.1000 RI with a quality factor of 24,861. For these four materials, the proposed biosensor quality has been calculated. The quality factor for MDCK is 21,214; blood is 22,250; water is 22,250; and glucose is 18,437, as shown in Fig. 9.23.

Multi-step graphene grating optical sensors are offered for highly sensing biomolecules based on IR surface plasmons (Sadeghi, Shojaeihagh, and Shirkani 2021). The graphene-gold gratings were structured and classified into three different sensors, such as single, double, and triple. This material can be detected by choosing a range of RI between 1.000 and 1.6000, including several gas alloys and biomolecules in the IR region. For the RI of 1.300, the best quality factor was 15.583, and for the three steps, the RI of 1.200, it was 42.944. The RI of 0.0005 for substances—G protein and guanine—can be measured by a double-step sensor and in the material order of water, glucose, blood, is 0.00001 for the tri-step sensor.

Three different structure types have been developed to assess the changes that occur for each step of the proposed sensor performance shown in (Fig. 9.24a). Initially, the impact of the first-stage structure was analyzed and significant values were obtained. Afterward, the graphene gold grating structures covered by two and three steps were examined and the structural parameters optimized. Between the gold and sensing media, a graphene monolayer has been inserted, gratings with the thickness (Ti) and width (Li) have been applied, and their index (i) shows different values for the number of steps such as 1–3.

Wu et al. (2019) proposed MXene-based SPR ultrasensitive carcinoembryonic antigen biosense, a 2D transition metal carbide, where different levels of PBS-diluted CEA (500 μL) of the flow cell was injected separately with an immobilized Ab_1-sensing film detection and incubation over a 30 min period. To remove

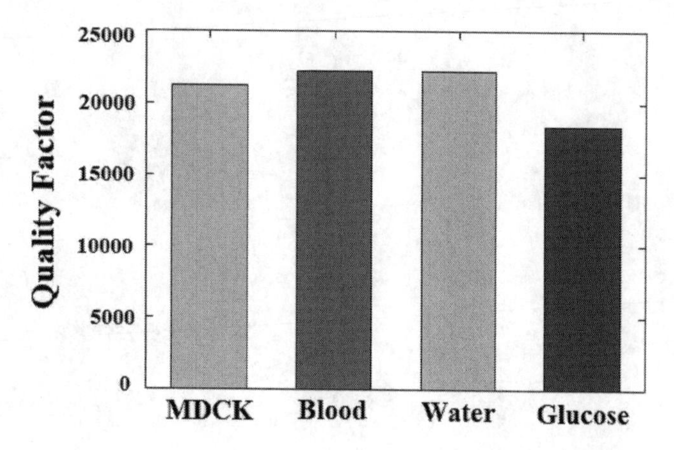

FIGURE 9.23 Calculated by the proposed biosensor the consistency factor of various materials. Reprinted with permission from Physica E: Low-Dimensional Systems and Nanostructures Journal Copyright, 2020, Elsevier (Sadeghi and Shirkani 2020).

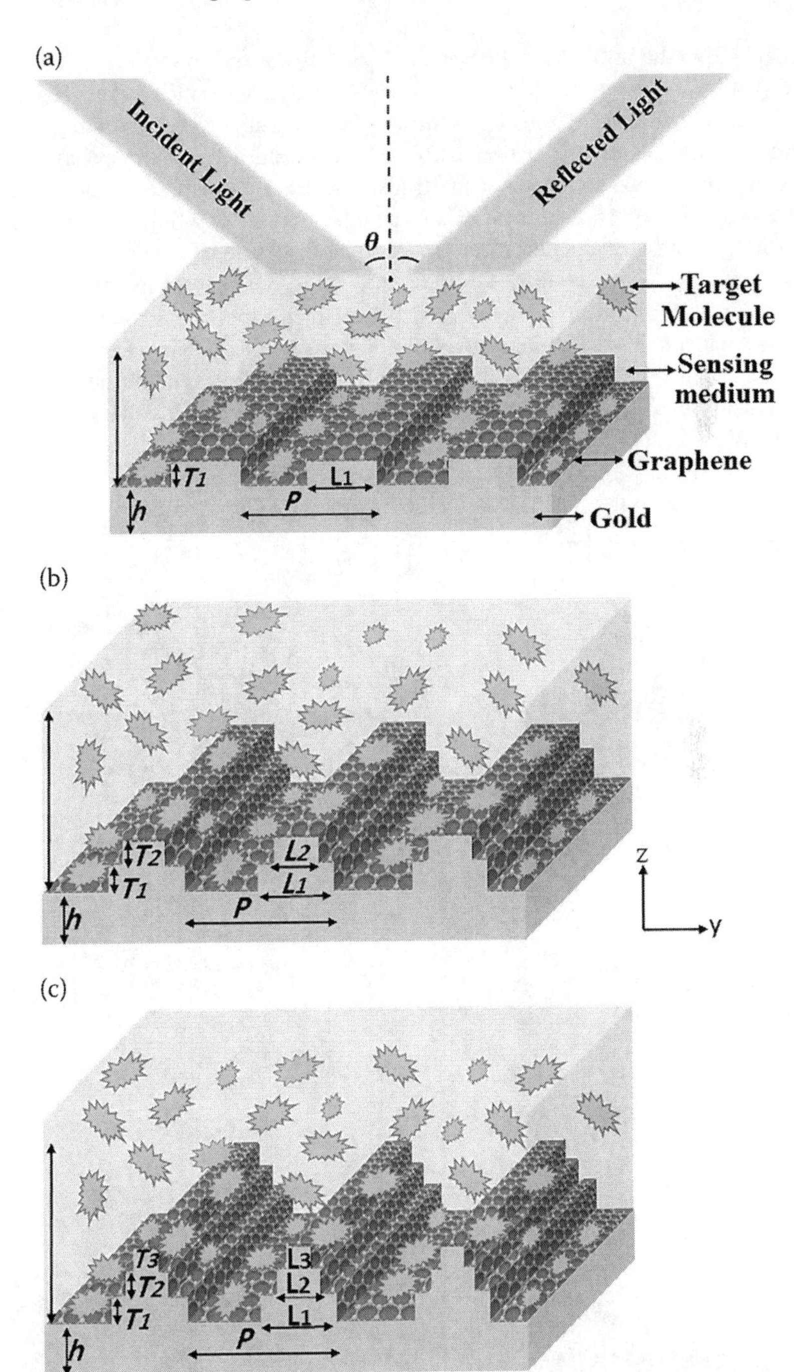

FIGURE 9.24 1D structures based on photonic crystal with gold-covered graphene with multi-step gratings: a) single-step, b) double-step, and c) triple-step. Reprinted with permission from Materials Science & Engineering B Journal Copyright, 2021, Elsevier (Sadeghi, Shojaeihagh, and Shirkani 2021).

unbound CEA, the injection of PBS in the flow cell was then completed. A resonant angle (about 5 min) of 0.4 mg mL^{-1} MWPAg-Ab_2 composites were used in the flow-zine and held for 20 min. The sensors were measured as resonant angle shifts, before and after sample injection with PBS. The entire test was performed at a standard temperature and stress, and all tests were repeated three times.

The proposed SPR biosensor indicates an ultra-low sensing limit of 0.07 fM. The Ti_3C_2-MXene-based SPR detection used for the detection of cancer biomarkers is an important tool for extending MXenes applications and shows its potential usage in biomedical applications. Pandey (2020) examined a plasmonic sampling sensor using MXene and $Ti_3C_2T_x$ glass substrates, with a significant RIU (RI = 1.33) FOM value of 55.596 RIU^{-1} for the three-layer BP–$Ti_3C_2T_x$ MXene-based monolayer SPR sensor. Sensing applications in which the extremely high angular sensitivity of 322.46 RIU^{-1} is achieved. This study analyzed SPR-based, five-layer sensors using monolayer BP and $Ti_3C_2T_x$ MXene. For gaseous analysis, gaseous analysis was calculated a sensitivity value of 64.40 RIU^{-1} and the FOM value of 111.226 RIU^{-1} (RI—1.0–1.005) (Fig. 9.25).

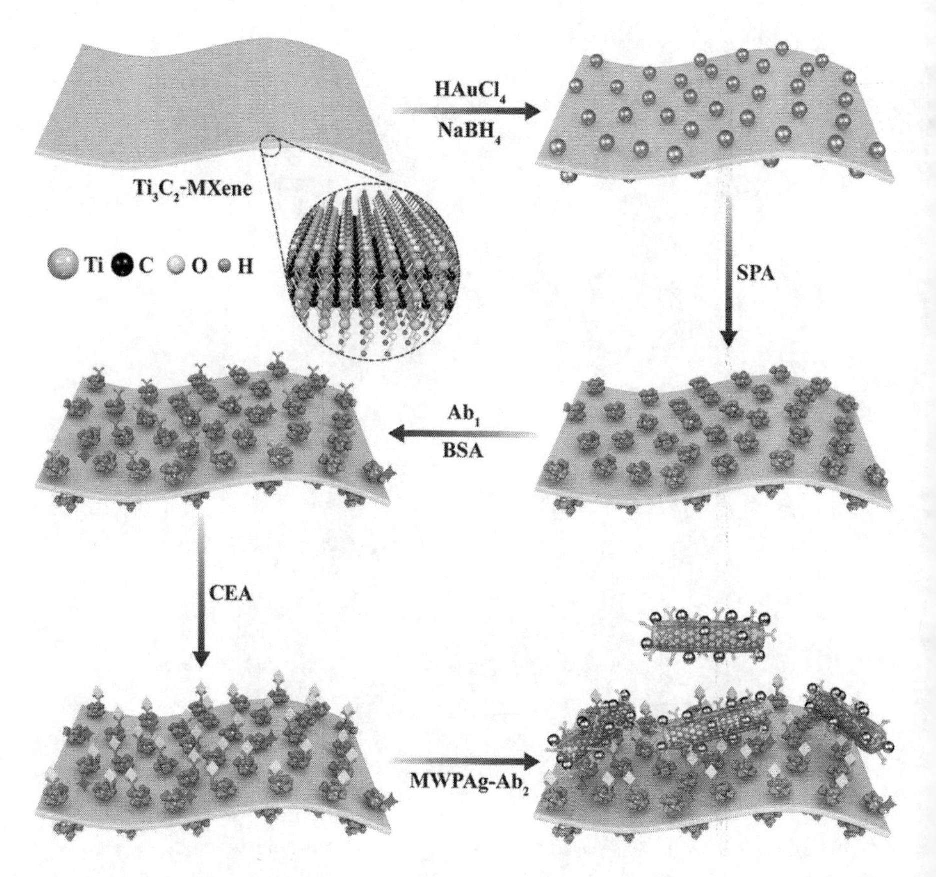

FIGURE 9.25 Scheme showing the biosensor detection protocol. Reprinted with permission from Biosensors and Bioelectronics Journal. Copyright, 2019, Elsevier (Wu et al. 2019).

Saifur Rahman, Rikta, Abdulrazak, and Anower (2020) have found an improvement in the hybrid S_nS_e-graphene photo-sensing surface biosensor for applications in biosensing. The sensitivity of this sensor was outstanding compared with a traditional SPR-based sensor with the value of 94.29 ° RIU^{-1} and QF up to 12.65 RIU^{-1}. Further, 2D titanium carbide MXenes were investigated by Zhu, Zhang, Liu, and Liu (2021) as emerging optical biosensing platforms. During the past few years, MXene's growth was rapid and its success had been unprecedented. Ti_3C_2 MXenes with different structures and morphologies and features can possibly be used as electrochemical and optical sensing platforms for multiple functions. Attractive electronic and optical properties are given by the Ti_3C_2 MXenes NSs and QDs in various optical sensor systems.

9.5 APPLICATIONS IN STRUCTURAL HEALTH MONITORING

Structural health monitoring (SHM) refers to automated methods for aerospace, civil, and mechanical damage monitoring. Early detection of defects, such as cracks or corrosion, can significantly decrease maintenance costs over a structure's lifetime and can prevent disasters. Classical time-based service has been transferred from classical maintenance to periodic large-scale structural systems control and condition-backed maintenance, combining low-cost sensor technology with digital twins—the precise virtual display of complex high industrial properties. A data-driven SHM approach uses the structure of interest to represent the physical structure. It is based on an approach based on data for SHM that uses the value structure physics-based representation. It derives damage-sensitive engineering features from raw discreet signals using the one-class support vector machine (OC-SVM) algorithm to equate safe exercise data with new blind test data (Bigoni and Hesthaven 2020). In a broad range of industries, carbon-fiber-resistant nanomaterial-coated sensors can be used for the on-site and structural health monitoring of composites. Irfan et al. (2021) investigated the monitoring of processes and SHM of carbon-coated piezoresistive fiber sensors. This work focused on the use and translation of carbon nanomaterials as fiber-reinforcement deposition materials in sensors for process surveillance and SHM during service time. The emphasis was also on various compound manufacturing parameters monitored, such as compaction reinforcement response, resin gelation, flow-front monitoring and cure, and damage detection by in-situ nano-material sensing.

Many techniques were used to test the structural performance of the composite using nanomaterial sensors to ensure that the structure is prepared for its entire use. However, it is an area for further development to use nanomaterials for process monitoring. In several key production processes of composites, compaction strengthening and resin impregnation are taken into account to be important steps (see Fig. 9.26). The computer vision of the SHM of civilian infrastructure was investigated by Feng and Feng (2018). From dynamic reaction to damage detection, this research is based on the recent development of computer-based vision sensors and their applications for the SHM structural displacement response. This research examines how pixel displacements can be converted to physical displacements and how subpixel resolution can be achieved, what causes measuring errors, and how

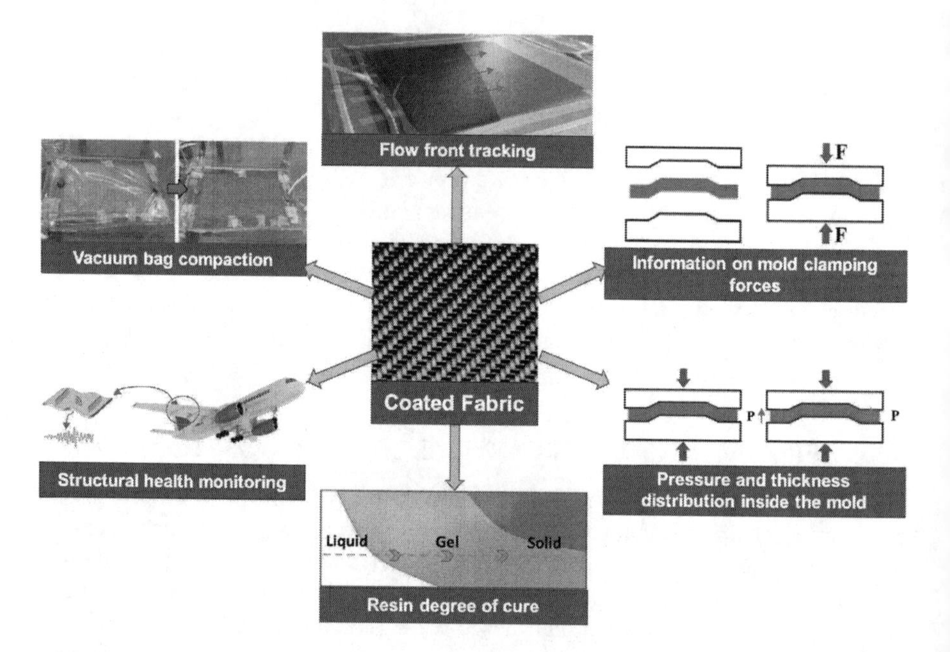

FIGURE 9.26 The application of process-coated carbon nanomaterials and structural health surveillance. Reprinted with permission from Composites A: Applied Science and Manufacturing Journal. Copyright, 2021, Elsevier (Irfan et al. 2021).

errors can be mitigated. The main focus of most recent field studies is measuring relatively large amplitude displacements such as bridges that are subject to moving train loads. Different noise sources in complex outside conditions, like heat haze, create difficulties in the accurate measuring of small-amplitude movements, such as the reaction to lightweight or environmental stimulation of short- and medium-sized concrete bridges. The proposed condition of monitoring composite overwrap vessels with MXene sensors was proposed by Wang, Lin, Lu, and Li (2021). A new MXene film sensor technology capable of monitoring composite structures in real time and locally. In a composite pressure vessel, the MXene sensor is built on a flexible printed circuit (FPC) and it is used during hydraulic pressure testing for monitoring arrow and axial pressure field shifts. The MXene sensor has a regular piezoresistive response. The max value of $\Delta R/R_0$ is kept constant at equal value as the peak hydraulic pressure and after the COPV pressure has been reduced to 0 MP, resistance may return to the starting resilience evaluation of shear crack depth found for 2D-distributed fiber-optic concrete structures proposed by Rodriguez, Casas, and Villalba (2019). This work bases its application on distributed optical fiber sensors (DOFS) to get shear capacity of concrete structures. The stress profile in the web of the DOFS beams was measured in two orthogonal directions using a 2D grid mesh. High-range resolution and OBR sensitivity (optical backscattered reflectometer) system allows a cracking pattern absolute mapping and provides the necessary data to calculate a crack width. De Luca, Perfetto, Fenza, and Petrone (2020) proposed a guided SHM damage detection system in complex composite compound structures, such as a mixed glass fiber reinforced polymer (GFRP) wing to monitor and detect damage in real time. This work is based on FE models for SHM systems to simulate guided wave propagation. The detection of damage by the wave-driven SHM system results in a comparison of the signals obtained during the actual lifetime of the structure monitored, in an uneven reference setting. Further, Yu, Saito, and Okabe (2021) developed a Bragg fiber-optic grating sensor ultrasonic visualization device that detects harm at 1000°C. It works with an FBG sensing configuration that is heat-resistant and fiber operated. This device allows laser irradiation to excite an ultrasonic wave on the surface of a material and it can be received from an FBG sensor that is mounted remotely due to the excellent heat resistance of both the wave excitement and the sensing elements. Also, a wave frequency test based on a 3D Fourier transformation was performed for the extraction of the wave elements corresponding to the reflection induced by an artificial defect in the panel. The laser ultrasonic visualization system based on the phase-shifted FBG (PSFBG) is able to a contribute efficient SHM high-temperature approach to the condition-based maintenance (CBM) strategies on heat-resistant structures (HRS). Ciminello et al. (2019) proposed a Stringer debonding edge detection using fiber-optic methods for non-model-based SHM by combining distributed strain and wave dispersion profiles. A structural, not linear induced by damage can be associated with the events of such a dispersed pattern. The waves operate at 60 kHz but the sensors operate at 250 Hz for optical fibers. Therefore, both methodologies are easy to implement a hybrid control system in real time, in which both of these systems can also be used for mutual evaluation. The design and characterization of a strong, low-cost, 2D deflection fiber-optic sensor was proposed by Bajić et al. (2017). A standard method of processing used in the quadrant photodetector simplifies the sensor application and is used to obtain 2D deflection information. The error boundary acquired in the magnitude of deflection and the direction of deflection measures is ±0.15 mm and ±2.5. The signal output of the proposed sensor corresponds with the known photo-quadrant detector, making it easier to process the signal and more realistic for the sensor. A temperature control and compensation thermistor is incorporated inside the beam. Error deflection and resolution limit of ±2.5° and 0.7° in 360°, respectively, are obtained (Fig. 9.27).

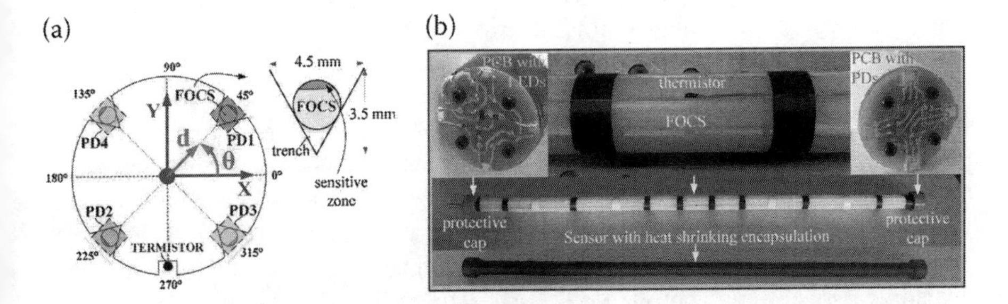

FIGURE 9.27 a) Beam cross-section arrangement of FOCSs and PDs with reference coordinate device; b) photograph of the sensor manufactured. Reprinted with permission from Sensors and Actuators A Journal. Copyright, 2017, Elsevier (Bajić et al. 2017).

In addition to four trenches with the fiber-optic curvature sensors (FOCs), a further trench is constructed in which a thermistor is installed to offset the temperature of a sensor system. At one end of the fiber, each FOC is joined to an LED and at the other, the photodetector (PD) is connected to a PD. An LED is used with a wavelength of 660 nm while a photodiode and a trans Impedance amplifier are also applied as a PD TSL250R-LF detector. The LEDs and PDs have been mounted on printed circuit boards in a circular shape (30 mm diameter), which are torn to the ends of the beam. Beams and FOCSs are also wrapped in thermally reduced protective encapsulation. Bigoni and Hesthaven (2020) are providing the simulation-based analysis and decision-making technique for the proactive preservation of complicated structures with a specific use for the SHM. It is based on an approach analysis of data for SHM that uses the physics-based representation of the interest structure. Damage detection and location are done by building synthetic training data to simulate the structure responsiveness from the sensor to active sources based on sensor by sensor. The replacement of the transform of the Laplace by a transform of the Fourier can incorporate the alternative passive regular sources that mimic the tide or wind effect. In that case, the characteristics used as indicators of damage must be adapted such that autoencoders are used to determine the characteristics behind healthy signals automatically. In the absence of damaged states, the systemic positioning of sensors was proposed by Bigoni, Zhang, and Hesthaven (2020) for structural detection anomalies. Simulated data can be produced to imitate monitoring phenomena under various natural and environmental conditions to discriminate relevant characteristics and thus identify potential anomalies. This work is based on an approach to sparse Gaussian processes, where a fixed number of sensors are consistently placed on an interesting structure. The healthy parametric structural differences are incorporated by the division of inducing inputs. Planar's configuration for extraordinary 2D-sub-material magnetic field sensors has been investigated by El-Ahmar et al. (2019). Magnetic resistors are especially characteristic of their simple nature and fast adaptation, coupled with their ability to detect small and medium magnetic fields. The broad range of magnetoresistance effects is unusual since a magnetic material cannot be needed as part of a structure, which can be useful in some applications. This work is focused on the use of epitaxial monolayer and biolayer graphene, of extraordinary magnetoresistance (EMR) flat configuration as construction material for sensitive substrate. The reliance on magnetic field induction of the EMR signal is consistent with the previous research in InSb/Ag structure-based EMR structures for Bi_2Se_3 EMR structures with two separate terminals. Butaud et al. (2020) researched to understand more effectively the potential for SHM applications of capacitive micromachined ultrasonic transducers (CMUTs). There are two dimensions to be considered: the multi-frequency dimension affecting the various circular membranes (50–250 m) and the bandwidth factor that influences the inherent capacity of a 100-m radius of an identical nine-membrane array. The frequency range for the CMUT is determined by an analytical approach: 80 kHz–2 MHz by radius (Madbouly, Mokhtar, and Morsy 2020). In this process, evaluating the performance assessment for SHM applications for rGO/cement composites. In this report, rGO explores the effect of the hardened cement paste on mechanical, pores, microstructural, and automatic sensing characteristics. The addition of rGO modified the pores structure by reducing the number of pores and making cement paste more homogeneous and compact. The net cement paste was applied to rGO by 0 wt, 0.01 wt, 0.02 wt, 0.03 wt, 0.04 wt, and 0.05 wt. The compression strength was assessed at 28 days of hydration. High-resolution transmission electron microscope HRTEM and Raman transmission electron microscopy characterized rGO sheets. Thermo-gravimetric analysis of the distribution of pores and electron scanning microscopy was the characteristic composites of rGO/cement. In addition to rGO, hydration products, particularly CH and CC phases, have also increased their thermal stability and can efficiently be used for the application of SHM as smart sensors.

9.6 SUMMARY

This chapter is based on the future trends of emerging materials in SPRs for different applications. The principle of sensors is discussed by expressing Maxwell equations and relations with evanescent wave-phenomena for SPR. Also discussed is the reason behind inflation in the research area of 2D metal oxide. Various methods for preparing 2D nanomaterials are also discussed under the top-down method (mechanical exfoliation, sonication exfoliation, intercalation exfoliation, chemical etching) and bottom-up method (CVD, hydrothermal, double decomposition reaction, hydrolysis, thermal decomposition). This chapter also briefly focused on the structures and the properties of emerging materials for analyzing their stability for their bandgap and charge mobility, such as graphene, h-BN monolayer, 2H-MoS_2 monolayer, antimonene, phosphorene, α-MoO_3, and $Ti_3C_2T_x$. Analysis of different composite structures for the gas-sensing mechanism and energy band was also done.

Various fields for sensing applications of 2D materials such as physical sensing, mining industries, biomedical industries, and structural health monitoring are discussed in this chapter. The sensing mechanism is somewhat different for different applications; in the physical sensing section, the sensing parameter of the different fields like strain sensor, alcohol sensor, humidity sensor, piezoresistive sensor, and RI sensor with different 2D materials have been explored. The exciting mechanical, physical, visual, and transport characteristics of the processing material methods and modern sensing methods and the flexibility of graphic piezoresistive physical sensors such as tension sensors, pressure sensors, and temperature sensors were quickly created by graphene. Further, different 2D materials are used in mining industries where standard gas sensors are designed to gather information in physical and chemical terms from gas analytics and give signals that can be obtained effectively. Some types of technology and their principles are discussed in this section.

The applications in biomedical industries of different types of fiber-optic SPR have been used because of their advantages, such as portability, size, ability to multiplex, remote sensing for psychological stress and cardiovascular disorders, and are essential to controlling metabolism, immune response, obesity, weakness, and bone fragility of the human body. The applications in structural health monitoring focused on the use and translation of carbon nanomaterials as fiber-reinforcement deposition materials in sensors for process surveillance and SHM, in monitoring aerospace, civil, and mechanical damage detection of defects, such as cracks or corrosion, can significantly decrease maintenance costs.

REFERENCES

Ashraf, Naveed, Muhammad Isa Khan, Abdul Majid, Muhammad Rafique, and Muhammad Bilal Tahir. 2020. "A Review of the Interfacial Properties of 2-D Materials for Energy Storage and Sensor Applications." *Chinese Journal of Physics* 66 (March): 246–57. 10.1016/j.cjph.2020.03.035.

Bajić, Jovan S., Marko Z. Marković, Ana Joža, Dejan D. Vasić, and Toša Ninkov. 2017. "Design Calibration and Characterization of a Robust Low-Cost Fiber-Optic 2D Deflection Sensor." *Sensors and Actuators, A: Physical* 267: 278–86. 10.1016/j.sna.2017.10.014.

Bigoni, Caterina, and Jan S. Hesthaven. 2020. "Simulation-Based Anomaly Detection and Damage Localization: An Application to Structural Health Monitoring." *Computer Methods in Applied Mechanics and Engineering* 363: 112896. 10.1016/j.cma.2020.112896.

Bigoni, Caterina, Zhenying Zhang, and Jan S. Hesthaven. 2020. "Systematic Sensor Placement for Structural Anomaly Detection in the Absence of Damaged States." *Computer Methods in Applied Mechanics and Engineering* 371: 113315. 10.1016/j.cma.2020.113315.

Butaud, Pauline, Patrice Le Moal, Gilles Bourbon, Vincent Placet, Emmanuel Ramasso, Benoit Verdin, and Eric Joseph. 2020. "Towards a Better Understanding of the CMUTs Potential for SHM Applications." *Sensors and Actuators, A: Physical* 313: 112212. 10.1016/j.sna.2020.112212.

Ciminello, Monica, Natalino Daniele Boffa, Antonio Concilio, Vittorio Memmolo, Ernesto Monaco, and Fabrizio Ricci. 2019. "Stringer Debonding Edge Detection Employing Fiber Optics by Combined Distributed Strain Profile and Wave Scattering Approaches for Non-Model Based SHM." *Composite Structures* 216 (November 2018): 58–66. 10.1016/j.compstruct.2019.02.088.

De Luca, A., D. Perfetto, A. De Fenza, G. Petrone, and F. Caputo. 2020. "Guided Wave SHM System for Damage Detection in Complex Composite Structure." *Theoretical and Applied Fracture Mechanics* 105 (September 2019): 102408. 10.1016/j.tafmec.2019.102408.

Dey, Biswajit, Md Sherajul Islam, and Jeongwon Park. 2021. "Numerical Design of High-Performance WS2/Metal/WS2/Graphene Heterostructure Based Surface Plasmon Resonance Refractive Index Sensor." *Results in Physics* 23 (February): 104021. 10.1016/j.rinp.2021.104021.

Dral, A. Petra, and Johan E. ten Elshof. 2018. "2D Metal Oxide Nanoflakes for Sensing Applications: Review and Perspective." *Sensors and Actuators, B: Chemical* 272 (May): 369–92. 10.1016/j.snb.2018.05.157.

El-Ahmar, S., W. Koczorowski, A. A. Poźniak, P. Kuświk, M. Przychodnia, J. Dembowiak, and W. Strupiński. 2019. "Planar Configuration of Extraordinary Magnetoresistance for 2D-Material-Based Magnetic Field Sensors." *Sensors and Actuators, A: Physical* 296: 249–53. 10.1016/j.sna.2019.07.016.

Feng, Dongming, and Maria Q. Feng. 2018. "Computer Vision for SHM of Civil Infrastructure: From Dynamic Response Measurement to Damage Detection—A Review." *Engineering Structures* 156 (November 2017): 105–17. 10.1016/j.engstruct.2017.11.018.

Irfan, M. S., T. Khan, T. Hussain, K. Liao, and R. Umer. 2021. "Carbon Coated Piezoresistive Fiber Sensors: From Process Monitoring to Structural Health Monitoring of Composites—A Review." *Composites Part A: Applied Science and Manufacturing* 141 (July 2020): 106236. 10.1016/j.compositesa.2020.106236.

Jia, Yue, Yunlong Liao, Zhongfu Li, and Houzhi Cai. 2021. "Niobium Disulfide Nanosheets Modified Surface Plasmon Resonance Sensors for Ultrasensitive Detection of Mercury Ion." *Journal of Alloys and Compounds* 869: 159328. 10.1016/j.jallcom.2021.159328.

Jiang, Fan, Wen Sheng Zhao, and Jun Zhang. 2020. "Mini-Review: Recent Progress in the Development of MoSe2 Based Chemical Sensors and Biosensors." *Microelectronic Engineering* 225 (February). 10.1016/j.mee.2020.111279.

Kanmani, R., N. A.M. Zainuddin, M. F.M. Rusdi, S. W. Harun, K. Ahmed, I. S. Amiri, and R. Zakaria. 2019. "Effects of TiO2 on the Performance of Silver Coated on Side-Polished Optical Fiber for Alcohol Sensing Applications." *Optical Fiber Technology* 50 (February): 183–7. 10.1016/j.yofte.2019.03.010.

Kumar, Anil, Awadhesh K. Yadav, Angad S. Kushwaha, and S.K. Srivastava. 2020. "A Comparative Study among WS2, MoS2 and Graphene Based Surface Plasmon Resonance (SPR) Sensor." *Sensors and Actuators Reports* 2 (1): 100015. 10.1016/j.snr.2020.100015.

Kumar, Sunil, Vladimir Pavelyev, Prabhash Mishra, Nishant Tripathi, Prachi Sharma, and Fernando Calle. 2020. "A Review on 2D Transition Metal Di-Chalcogenides and Metal Oxide Nanostructures Based NO2 Gas Sensors." *Materials Science in Semiconductor Processing* 107 (May 2019): 104865. 10.1016/j.mssp.2019.104865.

Leonardi, S. G., W. Wlodarski, Y. Li, N. Donato, Z. Sofer, M. Pumera, and G. Neri. 2018. "A Highly Sensitive Room Temperature Humidity Sensor Based on 2D-WS2 Nanosheets." *FlatChem* 9 (May): 21–6. 10.1016/j.flatc.2018.05.001.

Li, Jin, Hao Yan, Hongtao Dang, and Fanli Meng. 2021. "Structure Design and Application of Hollow Core Microstructured Optical Fiber Gas Sensor: A Review." *Optics and Laser Technology* 135 (May 2020): 106658. 10.1016/j.optlastec.2020.106658.

Madbouly, Ayman I., M. M. Mokhtar, and M. S. Morsy. 2020. "Evaluating the Performance of RGO/Cement Composites for SHM Applications." *Construction and Building Materials* 250: 118841. 10.1016/j.conbuildmat.2020.118841.

Pandey, Ankit Kumar. 2020. "Plasmonic Sensor Utilizing Ti3C2Tx MXene Layer and Fluoride Glass Substrate for Bio- and Gas-Sensing Applications: Performance Evaluation." *Photonics and Nanostructures—Fundamentals and Applications* 42 (January): 100863. 10.1016/j.photonics.2020.100863.

Patil, Pravin O., Gaurav R. Pandey, Ashwini G. Patil, Vivek B. Borse, Prashant K. Deshmukh, Dilip R. Patil, Rahul S. Tade, et al. 2019. "Graphene-Based Nanocomposites for Sensitivity Enhancement of Surface Plasmon Resonance Sensor for Biological and Chemical Sensing: A Review." *Biosensors and Bioelectronics* 139 (March): 111324. 10.1016/j.bios.2019.111324.

Rahman, Md Mahabubur, Md Masud Rana, Md Saifur Rahman, M. S. Anower, Md Aslam Mollah, and Alok Kumar Paul. 2020. "Sensitivity Enhancement of SPR Biosensors Employing Heterostructure of PtSe2 and 2D Materials." *Optical Materials* 107 (June): 110123. 10.1016/j.optmat.2020.110123.

Rodriguez, Gerardo, Joan R. Casas, and Sergi Villalba. 2019. "Shear Crack Width Assessment in Concrete Structures by 2D Distributed Optical Fiber." *Engineering Structures* 195 (September 2018): 508–23. 10.1016/j.engstruct.2019.05.079.

Sadeghi, Zeynab, and Hossein Shirkani. 2020. "Highly Sensitive Mid-Infrared SPR Biosensor for a Wide Range of Biomolecules and Biological Cells Based on Graphene-Gold Grating." *Physica E: Low-Dimensional Systems and Nanostructures* 119 (February): 114005. 10.1016/j.physe.2020.114005.

Sadeghi, Zeynab, Naser Shojaeihagh, and Hossein Shirkani. 2021. "Multiple-Step Graphene Grating Optical Sensors Based on Surface Plasmons in IR Range for Ultra-Sensing Biomolecules." *Materials Science and Engineering B: Solid-State Materials for Advanced Technology* 265 (November 2020): 114988. 10.1016/j.mseb.2020.114988.

Saifur Rahman, M., K. A. Rikta, Lway Faisal Abdulrazak, and M. S. Anower. 2020. "Enhanced Performance of SnSe-Graphene Hybrid Photonic Surface Plasmon Refractive Sensor for Biosensing Applications." *Photonics and Nanostructures—Fundamentals and Applications* 39 (September 2019): 100779. 10.1016/j.photonics.2020.100779.

Sharma, Anuj K., Baljinder Kaur, and Carlos Marques. 2020. "Simulation and Analysis of 2D Material/Metal Carbide Based Fiber Optic SPR Probe for Ultrasensitive Cortisol Detection." *Optik* 218 (June): 164891. 10.1016/j.ijleo.2020.164891.

Sharma, Anuj K., Baljinder Kaur, and Vasile A. Popescu. 2020. "On the Role of Different 2D Materials/Heterostructures in Fiber-Optic SPR Humidity Sensor in Visible Spectral Region." *Optical Materials* 102 (March): 109824. 10.1016/j.optmat.2020.109824.

Umar, Ahmad, H. Y. Ammar, Rajesh Kumar, Tubia Almas, Ahmed A. Ibrahim, M. S. AlAssiri, M. Abaker, and S. Baskoutas. 2020. "Efficient H2 Gas Sensor Based on 2D SnO2 Disks: Experimental and Theoretical Studies." *International Journal of Hydrogen Energy* 45 (50): 26388–401. 10.1016/j.ijhydene.2019.04.269.

Wang, Xiaoqiang, Lunyang Lin, Shaowei Lu, and Bohan Li. 2021. "Condition Monitoring of Composite Overwrap Pressure Vessels Using MXene Sensor." *International Journal of Pressure Vessels and Piping* 191 (February): 104349. 10.1016/j.ijpvp.2021.104349.

Wu, Qiong, Ningbo Li, Ying Wang, Ying Liu, Yanchao Xu, Shuting Wei, Jiandong Wu, et al. 2019. "A 2D Transition Metal Carbide MXene-Based SPR Biosensor for Ultrasensitive Carcinoembryonic Antigen Detection." *Biosensors and Bioelectronics* 144 (July): 111697. 10.1016/j.bios.2019.111697.

Yao, Ming Shui, Wen Hua Li, and Gang Xu. 2021. "Metal–Organic Frameworks and Their Derivatives for Electrically-Transduced Gas Sensors." *Coordination Chemistry Reviews* 426: 213479. 10.1016/j.ccr.2020.213479.

Yu, Haixia, Yang Chong, Penghao Zhang, Jiaming Ma, and Dachao Li. 2020. "A D-Shaped Fiber SPR Sensor with a Composite Nanostructure of MoS2-Graphene for Glucose Detection." *Talanta* 219 (June): 121324. 10.1016/j.talanta.2020.121324.

Yu, Fengming, Osamu Saito, and Yoji Okabe. 2021. "An Ultrasonic Visualization System Using a Fiber-Optic Bragg Grating Sensor and Its Application to Damage Detection at a Temperature of 1000°C." *Mechanical Systems and Signal Processing* 147: 107140. 10.1016/j.ymssp.2020.107140.

Zainuddin, N. A.M., M. M. Ariannejad, P. T. Arasu, S. W. Harun, and R. Zakaria. 2019. "Investigation of Cladding Thicknesses on Silver SPR Based Side-Polished Optical Fiber Refractive-Index Sensor." *Results in Physics* 13 (April): 102255. 10.1016/j.rinp.2019.102255.

Zhang, Changgeng, Yanming Zhang, Kaixiang Cao, Zehao Guo, Yaqin Han, Wei Hu, Yingjie Wu, Yin She, and Yong He. 2021. "Ultrasensitive and Reversible Room-Temperature Resistive Humidity Sensor Based on Layered Two-Dimensional Titanium Carbide." *Ceramics International* 47 (5): 6463–9. 10.1016/j.ceramint.2020.10.229.

Zheng, Qingbin, Jeng-hun Lee, Xi Shen, Xiaodong Chen, and Jang-Kyo Kim. 2020. "Graphene-Based Wearable Piezoresistive Physical Sensors." *Materials Today* 36 (June): 158–79. 10.1016/j.mattod.2019.12.004.

Zhou, Jie, Tingqiang Yang, Jiajie Chen, Cong Wang, Han Zhang, and Yonghong Shao. 2020. "Two-Dimensional Nanomaterial-Based Plasmonic Sensing Applications: Advances and Challenges." *Coordination Chemistry Reviews* 410: 213218. 10.1016/j.ccr.2020.213218.

Zhu, Xiaohua, Youyu Zhang, Meiling Liu, and Yang Liu. 2021. "2D Titanium Carbide MXenes as Emerging Optical Biosensing Platforms." *Biosensors and Bioelectronics* 171 (May 2020): 112730. 10.1016/j.bios.2020.112730.

Zrelli, Amira, and Tahar Ezzedine. 2018. "Design of Optical and Wireless Sensors for Underground Mining Monitoring System." *Optik* 170: 376–83. 10.1016/j.ijleo.2018.04.021.

Index

9781032041469